# Remote Sensing and Geospatial Technologies in Public Health

# Remote Sensing and Geospatial Technologies in Public Health

Special Issue Editor

**Fazlay S. Faruque**

MDPI • Basel • Beijing • Wuhan • Barcelona • Belgrade

**MDPI**

*Special Issue Editor*
Fazlay S. Faruque
University of Mississippi Medical Center
USA

*Editorial Office*
MDPI
St. Alban-Anlage 66
Basel, Switzerland

This is a reprint of articles from the Special Issue published online in the open access journal *ISPRS International Journal of Geo-Information* (ISSN 2220-9964) from 2014 to 2018 (available at: http://www.mdpi.com/journal/ijgi/special issues/geospatial-technologies)

For citation purposes, cite each article independently as indicated on the article page online and as indicated below:

LastName, A.A.; LastName, B.B.; LastName, C.C. Article Title. *Journal Name* **Year**, *Article Number*, Page Range.

**ISBN 978-3-03897-172-6 (Pbk)**
**ISBN 978-3-03897-173-3 (PDF)**

It depicts a NASA satellite image from January 14, 2002, shows a thick blanket of air pollution butting up against the Himalayas and stretching out into the Bay of Bengal. https://visibleearth.nasa.gov/view.php?id=57559.

# Contents

# About the Special Issue Editor

**Fazlay S. Faruque** is a Professor of Preventive Medicine in the John D. Bower School of Population Health at the University of Mississippi Medical Center (UMMC). He joined UMMC in 2000 as the Director of the GIS and Remote Sensing program with a faculty appointment as an Assistant Professor. Before joining UMMC, he worked for six years for a consulting firm responsible for managing their GIS division. Dr. Faruque has been teaching and conducting research in the areas of geospatial science and technology for twenty nine years with a focus on environmental health. He received funding from external sources including NIH and NASA. He organized several international meetings on the application of geospatial technology in health. Currently, Dr. Faruque is serving as the chair of the ISPRS Working Group on Environment and Health.

International Journal of
**Geo-Information**

MDPI

*Editorial*

# Remote Sensing and Geospatial Technologies in Public Health

Fazlay S. Faruque

Department of Preventive Medicine, John D. Bower School of Population Health, University of Mississippi
Medical Center, 2500 North State Street, Jackson, MS 39216-4505, USA; ffaruque@umc.edu

Received: 11 July 2018; Accepted: 20 July 2018; Published: 30 July 2018

---

The utilization of remote sensing and geospatial technologies has been instrumental in advancing our understanding of environmental factors affecting human health and well-being. Extreme weather and related phenomena appear to be rising in frequency and intensity which pose growing health risks to human populations. Earth-observing technologies and data are important elements of a comprehensive and multi-scaled public health response at both the micro and macro levels which can identify immediate and long-term impacts. Remote sensing and geospatial technologies have been successfully implemented for more than 50 years; they have examined the role of environmental factors in air-borne, vector-borne, soil-borne, and water-borne diseases. With the availability of new data and advanced technologies, more robust public health measures are being implemented to improve our health and well-being.

Through its regular and special issue publications, the ISPRS International Journal of Geo-Information (IJGI) has been active in providing an advanced forum for the science and technology of geographic information. The idea of a special issue on "Remote Sensing and Geospatial Technologies in Public Health" emerged from the 2nd Symposium on Advances in Geospatial Technologies for Health, which was organized by the ISPRS Working Group on Health in Arlington, Virginia, USA during 24–30 August 2013 [1]. The goals of that symposium were to bridge the geospatial science, Earth science, and health science communities as well as to explore interdisciplinary collaborations to improve our overall health and well-being. This symposium was attended by representatives from international organizations, government agencies, and academia who were actively engaged in advancing different aspects of Remote Sensing and geospatial technologies to benefit public health.

On the basis of the themes of this symposium, a call for papers for this special issue was announced. Submitted manuscripts went through multiple reviews by experts in each field. After intense evaluation, out of 26 submissions, 15 papers were accepted, which are the contents of this book. These papers cover a wide range of health-related issues, including air pollutions, vector-borne diseases, water quality, demographic factors and scale factors, along with emerging analytical techniques. Below are the summaries of the papers.

In "CALPUFF and CAFOs: Air Pollution Modeling and Environmental Justice Analysis in the North Carolina Hog Industry", Yelena Ogneva-Himmelberger and her colleagues address air pollution issues related to Concentrated Animal Feeding Operations (CAFOs) [2]. Here, they demonstrate the uses of meteorological and CAFO data in developing an air pollution dispersion model to estimate ammonia concentrations; they use their estimated ammonia concentrations to evaluate disproportionate exposure of children, elderly, whites, and minorities to this pollutant. They have found that their modelled ammonia concentration is comparable to other established methods. Hence, they feel comfortable to recommend their air pollution dispersion models for environmental justice studies to assess the impacts of the CAFOs and to address concerns regarding the health and quality of life of vulnerable populations.

The Public Health Agency of Canada (PHAC) predicts that, by the year 2020, 80% of Canadians will live in Lyme endemic areas. In the paper, "Analyzing the Correlation between Deer Habitat and the Component of the Risk for Lyme Disease in Eastern Ontario, Canada: A GIS-Based Approach", Chen et al. share their experience of developing a geospatial deer habitat suitability model for Eastern Ontario, Canada [3]. The authors expect their model will assist in the development of management strategies to prevent Lyme from becoming a threat to public health in Canada.

Geospatial technology has played a significant global role in the reduction of malaria. In their article, "Geospatial Technology: A Tool to Aid in the Elimination of Malaria in Bangladesh" Kirk et al. evaluate the success in malaria reduction in Bangladesh by effective utilization of geospatial technologies at various levels [4].

To make environmental health-related investigations useful in a clinical environment, it is necessary to incorporate personal level exposure to pollution [5]. In the article, "Examining Personal Air Pollution Exposure, Intake, and Health Danger Zone Using Time Geography and 3D Geovisualization", Yongmei Lu and Tianfang Fang from Texas State University examine personal exposure to air pollution and pollutant intake and define personal health danger zones by expanding traditional time geography and accounting for individual level space-time behavior [6]. Although this study focuses only on ozone, the authors believe that their 3D time-geography approach can be extended to other pollutants for evaluating personal health risks.

Many countries, including the U.S., make huge investments to gather and disseminate geospatial resources that are freely available for public use. However, researchers often cannot take full advantage of these resources due to a lack of basic information. James Acker and his colleagues diligently bring required information about the NASA Giovanni Data System and its potential use to the geospatial public health research community [7]. Since 2003, the Giovanni system has been providing access to a wide variety of NASA remote sensing data and other Earth science data sets, allowing researchers to apply it to a broad range of research topics. In this article, "Use of the NASA Giovanni Data System for Geospatial Public Health Research: Example of Weather-Influenza Connection", the authors describe the public health-related datasets in Giovanni, and inform how to access and utilize those resources. They also demonstrate how Giovanni resources can be used to study the relationship between influenza and meteorological parameters as well as to predict influenza activity.

In the paper, "Mapping Entomological Dengue Risk Levels in Martinique Using High-Resolution Remote-Sensing Environmental Data", Machault et al. discuss the development of entomological risk levels of dengue virus transmission in Martinique, French Antilles through the use of high spatial resolution remote-sensing environmental data along with field entomological and meteorological data [8]. In this paper, they demonstrate the use of the high resolution Geoeye-1 image to extract landscape elements which surrogate societal or biological information related to the life cycle of *Aedes* vectors.

Use of Remote Sensing data and techniques is not uncommon in water quality monitoring for large water bodies. However, this approach has limitations in smaller areas, such as small lakes. In "Improving Inland Water Quality Monitoring through Remote Sensing Techniques", Igor Ogashawara and Max Moreno-Madriñán document their proposed algorithm, which is specifically designed for monitoring the water quality of small lakes [9].

Although the impact of Modifiable Areal Unit Problem (MAUP) is well known, certain Census geographic unit is often used as the de facto data scale for health geography studies without considering other options. Different geographic scales may result in different results, which can bias public health-related decisions. Lee et al., in their article, "Impacts of Scale on Geographic Analysis of Health Data: An Example of Obesity Prevalence", demonstrate the differences in results from studying the prevalence of obesity using Census Tract and Block Group level data [10].

The location of disease clusters is often a matter of great interest among public health professionals and policymakers. However, the techniques of cluster identification can be quite diverse, resulting in varied results. Mahmoud Torabi and Katie Galloway from the University of Manitoba demonstrate five different cluster detection procedures to study geographic variation in the incidence of Chronic Obstructive Pulmonary Disease (COPD) in Manitoba, Canada [11]. In their paper, "Geographical

Variation of Incidence of Chronic Obstructive Pulmonary Disease in Manitoba, Canada", Torabi and Galloway discuss the methods, advantages and disadvantages, and the results of the various procedures, which provides a good reference for studying regional cluster analysis for other disease incidences.

With the availability of large amounts of multidimensional data, new techniques are emerging which apply these data to a wide range of applications, including those of health. In their article, "Holistics 3.0 for Health", Lary et al. present a holistic system suitable for complex health studies; this system combines multiple big datasets and multivariate computational techniques, also incorporating new geospatial techniques [12].

Haley Cleckner and Thomas Allen from East Carolina University apply a dasymetric mapping technique to map spatial patterns of vulnerable populations and to characterize potential exposure to mosquito vectors of West Nile Virus across the Chesapeake region of Virginia [13]. In their article, "Dasymetric Mapping and Spatial Modeling of Mosquito Vector Exposure, Chesapeake, Virginia, USA", they quantify mosquito vector abundance, which in combination with a population vulnerability index, they demonstrate an evaluative technique which examines the exposure of human populations to mosquitoes.

In the paper, "Modeling Properties of Influenza-Like Illness Peak Events with Crossing Theory", Wang and his colleagues from the University of Florida, Gainesville propose a methodology to redefine the "peak event" for influenza-like illness (ILI) [14]. As illustrated in this paper, their proposed "peak event" can aid public health professionals to improve the surveillance of influenza and to implement efficient intervention strategies.

In the article, "Correlating Remote Sensing Data with the Abundance of Pupae of the Dengue Virus Mosquito Vector, *Aedes aegypti*, in Central Mexico", Moreno-Madriñán et al. discuss the successful use of Remote Sensing products to correlate the abundance of the dengue virus mosquito vector, *Ae. aegypti*, with nighttime land surface temperature [15]. The authors note that, in their study area of central Mexico, the elevation and meteorological conditions significantly vary, but the socio-economic conditions remain relatively uniform, which simplifies the analysis. They demonstrate how their approach of substituting ground meteorological data by remote sensing products can be successfully implemented to predict the abundance of *Ae. aegypti*, the primary mosquito vector of dengue virus. This approach can be replicated in other regions where variables are less complicated, similar to the original study region of Central Mexico.

In their paper, "Canadian Forest Fires and the Effects of Long-Range Transboundary Air Pollution on Hospitalizations among the Elderly", George Le and his colleagues evaluate the transboundary movement of the forest fire-induced $PM_{2.5}$, which originated in Canada, and its impact on the increase of respiratory and cardiovascular hospital admissions of vulnerable population in the U.S. [16]. Lightning strikes often ignite forest fires, which can contribute to serious air pollution. The authors study the short-term increases in $PM_{2.5}$ concentrations resulting from forest fires in the province of Quebec, Canada and their impact on respiratory and cardiovascular hospital admissions for the elderly across the east coast of the U.S., as far south as Washington D.C.

In their paper, "Nexus of Health and Development: Modelling Crude Birth Rate and Maternal Mortality Ratio Using Nighttime Satellite Images", Koel Roychowdhury and Simon Jones combine development indicators, health indicators, and nighttime satellite images to propose models for predicting health determinants and development determinants [17].

Although the bridge between the broad areas of health and geospatial technology is not yet well established, there has been tremendous progress in the overarching disciplines as well as in their sub-disciplines. As a member of the geospatial health community, I would urge that we need to keep moving forward and be proactive in making our contributions useful to improve human health and well-being.

In recent years, there have been unprecedented advancements in the availability of geospatial data, analytical tools, and computational capacity. Access to these resources often tempts us to rush to utilize these resources without giving in-depth thought to the problems and the suitability of the tools for solutions. My hope is that young researchers will pay more attention to the theoretical background of these techniques and their applicability to investigate specific health issues. New generations of geospatial researchers can further advance geospatial technologies to make this world a better place.

I must express my gratitude to the reviewers who contributed their valuable time to provide detailed comments and suggestions to the authors. I also commend the authors for their patience as their manuscripts were going through multiple reviews as well as their diligence in addressing the comments. I also need to mention the professionalism of the IJGI publication team for their support throughout the entire process, from submission to publication, which has been exemplary for an academic publication.

**Funding**: This research received no external funding.

**Conflicts of Interest:** The author declares no conflict of interest.

## References

1. ISPRS ICWG III/IVc: Environment and Health. Past Events. 2nd Symposium on Advances in Geospatial Technologies for Health. Available online: http://www2.isprs.org/commissions/comm3/icwg-3-4c/2nd-symposium.html (accessed on 2 July 2018).
2. Ogneva-Himmelberger, Y.; Huang, L.; Xin, H. CALPUFF and CAFOs: Air Pollution Modeling and Environmental Justice Analysis in the North Carolina Hog Industry. *ISPRS Int. J. Geo-Inf.* **2015**, *4*, 150–171. [CrossRef]
3. Chen, D.; Wong, H.; Belanger, P.; Moore, K.; Peterson, M.; Cunningham, J. Analyzing the Correlation between Deer Habitat and the Component of the Risk for Lyme Disease in Eastern Ontario, Canada: A GIS-Based Approach. *ISPRS Int. J. Geo-Inf.* **2015**, *4*, 105–123. [CrossRef]
4. Kirk, K.; Haq, M.; Alam, M.; Haque, U. Geospatial Technology: A Tool to Aid in the Elimination of Malaria in Bangladesh. *ISPRS Int. J. Geo-Inf.* **2015**, *4*, 47–58. [CrossRef]
5. Faruque, F.S.; Finley, R.W. Geographic Medical History: Advances in Geospatial Technology Present New Potentials in Medical Practice. *Int. Arch. Photogramm. Remote Sens. Spat. Inf. Sci.* **2016**. [CrossRef]
6. Lu, Y.; Fang, T. Examining Personal Air Pollution Exposure, Intake, and Health Danger Zone Using Time Geography and 3D Geovisualization. *ISPRS Int. J. Geo-Inf.* **2015**, *4*, 32–46. [CrossRef]
7. Acker, J.; Soebiyanto, R.; Kiang, R.; Kempler, S. Use of the NASA Giovanni Data System for Geospatial Public Health Research: Example of Weather-Influenza Connection. *ISPRS Int. J. Geo-Inf.* **2014**, *3*, 1372–1386. [CrossRef]
8. Machault, V.; Yébakima, A.; Etienne, M.; Vignolles, C.; Palany, P.; Tourre, Y.; Guérécheau, M.; Lacaux, J.-P. Mapping Entomological Dengue Risk Levels in Martinique Using High-Resolution Remote-Sensing Environmental Data. *ISPRS Int. J. Geo-Inf.* **2014**, *3*, 1352–1371. [CrossRef]
9. Ogashawara, I.; Moreno-Madriñán, M. Improving Inland Water Quality Monitoring through Remote Sensing Techniques. *ISPRS Int. J. Geo-Inf.* **2014**, *3*, 1234–1255. [CrossRef]
10. Lee, J.; Alnasrallah, M.; Wong, D.; Beaird, H.; Logue, E. Impacts of Scale on Geographic Analysis of Health Data: An Example of Obesity Prevalence. *ISPRS Int. J. Geo-Inf.* **2014**, *3*, 1198–1210. [CrossRef]
11. Torabi, M.; Galloway, K. Geographical Variation of Incidence of Chronic Obstructive Pulmonary Disease in Manitoba, Canada. *ISPRS ISPRS Int. J. Geo-Inf.* **2014**, *3*, 1039–1057. [CrossRef]
12. Lary, D.; Woolf, S.; Faruque, F.; LePage, J. Holistics 3.0 for Health. *ISPRS Int. J. Geo-Inf.* **2014**, *3*, 1023–1038. [CrossRef]
13. Cleckner, H.; Allen, T. Dasymetric Mapping and Spatial Modeling of Mosquito Vector Exposure, Chesapeake, Virginia, USA. *ISPRS Int. J. Geo-Inf.* **2014**, *3*, 891–913. [CrossRef]

14. Wang, Y.; Waylen, P.; Mao, L. Modeling Properties of Influenza-Like Illness Peak Events with Crossing Theory. *ISPRS Int. J. Geo-Inf.* **2014**, *3*, 764–780. [CrossRef]

15. Moreno-Madriñán, M.; Crosson, W.; Eisen, L.; Estes, S.; Estes, M., Jr.; Hayden, M.; Hemmings, S.; Irwin, D.; Lozano-Fuentes, S.; Monaghan, A.; et al. Correlating Remote Sensing Data with the Abundance of Pupae of the Dengue Virus Mosquito Vector, Aedes aegypti, in Central Mexico. *ISPRS Int. J. Geo-Inf.* **2014**, *3*, 732–749. [CrossRef]

16. Le, G.; Breysse, P.; McDermott, A.; Eftim, S.; Geyh, A.; Berman, J.; Curriero, F. Canadian Forest Fires and the Effects of Long-Range Transboundary Air Pollution on Hospitalizations among the Elderly. *ISPRS Int. J. Geo-Inf.* **2014**, *3*, 713–731. [CrossRef]

17. Roychowdhury, K.; Jones, S. Nexus of Health and Development: Modelling Crude Birth Rate and Maternal Mortality Ratio Using Nighttime Satellite Images. *ISPRS Int. J. Geo-Inf.* **2014**, *3*, 693–712. [CrossRef]

International Journal of
**Geo-Information**

isprs

MDPI

*Article*

# CALPUFF and CAFOs: Air Pollution Modeling and Environmental Justice Analysis in the North Carolina Hog Industry

**Yelena Ogneva-Himmelberger \*, Liyao Huang and Hao Xin**

Department of International Development, Community and Environment, Clark University, 950 Main St., Worcester, MA 01610, USA; hlyao1108@gmail.com (L.H.); greenxinhao@gmail.com (H.X.)

\* Author to whom correspondence should be addressed; yogneva@clarku.edu; Tel.: +1-508-421-3805.

Academic Editors: Fazlay S. Faruque and Wolfgang Kainz

Received: 23 June 2014; Accepted: 9 January 2015; Published: 26 January 2015

**Abstract:** Concentrated animal feeding operations (CAFOs) produce large amounts of animal waste, which potentially pollutes air, soil and water and affects human health if not appropriately managed. This study uses meteorological and CAFO data and applies an air pollution dispersion model (CALPUFF) to estimate ammonia concentrations at locations downwind of hog CAFOs and to evaluate the disproportionate exposure of children, elderly, whites and minorities to the pollutant. Ammonia is one of the gases emitted by swine CAFOs and could affect human health. Local indicator of spatial autocorrelation (LISA) analysis uses census block demographic data to identify hot spots where both ammonia concentrations and the number of exposed vulnerable population are high. We limit our analysis to one watershed in North Carolina and compare environmental justice issues between 2000 and 2010. Our results show that the average ammonia concentrations in hot spots for 2000 and 2010 were 2.5–3-times higher than the average concentration in the entire watershed. The number of people living in the areas where ammonia concentrations exceeded the minimal risk level was 3647 people in 2000 and 3360 people in 2010. We recommend using air pollution dispersion models in future environmental justice studies to assess the impacts of the CAFOs and to address concerns regarding the health and quality of life of vulnerable populations.

**Keywords:** exposure to air pollutants; CALPUFF; ammonia; CAFO; environmental justice; hog industry

---

## 1. Introduction

Livestock farming has experienced significant changes in the last few decades: while the number of small, family-owned animal farms has been decreasing, the number of large, industrial animal farms has been increasing, similar to consolidation in other commercial operations, such as grocery and clothing stores. According to the National Agricultural Statistics Service, 86% of all hogs raised in the U.S. in 2010 were concentrated in just 12% of hog operations [1]. Proponents of industrial agriculture argue that concentrated animal feeding operations (CAFOs) provide a "low-cost source of meat, milk, and eggs, due to efficient feeding and housing of animals, increased facility size, and animal specialization" and "enhance the local economy and increase employment" [2]. However, numerous studies conducted in the last 15 years have shown that the rapid growth of CAFOs brought about a series of negative environmental and human health effects [3–6]. The main source of air and water pollution is animal manure. Manure contains a variety of nutrients and potential contaminants, such as nitrogen, phosphorous, pathogens (e.g., *E. coli*), growth hormones, antibiotics, animal blood and chemicals used to clean the equipment [2]. According to some estimates, livestock animals produce three- to 20-times more manure than people in the U.S. [7], and a hog farm with 1,000 animals produces

14,500 tons of manure each year [8]. It is channeled from animal houses into pits or storage lagoons and eventually sprayed untreated onto nearby fields, replacing commercial fertilizers. Regulations require that manure storage units be designed to not leak into the groundwater (using concrete, clay soil lining or a metal structures). In addition, units must not discharge to surface waters and must be inspected by state and, sometimes, federal regulatory agencies.

Manure storage facilities on livestock farms produce gaseous (ammonia, hydrogen sulfide, methane) and particulate substances proportionate to the number or mass of animals housed. Ammonia is formed during microbial decomposition of undigested organic nitrogen compounds in manure; hydrogen sulfide is produced during anaerobic bacterial decomposition of sulfur-containing organic matter; and methane is created during anaerobic microbial degradation of organic matter. Both ammonia and hydrogen sulfide pose serious risks to human health at elevated concentrations, and methane contributes to climate change. Ammonia irritates the respiratory tract and causes severe coughing, chronic lung disease and chemical burns to the respiratory tract, skin and eyes [2]. Hydrogen sulfide causes inflammation of the membranes of the eye and respiratory tract, as well as loss of smell [2].

A recent research found that odors produced by the CAFOs also have adverse effects on health and quality of life [9]. The odors contain a mixture of ammonia, hydrogen sulfide, carbon dioxide and volatile and semi-volatile organic compounds [6] and, according to a report [2], under certain atmospheric conditions (with wind and little or no thermal gradient), can be detected as far as three miles away, sometimes up to six miles away. Several studies have shown that intolerable odors prevent residents from opening windows, spending time outdoors or inviting visitors, causing tension, depression, anger and anxiety about deteriorating quality of life [10–13]. Other reports note that the growth of CAFOs has forced small family farms out of business and altered local economies and communities [14,15].

North Carolina experienced rapid changes in the livestock industry starting in the 1970s and now is the second largest state (after Iowa) in hog herd size, with 9–10 million animals [9]. Most hog CAFOs are located in the eastern counties of the state. Multiple incidences of swine lagoon overflows and water pollution caused by hurricanes in the 1990s led to public protests, and the state placed a moratorium of new hog farms housing more than 250 hogs. Despite this 10-year moratorium (1997–2007) [16], the number of hogs in the state "quadrupled between 1988 and 2010, while the number of farms fell by more than 80 percent" (http://www.factoryfarmmap.org/).

Several survey-based studies analyzed the health conditions of the residents [5,17,18]; other studies documented disproportional exposure of low-income, minority communities in North Carolina to CAFOs [15,16,19–22]. Disproportional exposure to environmental pollution is an environmental justice issue. EPA defines environmental justice as "the fair treatment and meaningful involvement of all people regardless of race, color, national origin, or income with respect to the development, implementation, and enforcement of environmental laws, regulations, and policies" (http://www.epa. gov/environmentaljustice/). All environmental justice studies related to CAFOs in North Carolina were conducted for the entire state and used the characteristics of CAFOs to represent potential pollution exposure. These studies used census-based units of analysis (county, census tract or block group) and socio-economic data from the census as analytical variables. Wing *et al.* (2000) used Poisson regression and the number of swine operations in the census block group as the dependent variable and the socio-economic characteristics of the block group as independent variables. They found that areas with the highest poverty and the highest percentage of minorities have the highest number of hog CAFOs per block group [17]. Wing *et al.* (2002) calculated ratios of the proportion of blacks to the proportion of whites living in areas with CAFOs that could be potentially flooded *vs.* areas not likely to be flooded [22]. They found that blacks were more likely than whites to live in areas with CAFOs that could be potentially flooded. Edward and Ladd used hog population per county as the dependent variable and county socio-demographics as independent variables [20] and found that minority communities are disproportionately exposed to high hog populations and that the relationship between income and hog population varies by region. A more recent study in eastern North Carolina [16] compared demographics of census tracts within one and three miles of CAFOs in 1990 and 2000 to random points within the same region. The results of this study showed that areas near CAFOs have higher percentages of minorities, low-income and low education level residents.

One of the limitations of these studies is that CAFO characteristics (the number of facilities or the number of hogs) are used as a surrogate measure of potential pollution produced by CAFOs. Our study tries to address this gap in the literature and uses modeled pollutant concentrations in the air as a measure of population exposure. We also use the smallest census-based unit of analysis, the census block, to analyze environmental justice at a finer spatial scale than previous studies. We limit our analysis to one watershed in eastern North Carolina and use longitudinal analysis to compare environmental justice issues between 2000 and 2010 in the context of ammonia pollution exposure. We chose ammonia because it is one of the most prevalent gases emitted by swine CAFOs.

None of the previous environmental justice studies in this area have analyzed the disproportionate exposure of children and the elderly. Children take in 20%–50% more air than adults and therefore are more susceptible to the health effects of air pollution [23]. Elderly people are more susceptible to air pollution due to ageing [24] and because air pollution can aggravate existing health conditions [25]. We include these populations in our analysis. Specifically, our study tries to answer the following question: Are children, the elderly, white and minority populations disproportionately exposed to ammonia emitted by CAFOs in Contentnea Creek Watershed in North Carolina?

## 2. Study Area

Stretching across nearly 275 miles, the Neuse River is the longest river entirely contained in North Carolina. In 1995, 1996 and 1997, it was continuously designated as one of North America's most threatened rivers, and in 2007, it was designated as one of the most endangered rivers in the U.S. [26]. CAFO pollution was named one of the leading causes of the river's continuing pollution problems [26]. There are approximately 500 CAFO facilities housing about 1.8 million animals in the Neuse River Watershed [27].

Our study is focused on the Contentnea Creek Watershed (4274.85 km$^2$), a sub-basin of the Neuse River. The watershed contains several counties (Figure 1) and has one of the highest concentrations of CAFO facilities in North Carolina.

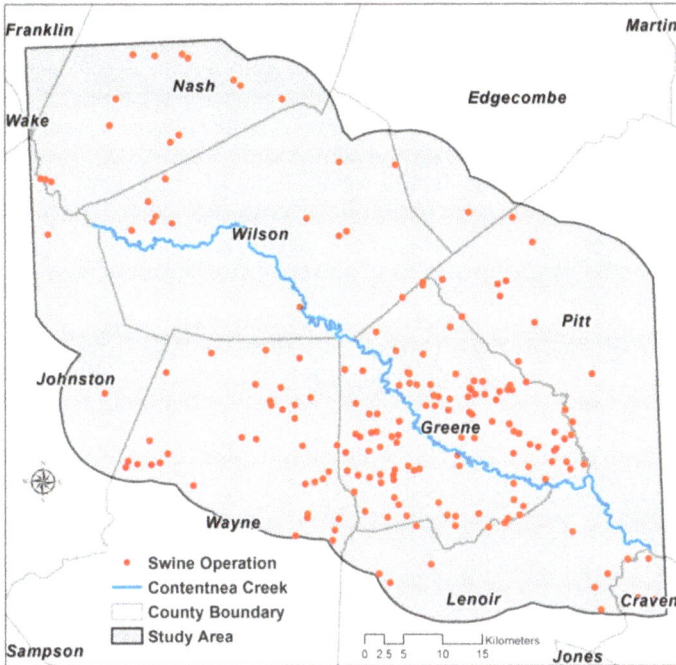

**Figure 1.** Study area with swine CAFO operations.

Pork production historically has been an important part of agriculture in this part of North Carolina, but it experienced an exponential growth in the 1990s. Population characteristics within this watershed are similar to the population in North Carolina as a whole: both have about a 32% minority population, about 20% of population below 15 years of age and about 13% of population over 65 years of age [28].

## 3. Data

Animal operations data were downloaded online from the NC Department of Environment and Natural Resources (NC DENR) Division of Water Quality website for 2010 (http://portal.ncdenr.org/web/wq/animal-facility-map). The data include the type of animal operation (swine, cattle, poultry and horse), capacity, geographic coordinates, total animal weight in kg and the description of each operation. Latitude/longitude information was used to map 195 swine CAFOs located in our study area (Figure 1).

Our goal was to use a spatial unit of analysis that would allow us to take full advantage of the spatial resolution of the ammonia concentration data (1 km × 1 km pixels) and detailed demographic information; the census block was the best option. We downloaded census block data from the U.S. Census Bureau website for the entire state (census block boundaries, age and racial composition) for 2000 and 2010. Poverty or income data were not included in the analysis, because these data are not publicly available at the census block level.

Both CAFO and census data were projected into the NAD 1983 State Plane North Carolina coordinate system. Census blocks located inside Contentnea Creek Watershed and within five miles outside its boundary were selected for the analysis. Census blocks within urban areas were removed from the analysis because CAFOs are located in rural areas. Boundaries of urban areas were obtained from the Census Bureau (http://www.Census.gov/geo/maps-data/data/cbf/cbf_ua.html).

Uninhabited census blocks were excluded, because our focus was on human exposure to air pollution. The final dataset included 3290 census blocks for 2000 and 3685 census blocks for 2010. The number of blocks is different between the two years, because some census boundaries had changed.

Using 2000 and 2010 U.S. Census data, we calculated the number of people aged 65 and older, the number of people aged 15 and younger and the number of white and minority people. Table 1 shows the demographic data, and Figure 2 shows their spatial distribution. We used the actual number of people within each population group in the environmental justice analysis (instead of percent per census block), because it is a more relevant measure in the context of human exposure to air pollution.

**Table 1.** Demographic characteristics of census blocks in 2000 and 2010.

| Year (# of Census Blocks) | Statistics per Census Block | # of People under 15 Years of Age | # of People over 65 Years of Age | # of Minority Population | # White Population |
|---|---|---|---|---|---|
| 2000 (*3290*) | Min | 0 | 0 | 0 | 0 |
| | Max | 263 | 93 | 815 | 348 |
| | Mean | 8 | 4 | 12 | 28 |
| | Median | 4 | 3 | 3 | 15 |
| | SD | 13 | 6 | 29 | 38 |
| 2010 (*3685*) | Min | 0 | 0 | 0 | 0 |
| | Max | 263 | 149 | 725 | 482 |
| | Mean | 8 | 5 | 13 | 28 |
| | Median | 4 | 3 | 4 | 15 |
| | SD | 14 | 7 | 31 | 40 |

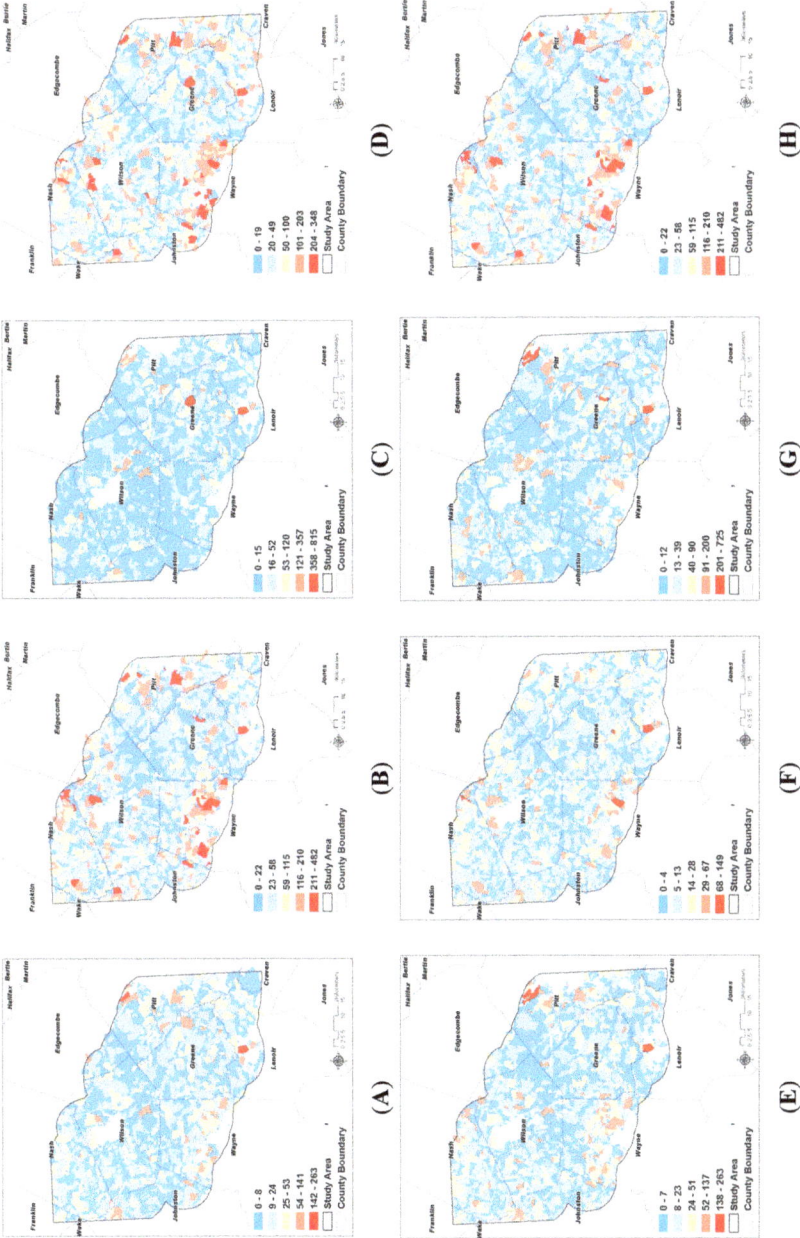

**Figure 2.** Demographic variables from 2000 and 2010 Census: (**A**) population of minorities in 2000; (**B**) population of children under 15 in 2000; (**C**) population of whites in 2000; (**D**) population of elderly over 65 in 2000; (**E**) population of minorities in 2010; (**F**) population of children under 15 in 2010; (**G**) population of elderly over 65 in 2010; (**H**) population of whites in 2010.

## 4. Methodology

### 4.1. Modeling Ammonia Concentrations

Air dispersion models have been used to estimate concentrations of pollutants emitted by CAFOs. Some studies modeled the dispersion of odor to define setback distances between CAFOs and residential areas [29]. Other studies attempted to model ammonia and hydrogen sulfide near CAFOs [30,31]. One study used the Industrial Source Complex Short-Term Model, version 3 (ISC-ST3), to model ammonia dispersion and deposition from CAFOs in North Carolina by hydrologic unit and county [30]. This model operates under the assumption that the concentration of the contaminant is defined by a normal, or Gaussian, curve and has some known deficiencies; it does not operate well during stable or near-calm conditions, and it cannot account for the effects of vegetation on concentrations, the effect of elevation nor wind distribution [29].

Another study used the CALPUFF model to model ammonia and hydrogen sulfide emitted by CAFOs in Minnesota [31]. The CALPUFF model accounts for variable wind directions and land cover pattern, includes a calm-wind algorithm and can model dispersion from multiple facilities and over a complex terrain. Due to these advantages, the "U.S. EPA has adopted CALPUFF as the preferred model for assessing long-range transport of pollutants" [29,32].

Following the U.S. EPA's recommendation [29,32], we selected CALPUFF to model ammonia dispersion in our study. The CALPUFF Modeling System includes three main components: CALMET, CALPUFF and CALPOST. CALMET is a meteorological model that develops wind and temperature models as three-dimensional grids. CALPUFF is a transport and dispersion model that emits "puffs" of material from modeled sources, simulating dispersion and transformation processes along the way. It uses the information generated by CALMET, and temporal and spatial variations in the meteorological grids are explicitly incorporated with the resulting distribution of puffs throughout a simulated period. The output files from CALPUFF contain concentrations evaluated at selected locations, called receptors. CALPOST is used to process these files and produce the summarized results of the simulation [33].

Three main types of data are required to run the model: hourly average meteorological data, facility layout and dimensions and emission data. We purchased meteorological data (MM5 file) from the company that distributes the CALPUFF model (http://www.src.com/calpuff/calpuff1.htm) and used it as input for the CALMET meteorological model. Each MM5 file covers a 120 km × 120 km area and contains a 10 × 10 grid with a spatial resolution of 12 km × 12 km. Our study area (Contentnea Creek Watershed and areas within five miles) is covered by one MM5 file. This grid contains hourly information for an entire year, including wind speed and direction, temperature, relative humidity, pressure, mixing ratios of water vapors, precipitation amount, solar radiation, snow cover, 2-m temperature and specific humidity, 10-m wind speed and direction and sea surface temperature. Only one MM5 file was available for our area of interest and time period, for 2006, so we used it to model ammonia emissions from CAFOs. To check whether 2006 data are representative of weather conditions in the area, we obtained weather observation data for three meteorological stations for 2000–2010: Goldsboro (latitude: 35.37935; longitude: −78.0448), Greenville (latitude: 35.6352389; longitude: −77.3853194) and Rocky Mount (latitude: 35.89295; longitude: −77.67996). Data included monthly average daily temperature, average wind direction, average wind speed and total monthly precipitation. We compared monthly 2006 data with averages for the other 9 years for each station separately (2000–2005; 2006–2010). Our comparisons showed that there was little difference between each month in 2006 and the corresponding values for the other 9 years for this month. For example, wind direction for Greenville was west (265°) in March 2006, and in six out of nine years it had a similar direction (ranging from 219° to 285°). Table 2 shows that there was little difference in wind speed between 2006 and the average for the other 9 years.

Using MM5 data, CALMET calculates 3D wind fields as 1 km × 1 km grids that are used as input for the CALPUFF model.

**Table 2.** Wind speed (km/hour) averaged for 2000–2005; 2007–2010 and its difference with 2006 wind speed for three meteorological stations in North Carolina [34].

| Month | Goldsboro | | | Greenville | | | Rocky Mount | | |
|---|---|---|---|---|---|---|---|---|---|
| | Average | 2006 | Difference | Average | 2006 | Difference | Average | 2006 | Difference |
| January | 5.72 | 5.31 | 0.41 | 3.60 | 4.35 | −0.74 | 3.92 | 4.35 | −0.42 |
| February | 6.37 | 4.67 | 1.70 | 4.20 | 4.51 | −0.30 | 4.18 | 4.51 | −0.32 |
| March | 3.36 | 4.67 | −1.31 | 2.37 | 3.54 | −1.17 | 1.97 | 3.54 | −1.57 |
| April | 4.05 | 2.57 | 1.47 | 3.38 | 2.09 | 1.29 | 2.96 | 2.25 | 0.70 |
| May | 3.04 | 4.02 | −0.99 | 3.02 | 2.90 | 0.13 | 2.92 | 3.54 | −0.62 |
| June | 3.04 | 1.93 | 1.11 | 3.54 | 3.38 | 0.16 | 2.72 | 2.41 | 0.30 |
| July | 3.54 | 4.02 | −0.48 | 3.38 | 4.35 | −0.97 | 3.08 | 3.70 | −0.62 |
| August | 2.45 | 1.45 | 1.01 | 2.50 | 1.29 | 1.22 | 2.04 | 1.77 | 0.27 |
| September | 4.87 | 1.29 | 3.58 | 3.38 | 0.16 | 3.22 | 3.50 | 0.64 | 2.86 |
| October | 3.32 | 3.22 | 0.10 | 1.97 | 1.77 | 0.20 | 1.95 | 2.09 | −0.14 |
| November | 3.72 | 2.90 | 0.82 | 2.25 | 2.09 | 0.16 | 2.29 | 2.09 | 0.20 |
| December | 4.18 | 1.45 | 2.74 | 2.75 | 1.93 | 0.82 | 3.17 | 1.77 | 1.39 |

CAFOs emissions were modeled as area emissions (*vs.* point or line), because hog houses and lagoons are more accurately represented as areas. We used the following procedure to calculate the dimensions of each operation. First, we randomly selected ten CAFO operations in the area and used Google Earth to draw their boundaries. Then, we calculated the total area of each CAFO (hog houses and lagoons together) and explored the relationship between the total area and the corresponding total weight of animals at each sampled operation. Our calculations showed that, on average, one kg of hog weight takes up about $0.1 \text{ m}^2$ of CAFO area. Using this conversion factor, we calculated the areal extent of each CAFO operation, representing it as a square of a certain size. Each square was centered on the latitude/longitude coordinates of the corresponding CAFO. Each CAFO's area size information, along with its elevation above sea level, was put into the CALPUFF model.

Finally, to calculate ammonia emissions from each CAFO, we used ammonia emission rates reported in the literature. A study by Aneja *et al.* [35] reported that emissions from barns ranged from 0.89 to 1.05 kg-N/week/1000 kg $\times$ lm for the cold and warm season, respectively (lm = live animal mass). When recalculated in different units (g/year), these emissions amount to 46.28–54.6 g-N/year/kg live mass. This study also reported ammonia flux rates per minute per square meter of waste lagoon surface. Since we did not have information about the size of each waste lagoon, we could not use the rates reported by this study and instead applied emission rates that were calculated per year per live animal weight or per animal, as reported in a study by Doorn *et al.* [36]. This study, conducted by the EPA, recommended a general emission factor for hog houses of $59 \pm 10$ g $NH_3$/kg live weight/year and for swine lagoons, of 2.4 kg $NH_3$/year/hog. We multiplied these rates by the corresponding numbers from the CAFOs table (total live weight and the number of hogs at each operation, respectively) and added two emissions together to get the total emissions amount per CAFO per year. We accepted CALPUFF's default settings and did not model the removal of the ammonia (wet deposition, dry deposition) or its chemical transformation. The output from CALPUFF model was input in CALPOST to produce daily average ammonia concentrations for the entire study area. It was then imported into ArcGIS software as a raster grid with a pixel size of 1 km $\times$ 1 km. To calculate the average daily ammonia concentrations per census block, we used the zonal statistics operation in ArcGIS 10.

*4.2. Assessing Disproportionate Exposure*

To assess disproportionate population exposure to ammonia concentrations, we used techniques of traditional and spatial statistics. First, we calculated the correlation coefficient between ammonia concentrations and each demographic variable for 2000 and 2010. Since North Carolina imposed a CAFO moratorium in 1997–2007, we assumed that the number of CAFOs did not change between 2000 and 2010 and used CALPUFF results for both 2000 and 2010 analyses.

13

ISPRS Int. J. Geo-Inf. **2015**, 4, 150–171

We utilized a Spearman's correlation coefficient in IBM SPSS Statistics [37] to measure the associations between the average $NH_3$ concentrations and the specific sociodemographic characteristics of each census block. Spearman's rank correlation is a nonparametric statistic that first converts the values of the variables to ranks and then calculates the correlations as follows [38]

$$r_s = 1 - \frac{6 \sum D^2}{n^3 - n} \tag{1}$$

where:

$r_s$ = Spearman's rank correlation coefficient;
$D$ = the differences between the rank values for each feature on the two variables;
$n$ = the number of features.

The value of $r_s$ in the result is constrained from 1 (a perfect direct correlation) to −1 (a perfect inverse correlation). The closer $r_s$ is to ±1, the stronger the monotonic relationship, while a $r_s$ near 0 indicates no relationship between the two variables [39].

In order to examine the correlation between ammonia concentrations and sociodemographic variables spatially, a bivariate local indicator of spatial association (LISA) analysis was performed using GeoDa software [40]. Developed by Luc Anselin [41], a local indicator of spatial association, also known as a univariate LISA, tests whether local correlations between values of a feature and values of its neighbors are significantly different from what would be expected from a complete spatial randomization. It identifies significant spatial clusters by involving the cross product between the standardized value of a variable for feature $i$ and that of the average of the neighboring values:

$$I_i = \frac{x_i - \overline{X}}{S_i^2} \sum_{j=1, j \neq i}^{n} w_{i,j} \left( x_j - \overline{X} \right) \tag{2}$$

where:

$x_i$ = an attribute value for feature $i$;
$\overline{X}$ = the mean of the corresponding attribute;
$w_{i,j}$ = the spatial weight between features $i$ and $j$, and:

$$S_i^2 = \frac{\sum_{j=1, j \neq i}^{n} \left( x_j - \overline{X} \right)^2}{n - 1} - \overline{X}^2 \tag{3}$$

$n$ = the total number of features.

As a simple extension of the univariate LISA, the bivariate LISA identifies the extent of spatial clusters by involving the cross product of the standardized values of one variable at location $i$ with that of the average neighboring values of the other variable. The statistical significance of these spatial clusters is evaluated using Monte-Carlo spatial randomization [41].

Bivariate LISA produces four clusters: high-high, high-low, low-high and low-low. In the context of our study, a high-high cluster indicates areas with a higher than average concentration of ammonia surrounded by neighbors with more than the average number of people from a vulnerable population group (children, the elderly, whites or minorities). A high-low cluster indicates areas with a higher than average concentration of ammonia surrounded by neighbors with less than the average number of people from a vulnerable population group. A low-high cluster indicates areas with a lower than average concentration of ammonia surrounded by neighbors with more than the average number of people from a vulnerable population group, and a low-low cluster indicates areas with a lower than average concentration of ammonia surrounded by neighbors with less than the average number of people from a vulnerable population group.

14

To conduct the LISA analysis, a weights matrix is created to conceptualize the spatial relationships. Considering that the areal units are irregular, this study used a distance-based spatial weight matrix, selecting polygons located within a particular distance as neighbors of the target polygon. Since this particular distance is an important parameter for modeling spatial relationships, we selected an appropriate distance threshold with the assistance of the Incremental Spatial Autocorrelation tool in ArcGIS. Specifically, this tool examines spatial autocorrelation at various distances and provides the associated z-scores reflecting the intensity of spatial clustering [42]. The distance associated with the highest z-score is chosen as the threshold distance for the weight matrix. Incremental Spatial Autocorrelation identified 4 kilometers as the distance that resulted in the highest z-score, so we used it as the threshold for the calculation of the spatial weights matrices for both 2000 and 2010.

## 5. Results

The raster grid of modeled ammonia concentrations is shown in Figure 3. The average modeled concentration is 14.773 $\mu g/m^3$, (min = 0.549; max = 540.837 $\mu g/m^3$). The maximum value is observed in areas where multiple CAFOs are located next to each other, mostly in Green County. To validate our model, we compared our results with a study by Wilson and Serre [43], who used passive samplers to measure ammonia in eastern North Carolina. They found that, at sites within 2 km from a hog CAFO, the ammonia concentration averaged 19.872 $\mu g/m^3$, reaching as high as 115.2 $\mu g/m^3$. While these measurements are very similar to our modeled concentrations, it is important to note that we only modeled emissions from swine CAFOs and did not account for other sources of ammonia. According to Battye *et al.* [44], livestock waste accounts for about 80% of ammonia emissions in North Carolina, and other sources include fertilizer application, forests, non-agricultural vegetation and motor vehicles.

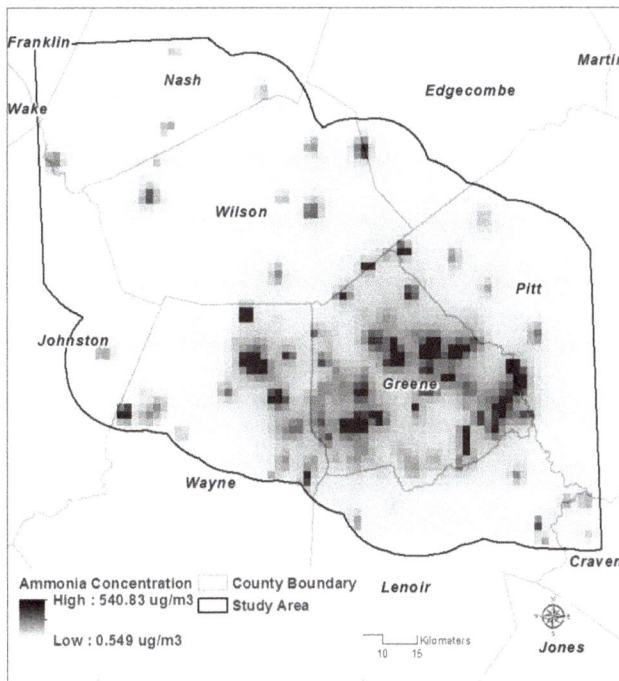

**Figure 3.** CALPUFF output: modeled ammonia concentrations ($\mu g/m^3$).

We also compared our results with concentrations measured at an Ambient Ammonia Monitoring Network (AMoN; http://nadp.sws.uiuc.edu/amon/) site located about 50 km outside our study area (Clinton Crops Research Station; latitude 35.0258; longitude −78.2783). This AMoN site is the closest to our study area. Based on biweekly samples from February 2009 to February 2010, ammonia concentrations at this site ranged from 1.11 to 8.3 $\mu g/m^3$, with a mean concentration of 4.191 $\mu g/m^3$. These measurements are comparable to our modeled concentrations (mean values of 4.191 *vs.* 14.773 $\mu g/m^3$, respectively). Lower values at the monitoring station could be explained by the fact that it is located 1.4 km from the nearest CAFO, while our model predicts concentrations for the entire study area, including areas in the immediate vicinity of CAFOs.

The Spearman's correlation between the average ammonia concentrations and each demographic variable for 2000 and 2010 is shown in Table 3. In 2000, statistically significant, but weak, relationships were identified between the average ammonia concentration and three demographic variables. Specifically, population under 15 years of age and minority population have a significant positive correlation with the ammonia concentration at the 99% confidence level; population over 65 years of age and the ammonia concentration are significantly correlated at the 95% confidence level. However, in 2010 only, minority population had a significant, but weak, positive correlation with the average concentration of ammonia. No significant relationship was found for the other three variables.

**Table 3.** Spearman's rank correlation coefficients (* $p < 0.05$; ** $p < 0.01$).

| Year | Demographic Variable | Ammonia Concentration |
|---|---|---|
| 2000 Census Data | Under 15 years of age | 0.046 ** |
| | Over 65 years of age | 0.038 * |
| | Minority population | 0.162 ** |
| | White population | −0.034 |
| 2010 Census Data | Under 15 years of age | 0.024 |
| | Over 65 years of age | 0.008 |
| | Minority population | 0.104 ** |
| | White population | −0.031 |

Bivariate LISA examined the correlation between the ammonia exposure and demographics and identified significant spatial clusters ($p = 0.05$). For the purposes of this study, we focus our attention on high-high clusters, because they represent areas where high numbers of vulnerable people are exposed to higher than the average level of pollutant concentrations. These high-high clusters are also often referred to as "hot spots". We calculated the average ammonia concentrations in hot spots for 2000 and 2010 and compared them with ammonia concentrations for the whole study area (Table 4). Our calculations show that average ammonia concentrations in hot spots for 2000 and 2010 are 2.5–3-times higher than the average concentration in the entire watershed. Figures 4–7 show the locations of high-high clusters for both years for four vulnerable population groups.

**Table 4.** Statistics for ammonia concentrations ($\mu g/m^3$) in the entire study area and in the 2000 and 2010 hot spots for minorities, whites, children (under 15 years old) and the elderly (over 65 years old).

| Statistics | Entire Study Area | 2000 Hot Spots | | | | 2010 Hot Spots | | | |
|---|---|---|---|---|---|---|---|---|---|
| | | Non-White | White | Under 15 | Over 65 | Non-White | White | Under 15 | Over 65 |
| Min | 0 | 14 | 4 | 14 | 14 | 14 | 6 | 14 | 0 |
| Max | 530 | 316 | 530 | 530 | 138 | 87 | 530 | 530 | 530 |
| Mean | 12 | 37 | 34 | 33 | 30 | 34 | 32 | 32 | 31 |

In both years, high-high clusters indicating a high ammonia concentration and a high population under 15 years of age are mainly located in Wayne County, close to the boundary of Greene County. Several small clusters can also be found in Greene, Lenoir and Pitt counties. In 2000, 148 census blocks were included in high-high clusters; in 2010, that number increased to 160. This change is also reflected in the number of children who lived in these hot spots: it increased by 15 children.

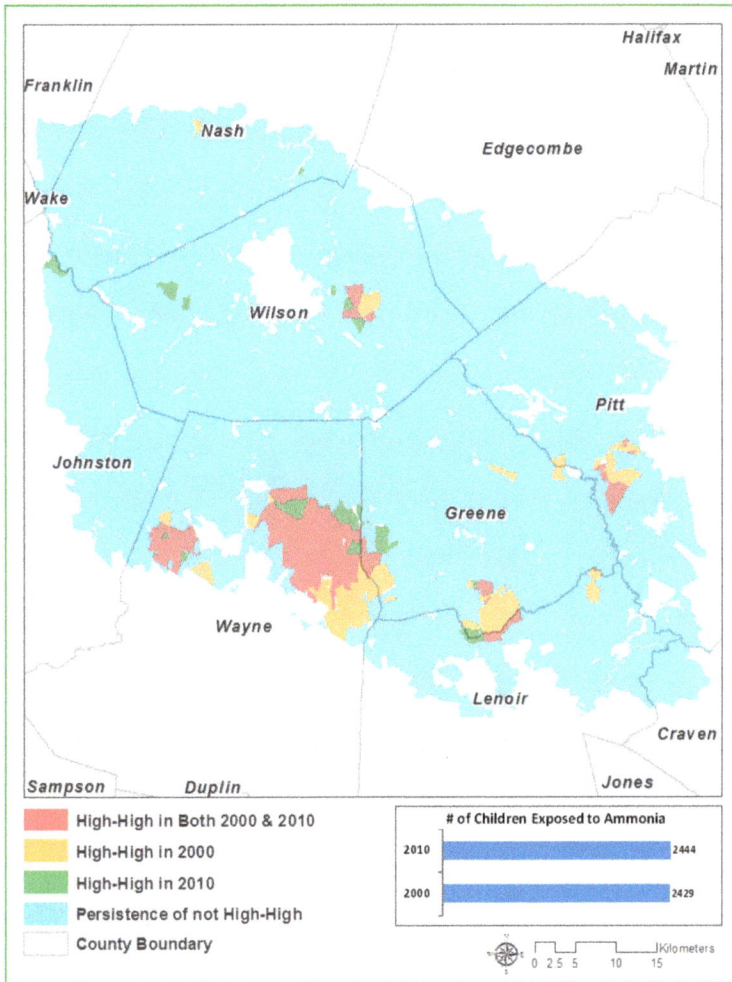

**Figure 4.** Bivariate local indicator of spatial association (LISA) hot spots in 2000 and 2010: average ammonia concentration with population under 15.

When analyzing the spatial association between ammonia concentration and population over 65 years of age, large high-high clusters for both years can be found in Wayne and Greene counties. In 2000, 190 census blocks were included in high-high clusters; in 2010, that number was 189. The number of elderly living in hot spots decreased by 73 people between 2000 and 2010.

A different pattern is observed for minority population, and most of the high-high areas showing persistence in both years are located in Greene County. Wayne and Wilson Counties also contain a small number of high-high clusters in both years. In 2000, 188 census blocks were included in high-high clusters; in 2010, that number decreased to 124. One hundred fewer minority people were living in these hot spots in 2010 as compared to 2000.

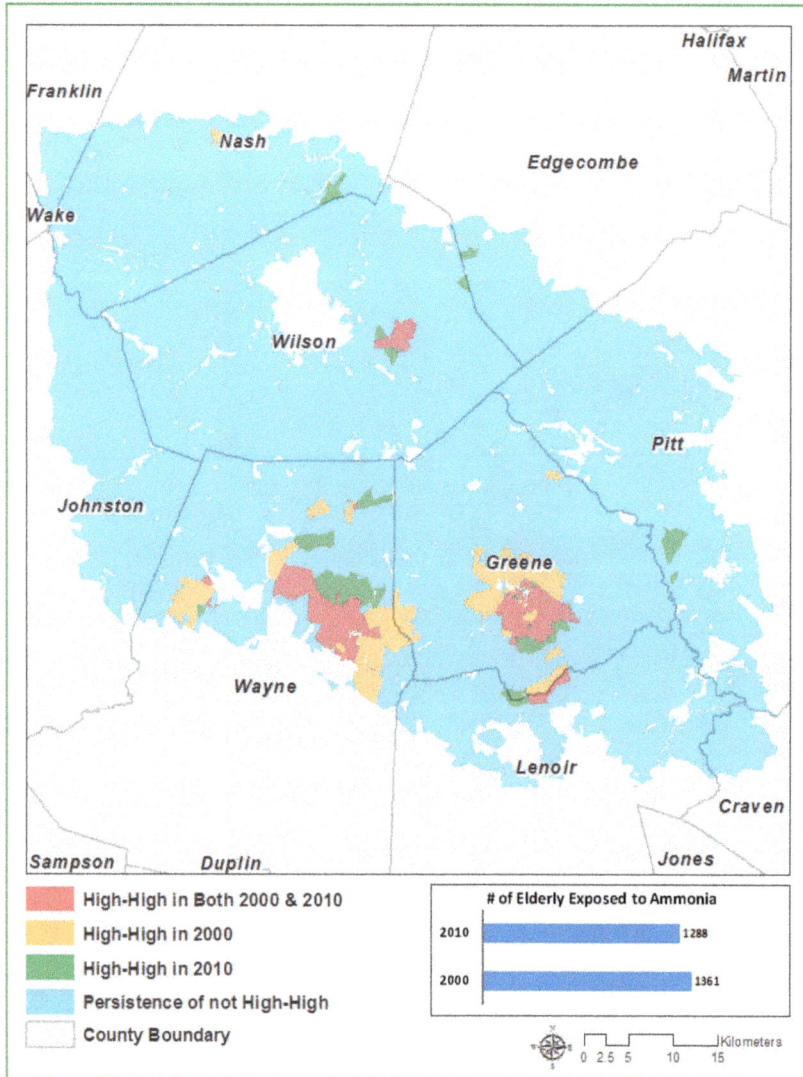

**Figure 5.** Bivariate LISA hot spots in 2000 and 2010: average ammonia concentration with population over 65.

In both years, high-high clusters indicating a high ammonia concentration and high white population are mainly located in Wayne and Pitt Counties. In 2000, 132 census blocks were included in high-high clusters; in 2010, that number increased to 182. The number of white people who lived in these hot spots increased by 934.

18

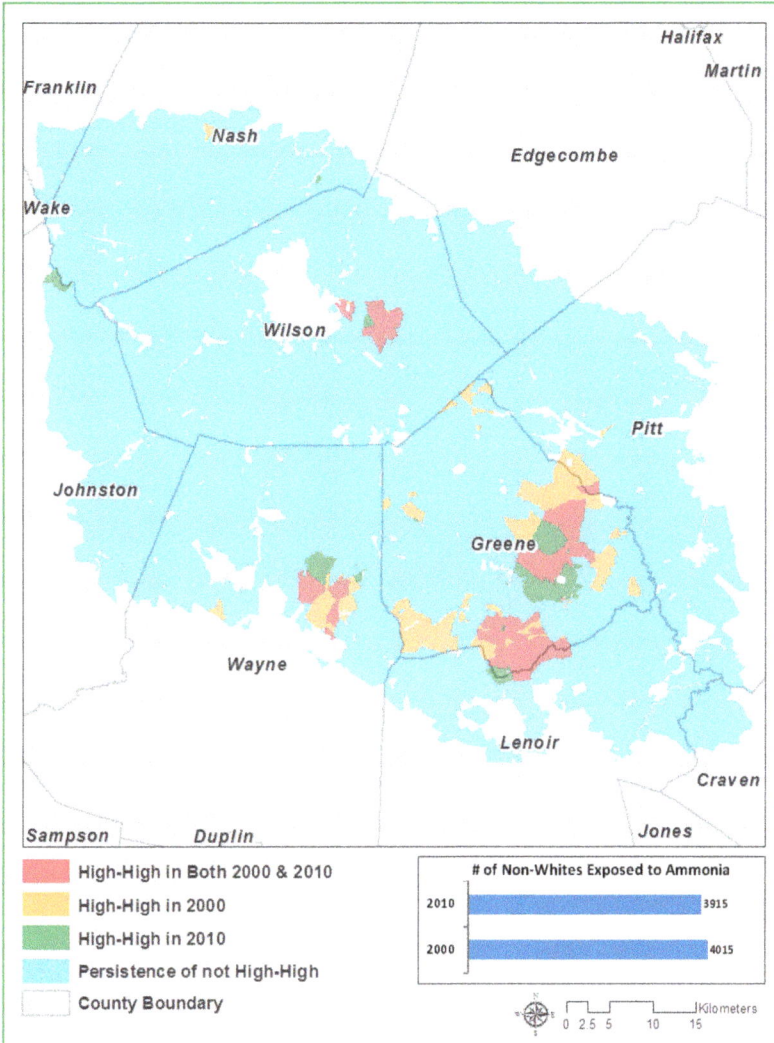

**Figure 6.** Bivariate LISA hot spots in 2000 and 2010: average ammonia concentration *vs.* population of minorities.

To compare the spatial locations of persisting hot spots of four vulnerable populations, we overlaid four maps. Spatial coincidence analysis showed 12 census blocks that belonged to a persisting hot spot in all four maps (Figure 7). These areas are located in Wayne County and represent an extreme case of potential environmental injustice, where disproportionately high numbers of children, elderly, whites and minorities could have been exposed to high ammonia concentrations in 2000 and 2010.

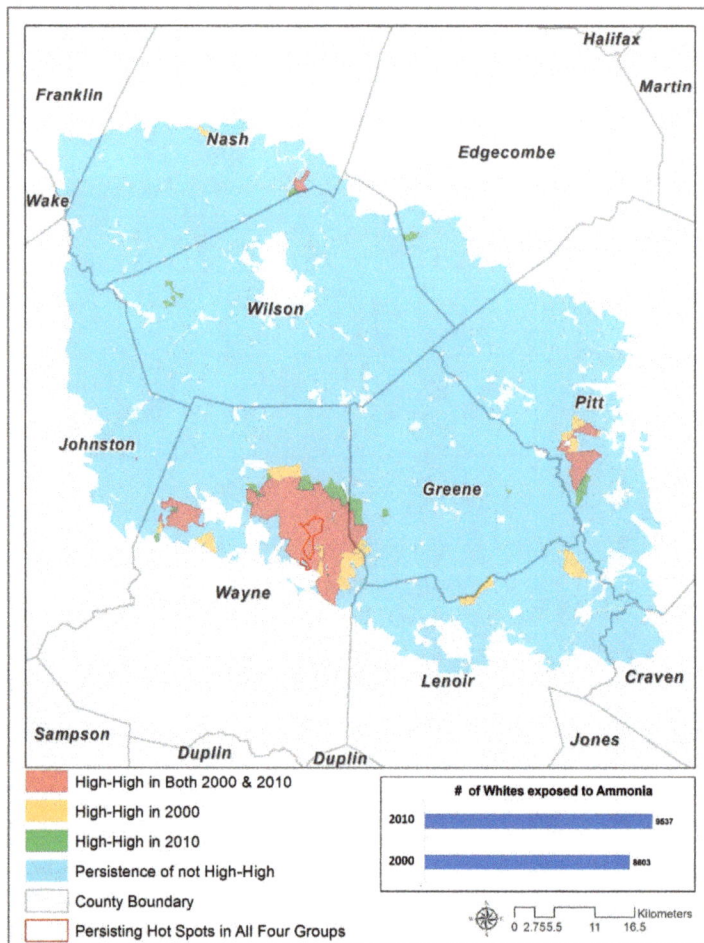

**Figure 7.** Bivariate LISA hot spots in 2000 and 2010: average ammonia concentration *vs.* population of whites and persisting hot spots of all four population groups.

## 6. Discussion and Conclusions

The main research question of this study asked if children, the elderly, white and minority populations are disproportionately exposed to ammonia emitted by CAFOs in Contentnea Creek Watershed in North Carolina. To answer this question, the study used the CALPUFF software to model ammonia emission and dispersion from CAFOs and applied the Spearman correlation and LISA analysis to examine the relationship between ammonia concentrations and demographic characteristics in 2000 and 2010.

The CALPUFF model used very detailed data about meteorological conditions and the characteristics of CAFO facilities (location, dimensions and ammonia emissions per hog or kg of live weight) to produce ammonia emission estimates as a continuous surface with 1-km$^2$ pixels. The fine spatial scale of the modeled ammonia output matched the spatial dimensions of census data very well: the average census block in the study areas is 0.78 km$^2$. Areas with the highest concentration of ammonia were found in Greene County and Wayne County, where the concentration of CAFOs is the highest.

The Spearman's correlation analysis showed that a weak positive relationship existed between average ammonia concentration and some of the demographic variables. Most of the correlation coefficients were statistically significant, probably due to the large sample size (the number of census blocks in each year was over 3000). While these findings indicate that higher numbers of vulnerable people are associated with higher ammonia concentrations in the area, they do not provide any insight into the spatial pattern of these relationships.

Bivariate LISA analysis identified hot spots of environmental injustice: areas with high ammonia concentrations are surrounded by areas with high numbers of vulnerable population. Although the population in three vulnerable groups within hot spots has slightly decreased between the two years, the number of people disproportionately exposed to ammonia concentrations was still large in 2010: 2444 children, 1288 elderly people, 9537 whites and 3915 minorities.

Using spatial overlay in GIS, we identified areas that could have experienced an extreme case of environmental injustice, because they were included in hot spots for all four population groups in both years. A spatial query showed that just within three miles from these areas, there were 12 CAFOs. The results of air pollution modeling suggest that these areas should be prioritized for ambient air quality monitoring.

It is beneficial to discuss these findings in the context of existing air quality regulations. Unfortunately, there is no federal-level standard regarding ammonia, because it is not one of the six criteria air pollutants covered by the Clean Air Act (http://scorecard.goodguide.com/env-releases/def/cap_naaqs.html). EPA requires some CAFOs to report estimated ammonia and hydrogen sulfide emissions based on the number of confined animals. For example, swine CAFOs that have more than 2500 swine, each weighing 55 pounds or more, or 10000 swine, each weighing less than 55 pounds, are required to report their emission estimates (http://www.ecy.wa.gov/epcra/CERCLA_CAFOairexempt.pdf). The Occupational Safety and Health Administration (OSHA) has set an acceptable eight-hour exposure and a short-term (15 min) exposure level for ammonia (http://www.atsdr.cdc.gov/toxfaqs/tf.asp?id=10&tid=2), but these standards mainly concern CAFO workers and are not directly applicable to the general population living in the area. State-level air quality regulations vary, and the majority of states do not have a comprehensive air quality regulatory system. For example, Missouri has an ambient acceptable level of ammonia, and North Carolina and Colorado have regulations concerning odor emissions from CAFOs, but no emission standards for hydrogen sulfide or ammonia [45].

The Agency for Toxic Substances and Disease Registry (ATSDR) includes ammonia in its toxic substances list and provides minimal risk level concentrations for it. According to the ATSDR, the minimal risk level (MRL) "is an estimate of the daily human exposure to a hazardous substance that is likely to be without appreciable risk of adverse, non-cancer health effects over a specified duration of exposure. The information in this MRL serves as a screening tool to help public health professionals decide where to look more closely to evaluate possible risk of adverse health effects from human exposure." (http://www.atsdr.cdc.gov/mrls/index.asp). Ammonia's MRL for chronic exposure (meaning exposure for one year or longer) is set at 0.1 ppm, or 75 $\mu g/m^3$ (http://www.atsdr.cdc.gov/mrls/mrllist.asp#2tag). Applying this threshold to our modeled concentrations, we identified areas where chronic exposure exceeds MRL. Most of these areas overlap with our identified hot spots and correspond to high CAFO density areas in Greene and Wayne Counties. While the spatial extent of these areas is small (2.5% of the watershed), they correspond to densely populated areas with total population ranging from 3647 people in 2000 to 3360 people in 2010. Following ATSDR's suggestion, these areas deserve further attention of public health professionals to examine possible adverse health effects due to chronic exposure to higher than the minimal risk levels of ammonia.

This study contributes to the body of research on CAFOs and environmental justice, because no other study has yet analyzed environmental justice on the basis of unequal exposure to CAFO-related emissions. Previous environmental justice studies of CAFOs used proximity to CAFOs or their density as the proxy for air pollution exposure. Our study is the first one to directly link modeled pollutant concentrations and demographic characteristics. The proposed methodology is also the first one of its kind to analyze CAFOs-related environmental justice at the finest possible spatial scale, the census block. Previous studies conducted their analysis at the county, census track or census block group levels.

Our study also contributes to a broader literature on environmental justice. Traditionally, environmental justice research has analyzed unequal exposure in the context of race and poverty; including age-based vulnerable population groups in the analysis is an advantage of our study. Both children and elderly are recognized as the most vulnerable age groups, but they are rarely included in environmental justice studies.

There are several limitations to this study, mostly associated with the CALPUFF model and data availability. First, to run the CALPUFF model, we needed very detailed meteorological data, and the only available data compatible with CALPUFF was for 2006. Therefore, we assumed that the 2006 meteorological data represents the average meteorological situation and used CALPUFF output based on 2006 to analyze environmental justice in 2000 and 2010. Since the entire year of data was used to model average daily ammonia concentrations, this assumption seems reasonable. The second limitation relates to CALPUFF model parameters (for example, the user's choice for background pollution concentrations, the incorporation of wet and dry deposition or the choice of chemical conversion mechanism). We accepted the default settings within the model, and future studies should assess the sensitivity of modeling results to these parameter settings. The third limitation is related to CAFO data availability, including the type and individual size of manure storage facilities. We used 2010 CAFO data to analyze unequal exposure in 2000 and 2010, assuming that the CAFO size did not change during this time to an extent that it would have an effect on the modeled ammonia concentrations. We also did not have data on the type of facility (breeder *versus* finisher facility) nor the management practices and assumed that the emissions rate is the same for all facilities and remains constant thought the year.

Another limitation of this study is that our findings are only valid at the census block level and cannot be extrapolated to other spatial scales. The reason for that is that the relationships between hazardous facilities and socioeconomic variables may change or become more or less significant when the spatial scale changes [46]. This issue is often referred to as the modifiable area unit problem [47].

Our modeled ammonia concentration was comparable to ammonia concentrations measured in the field by other researchers [43] and by the Ambient Ammonia Monitoring Network. In the future, it would be important to test other atmospheric dispersion models and to compare their results with CALPUFF and field measurement for various pollutants and different geographical regions. These studies should include up-to-date characteristics of polluting facilities, such as individual CAFO operations (e.g., exact size of animal houses and lagoons, number of animals, total animal weight). More studies of this kind (based on pollution dispersion models and using reliable fine-scale demographic data) will allow one to assess the impacts of the CAFOs and to address the concerns regarding the health and quality of life of vulnerable populations.

**Acknowledgments:** The authors would like to thank Joseph S. Scire and Zhong-Xiang Wu, from TRC Atmospheric Studies Group, for their invaluable help with the CALPUFF model and Ksenia Varlyguina for her editorial assistance. We are also grateful to the anonymous reviewers for their insightful comments.

**Author Contributions:** Yelena Ogneva-Himmelberger conceived of and designed the study and drafted the manuscript. Liyao Huang performed census data collection and analyses, contributed to some sections of the manuscript and created maps. Hao Xin collected CAFO data and ran CALPUFF to model ammonia concentrations.

**Conflicts of Interest:** The authors declare no conflict of interest.

## References

1. USDA National Agricultural Statistics Service. Farms, Land in Farms, and Livestock Operations 2010 Summary, 2011 Edition. 2011. Available online: http://usda.mannlib.cornell.edu/usda/nass/FarmLandIn/ /2010s/2011/FarmLandIn-02--11--2011_revision.pdf (accessed on 15 November 2014).
2. Hribar, C.; Schultz, M. *Understanding Concentrated Animal Feeding Operations and Their Impact on Communities*; National Association of Local Boards of Health: Bowling Green, KY, USA, 2010. Available online: http://www.cdc.gov/nceh/ehs/Docs/Understanding_CAFOs_NALBOH.pdf (accessed on 15 November 2014).
3. UCS. *CAFOs Uncovered. The Untold Costs of Confined Animal Feeding Operations*; Union of Concerned Scientists: Cambridge, MA, USA, 2008. Available online: http://www.ucsusa.org/assets/documents/food_and_ agriculture/cafos-uncovered.pdf (accessed on 15 November 2014).
4. Greger, M.; Koneswaran, G. The public health impacts of concentrated animal feeding operations on local communities. *Fam Community Health* **2010**, *33*, 11–20. [CrossRef] [PubMed]
5. Bullers, S. Environmental stressors, perceived control, and health: The case of residents near large-scale hog farms in eastern North Carolina. *Hum Ecol.* **2005**, *33*, 1–16. [CrossRef]
6. Heederik, D.; Sigsgaard, T.; Thorne, P.S.; Kline, J.N.; Avery, R.; Bønløkke, J.H.; Chrischilles, E.A.; Dosman, J.A.; Duchaine, C.; Kirkhorn, S.R.; *et al.* Health effects of airborne exposures from concentrated animal feeding operations. *Environ. Health Perspect.* **2006**, *115*, 298–302. [CrossRef]
7. Rogers, S.; Haines, J.R. *Detecting and Mitigating the Environmental Impact of Fecal Pathogens Originating from Confined Animal Feeding Operations: Review. U.S.*; epa/600/r-06/021 (NNIS pb2006–109780); Environmental Protection Agency: Washington, DC, USA, 2005.
8. EPA. Risk Assessment Evaluation for Concentrated Animal Feeding Operations. 2004. Available online: http://www.epa.gov/nrmrl/pubs/600r04042.html (accessed on 15 November 2014).
9. Nicole, W. CAFOs and environmental justice: The case of North Carolina. *Environ. Health Perspect.* **2013**, *121*, a182–a189. [CrossRef] [PubMed]
10. Schiffman, S.S.; Miller, E.A.; Suggs, M.S.; Graham, B.G. The effect of environmental odors emanating from commercial swine operations on the mood of nearby residents. *Brain Res. Bull.* **1995**, *37*, 369–375. [CrossRef] [PubMed]
11. Tajik, M.; Muhammad, N.; Lowman, A.; Thu, K.; Wing, S.; Grant, G. Impact of odor from industrial hog operations on daily living activities. *New Solut.* **2008**, *18*, 193–205. [CrossRef] [PubMed]
12. Wing, S.; Horton, R.A.; Marshall, S.W.; Thu, K.; Tajik, M.; Schinasi, L.; Schiffman, S.S. Air pollution and odor in communities near industrial swine operations. *Environ. Health Perspect.* **2008**, *116*, 1362–1368. [CrossRef] [PubMed]
13. Horton, R.A.; Wing, S.; Marshall, S.W.; Brownley, K.A. Malodor as a trigger of stress and negative mood in neighbors of industrial hog operations. *Am. J. Public Health* **2009**, *99*, S610–S615. [CrossRef] [PubMed]
14. FWW. *Turning Farms into Factories: How the Concentration of Animal Agriculture Threatens Human Health, the Environment, and Rural Communities*; Food and Water Watch: Washington, DC, USA, 2007. Available online: http://documents.foodandwaterwatch.org/doc/FarmsToFactories.pdf (accessed on 15 November 2014).
15. Edwards, B.; Ladd, A.E. Environmental justice, swine prod uction and farm loss in North Carolina. *Sociol. Spectr.* **2000**, *20*, 263–290. [CrossRef]
16. Horton, J. The Siting of Hog CAFOs in Eastern North Carolina: A Case of Environmental Justice? Ph.D. Thesis, Univeristy of Michigan, Ann Arbor, MI, USA, 2012.
17. Wing, S.; Wolf, S. Intensive livestock operations, health, and quality of life among eastern North Carolina residents. *Environ. Health Perspect.* **2000**, *108*, 233–238. [CrossRef] [PubMed]
18. Mirabelli, M.C.; Wing, S.; Marshall, S.W.; Wilcosky, T.C. Race, poverty, and potential exposure of middle-school students to air emissions from confined swine feeding operations. *Environ. Health Perspect.* **2006**, *114*, 591–596. [CrossRef] [PubMed]
19. Wing, S.; Cole, D.; Grant, G. Environmental injustice in North Carolina's hog industry. *Environ. Health Perspect.* **2000**, *108*, 225–231. [CrossRef] [PubMed]
20. Edwards, B.; Ladd, A.E. Race, poverty, political capacity and the spatial distribution of swine waste in North Carolina, 1982–1997. *North Carolina Geographer.* **2001**, *9*, 55–77.
21. Ladd, A.E.; Edward, B. Corporate swine and capitalist pigs: A decade of environmental injustice and protest in North Carolina. *Soc. Justice* **2002**, *29*, 26–46.

23

22. Wing, S.; Freedman, S.; Band, L. The potential impact of flooding on confined animal feeding operations in eastern North Carolina. *Environ. Health Perspect.* **2002**, *110*, 387–391. [CrossRef] [PubMed]

23. Kleinman, M. The Health Effects of Air Pollution on Children. 2000. Available online: http://www.aqmd.gov/docs/default-source/students/health-effects.pdf?sfvrsn=0 (accessed on 15 November 2014).

24. Bentayeb, M.; Simoni, M.; Baiz, N.; Norback, D.; Baldacci, S.; Maio, S.; Viegi, G.; Annesi-Maesano, I. Adverse respiratory effects of outdoor air pollution in the elderly. *Int. J. Tuberc. Lung Dis.* **2012**, *16*, 1149–1161. [CrossRef] [PubMed]

25. EPA. *Age Healthier, Breathe Easier*; Publication Number EPA 100-F-09–045; EPA: Washington, DC, USA, 2009.

26. Spruill, T.B.; Tesoriero, A.J.; Mew, H.E., Jr.; Farrell, K.M.; Harden, S.L.; Colosimo, A.B.; Kraemer, S.R. *Geochemistry and Characteristics of Nitrogen Transport at a Confined Animal Feeding Operation in a Coastal Plain Agricultural Watershed, and Implications for Nutrient Loading in the Neuse River Basin, North Carolina, 1999–2002*; USGS: Reston, VA, USA, 2005.

27. Deike, J. Waterkeeper Alliance: Factory Farm Swine Operation Violates Clean Water Act. Available online: http://ecowatch.com/2014/03/14/waterkeeper-alliance-factory-farm-swine-operation-violates-clean-water-act (accessed on 15 November 2014).

28. Census Bureau, U.S. State & County Quickfacts—North Carolina. Available online: http://quickfacts.census.gov/qfd/states/37000.html (accessed on 15 November 2014).

29. Bunton, B.; O'Shaughnessy, P.; Fitzsimmons, S.; Gering, J.; Hoff, S.; Lyngbye, M.; Thorne, P.S.; Wasson, J.; Werner, M. Monitoring and modeling of emissions from concentrated animal feeding operations: Overview of methods. *Environ. Health Perspect.* **2007**, *115*, 303–307. [CrossRef] [PubMed]

30. Cajka, J.; Deerhake, M.; Yao, C. Modeling ammonia dispersion from multiple CAFOs using GIS. In Proceedings of the Esri International User Conference, San Diego, CA, USA, 9–13 August 2004.

31. MPCA. *Draft Environmental Impact Statement on the Hancock Pro Pork Feedlot Project, Stevens and Pope Counties*; Minnesota Pollution Control Agency: St. Paul, MN, USA, 2003. Available online: http://www.pca.state.mn.us/index.php/view-document.html?gid=9487 (accessed on 20 January 2015).

32. EPA. *Revision to the Guideline on Air Quality Models: Adoption of a Preferred General Purpose (Flat and Complex Terrain) Dispersion Model and Other Revisions*; Final Rule. 40 CFR Part 51; U.S. Environmental Protection Agency: Washington, DC, USA, 2005. Available online: http://www.epa.gov/ttn/scram/guidance/guide/appw_05.pdf (accessed on 15 November 2014).

33. Scire, J.S.; Strimaitis, D.G.; Yamartino, R.J. *A User's Guide for the Calpuff Dispersion Model (Version 5)*; Earth Tech: Concord, MA, USA, 2000.

34. The State Climate Office of North Carolina. Available online: http://www.nc-climate.ncsu.edu/cronos (accessed on 15 November 2014).

35. Aneja, V.P.; Arya, S.P.; Kim, D.S.; Rumsey, I.C.; Arkinson, H.L.; Semunegus, H.; Bajwa, K.S.; Dickey, D.A.; Stefanski, L.A.; Todd, L.; *et al.* Characterizing ammonia emissions from swine farms in eastern North Carolina: Part 1—Conventional lagoon and spray technology for waste treatment. *J. Air Waste Manag. Assoc.* **2008**, *58*, 1130–1144.

36. Doorn, M.R.J.; Natschke, D.F.; Meeuwissen, P.C. Review of Emission Factors and Methodologies to Estimate Ammonia Emissions from Animal Waste Handling. Available online: http://www.epa.gov/nrmrl/pubs/600sr02017.html (accessed on 15 November 2014).

37. Norusis, M. *Spss Statistics 17.0 Guide to Data Analysis*; Prentice Hall: Upper Saddle River, NJ, USA, 2008.

38. Mitchell, A. *The ESRI Guide to GIS Analysis: Spatial Measurements and Statistics*; ESRI Press: Redlands, CA, USA, 2005; Volume 2.

39. Ebdon, D. *Statistics in Geography*; Wiley-Blackwell: Hoboken, NJ, USA, 1985.

40. Anselin, L.; Syabri, I.; Kho, Y. Geoda: An introduction to spatial data analysis. *Geogr. Anal.* **2006**, *38*, 5–22. [CrossRef]

41. Anselin, L. Local indicators of spatial association—LISA. *Geogr. Anal.* **1995**, *27*, 93–115. [CrossRef]

42. ESRI. *ArcGIS 9.2 Desktop Help*; ESRI: Redlands, CA, USA, 2008. Available online: http://webhelp.esri.com/arcgisdesktop/9.2/index.cfm?TopicName=welcome (accessed on 15 November 2014).

43. Wilson, S.M.; Serre, M.L. Use of passive samplers to measure atmospheric ammonia levels in a high-density industrial hog farm area of eastern North Carolina. *Atmos. Environ.* **2007**, *41*, 6074–6086. [CrossRef]

44. Battye, W.; Aneja, V.P.; Roelle, P.A. Evaluation and improvement of ammonia emissions inventories. *Atmos. Environ.* **2003**, *37*, 3873–3883. [CrossRef]

45. Copeland, C. Air quality issues and animal agriculture: A primer. Congressional Research Service. 2014. Available online: http://nationalaglawcenter.org/wp-content/uploads/assets/crs/RL32948.pdf (accessed on 15 November 2014).

46. Sheppard, E.; Leitner, H.; McMaster, R.B.; Tian, H. GIS-based measures of environmental equity: Exploring their sensitivity and significance. *J. Expo. Anal. Environ. Epidemiol.* **1999**, *9*, 18–28. [CrossRef] [PubMed]

47. Wong, D.S. The modifiable areal unit problem (MAUP). In *Worldminds: Geographical Perspectives on 100 Problems*; Janelle, D., Warf, B., Hansen, K., Eds.; Springer: Dordrecht, The Netherlands, 2004; pp. 571–575.

isprs International Journal of
Geo-Information

MDPI

*Article*

# Analyzing the Correlation between Deer Habitat and the Component of the Risk for Lyme Disease in Eastern Ontario, Canada: A GIS-Based Approach

Dongmei Chen [1,*], Haydi Wong [1], Paul Belanger [1,2], Kieran Moore [2], Mary Peterson [2] and John Cunningham [3]

[1]  Department of Geography, Queen's University, Kingston, ON K7L 3N6, Canada;
     chendm@queensu.ca (D.C.); 7hyhw@queensu.ca (H.W.)
[2]  KFL & A Public Health, Kingston, ON K7M 1V5, Canada; Paul.Belanger@kflapublichealth.ca (P.B.);
     kieran.moore@kflapublichealth.ca (K.M.); Mary.Peterson@kflapublichealth.ca (M.P.)
[3]  Leeds, Grenville & Lanark District Health Unit, Brockville, ON K6V 7A3, Canada;
     John.Cunningham@healthunit.org
*   Author to whom correspondence should be addressed; chendm@queensu.ca.

Academic Editors: Fazlay S. Faruque and Wolfgang Kainz
Received: 7 August 2014; Accepted: 7 January 2015; Published: 15 January 2015

**Abstract:** Lyme borreliosis, caused by the bacterium, *Borrelia burgdorferi*, is an emerging vector-borne infectious disease in Canada. According to the Public Health Agency of Canada (PHAC), by the year 2020, 80% of Canadians will live in Lyme endemic areas. An understanding of the association of *Ixodes scapularis*, the main vector of Lyme disease, with it hosts is a fundamental component in assessing changes in the spatial distribution of human risk for Lyme disease. Through the application of Geographic Information System (GIS) mapping methods and spatial analysis techniques, this study examines the population dynamics of the black-legged Lyme tick and its primary host, the white-tailed deer, in eastern Ontario, Canada. By developing a habitat suitability model through a GIS-based multi-criteria decision making (MCDM) analysis, the relationship of the deer habitat suitability map was generated and the results were compared with deer harvest data. Tick submission data collected from two public health units between 2006 and 2012 were used to explore the relationship between endemic ticks and deer habitat suitability in eastern Ontario. The positive correlation demonstrated between the deer habitat suitability model and deer harvest data allows us to further analyze the association between deer habitat and black-legged ticks in our study area. Our results revealed that the high tick submission number corresponds with the high suitability. These results are useful for developing management strategies that aim to prevent Lyme from becoming a threat to public health in Canada. Further studies are required to investigate how tick survival, behaviour and seasonal activity may change with projected climate change.

**Keywords:** GIS; Lyme disease; habitat suitability; multi-criteria decision making

## 1. Introduction

The incidence of Lyme disease-carrying ticks has been increasing across different regions of North America over recent years. Research studies have shown that the distribution and abundance of tick populations has increased and migrated northwards from endemic areas in the United States towards new regions in southern Canada [1]. At present, endemic tick populations are found in various regions across Canada, including Nova Scotia, New Brunswick, Quebec, Ontario and southeastern Manitoba [2,3]. In 1980, the first reported endemic tick population in Ontario occurred at Long Point, Ontario [4]. More recently, populations of ticks infected with *Borrelia burgdorferi* (*B. burgdorferi*) were discovered in the eastern Ontario area. The first endemic population was found on Thwartway Island of the Thousand Islands in 2006 [5]. Between 2002 and 2008, there have been a total of 37 human Lyme disease cases identified within the health unit boundaries of Kingston, Frontenac, Lennox and Addington Public Health and the Leeds, Grenville and Lanark Health Unit.

The black-legged tick has three instars, consisting of larvae, nymph and adult stages, which all feed on the animal hosts. The ticks will fall off their host when fully engorged and will then progress into the next developmental stage. Ticks acquire *B. burgdorferi* when feeding on infected host species and will then infect any subsequent animal upon which they feed [6]. Due to their small size, ixodid ticks have very limited capacity for moving into new territory, and as a result, long-distance dispersal of the tick is typically attributed to a wide range of hosts, including rodents, white-tailed deer and migratory birds [7].

While birds migrating northwards in the spring can carry ticks long distances into Canada [8], research studies have suggested that the primary host of tick populations in Canada is the *Odocoileus virginianus*, or more commonly known as the white-tailed deer [9]. In Canada, the geographic range of the white-tailed deer is quite extensive, as populations of the deer had been identified in the east coast of British Columbia to the Maritime Provinces in the west and extending from the northern USA border to southeastern Canada [1]. The white-tailed deer play a critical role in the reproduction and life cycle of the tick and is often considered to be an important host to all tick stages. During the fall and early winter season, adult ticks will feed and mate on the white-tailed deer before dropping onto the ground to lay eggs in the following spring [10]. Since transovarial transmission of generalized infection to filial ticks is highly unlikely, larvae that hatch during midsummer are free from *B. burgdorferi* infection [11]. During the larvae blood meal, ticks will feed on deer and rodent hosts, but studies have found that white-footed mice are primarily responsible for infecting ticks with *B. burgdorferi* at this stage of the tick life cycle. Larvae will then molt into nymphs and then into adults in the following months, which will then seek a deer host on which to feed. The reproduction cycle begins again as these adult ticks mate on their deer host during the autumn season [10]. Thus, the location of the deer is an important determinate of the location of egg-laying adults during this time of the year and, hence, where larvae occur during their blood meal in the summer.

Recent studies have demonstrated that white-tailed deer population density is positively correlated with tick abundance in several regions across North America [12]. In 2001, a study analyzing *Ixodes* ticks found that the emergence of Lyme disease has been associated with land use changes that increased the density of the white-tailed deer [13]. Another study that conducted an experimental addition of acorn in eastern U.S. oak forests, a highly preferred habitat of the white-tailed deer, found a series of chain events linking tick populations and tick hosts. By attracting deer into the oak forests through the addition of acorns, their research showed that the amount of time that the deer spent feeding on acorns was longer during the autumn season and that the densities of black-legged ticks were also higher, suggesting greater Lyme disease risk [10]. Furthermore, researchers analyzing the relationship between tick stages and hosts demonstrated that over 95% of adult female ticks feed on the white-tailed deer. While ticks at the larvae and nymphal stage rely primarily on smaller hosts, such as rodents or birds, the study suggested that nymphal ticks will also feed on the deer whenever available [14]. In New England, field studies investigating the ecology of *Ixodes scapularis* (*I. scapularis*) quickly revealed the ticks' dependence on the white-tailed deer, with the deer feeding mostly on ticks at the adult stage [15]. White-tailed deer are considered incompetent reservoirs of *B. burgdorferi*, as deer serum contains a borreliacidial component, which is an antibody against the manifestations of Lyme disease [16]. Hence, the development of white-tailed deer populations in eastern North America over the past century may be responsible for the maintenance and establishment of tick populations. However, no similar studies have been conducted in Canada.

Global climate change has caused much speculation with the potential of causing serious effects on the future spatial and temporal distribution of vector-borne diseases. Since climate plays an important role in the survival of vectors and pathogens, studies have predicted that climate change may open up previously uninhabitable territory for vectors, increase reproduction rates and shorten pathogen incubation periods [17]. Furthermore, analyses were conducted using statistical models to explain the northern expansion of the tick's geographic range with climate change [18]. While many previous studies have analyzed the impact of climate change on the northwards expansion of endemic tick populations in the U.S. or Canada [1,19–21], few research studies have explored the relationship of the black-legged ticks and its primary host in Canada, as the impact of the white-tailed deer on the spread of *B. burgdorferi* at the local scale is not entirely understood. The use of passive surveillance to monitor the spatial and temporal occurrence of ticks and their infection with *B. burgdorferi* tick populations has occurred in Canada as early as the 1990s. Ticks that were found by the public attached to themselves or their pets were submitted either directly or via physicians and veterinarians to provincial and federal health organizations and Canadian universities [22]. To date, there have been a limited number of research studies focusing on the occurrence of tick populations in the eastern Ontario region at the local area level using passive surveillance.

The integration of multi-criteria decision making (MCDM) with Geographic Information Systems (GIS) and spatial analysis had been used in various applications as a practical method to solve complex multi-faceted environmental problems. MCDM is a powerful technique, as it enables decision makers to evaluate relative priorities by developing a framework based on a set of preferences or criteria combined into a single model. Since the 1990s, the use of MCDM and GIS had been applied in various fields and was shown in studies related to urban planning, land use allocation, forest conservation and site determination [23]. A GIS-based MCDM model was proven to be a useful tool in developing habitat suitability models to predict species presence and to support conservation planning [23–25]. Since evaluating habitat suitability often requires the use of data from different sources and at different spatial scales, an MCDM approach will allow us to find the best combination of habitat features based on the environmental preferences of the white-tailed deer, which will then be used to predict species spatial distribution across our study area.

The objective of this study is to determine the applicability of a deer habitat suitability model used as a proxy for determining black-legged tick distribution in eastern Ontario. A habitat suitability model of the white-tailed deer using a MCDM- and GIS-weighted sum analysis is developed to explore the relationship between deer habitat suitability and endemic tick populations in eastern Ontario. The results from habitat suitability are compared with deer harvest data. Through the application of GIS methods and spatial analysis, we will strive to gain greater insight for the first time into the spatial distribution of endemic tick populations at the dissemination area (DA) level and the potential role of white-tailed deer in the spatial expansion of Lyme ticks. We also aim to determine the relationship between the tick *B. burgdorferi* bacterium and deer establishment, while predicting future trends, which can be used to guide local health in developing disease prevention strategies and generating greater public awareness.

## 2. Materials and Methods

### 2.1. Study Area

The study of the deer habitat was conducted in the eastern Ontario region at the wildlife management unit (WMU) geographic level, which is an administrative coverage boundary developed by the Ontario Ministry of Natural Resources (MNR) based on a number of environmental requirements of wildlife species. In Ontario, there are a total of 95 WMUs. Of the 95, our study area consisted of 35 units in the southeastern Ontario region and includes the following units: 48, 50, 51, 53A, 54, 55A, 55B, 56, 57, 58, 59, 60, 61, 62, 63A, 63B, 64A, 64B, 65, 66A, 66B, 67, 68A, 68B, 69A, 69B, 70, 71, 72A, 73, 74A, 74B, 75, 76A and 78A. Figure 1 shows the extent of our deer study area, with the label representing each WMU.

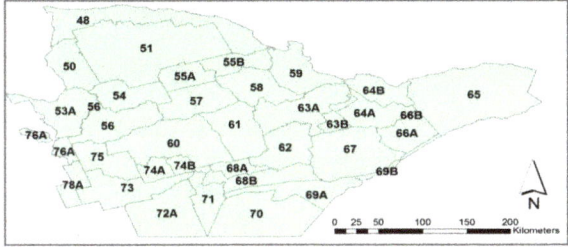

**Figure 1.** Wildlife management units in southeastern Ontario.

The tick-validation area is based on the boundaries of two public health units located in eastern Ontario, consisting of the Kingston, Frontenac, Lexington and Addington Public Health (KFL&A) and the Leeds, Grenville and Lanark District Health Unit (LGL). Since the study area of the tick is conducted within the southeastern region of the deer study area (Figure 1), Figure 2 outlines the extent of the tick-validation study area relative to the deer study area. As shown in Figure 2, Image A displays the study area of the deer within the province of Ontario, Image B shows the boundaries of the two public health units with respect to the deer study area and Image C shows the dissemination area (DA) boundaries of the two public health units.

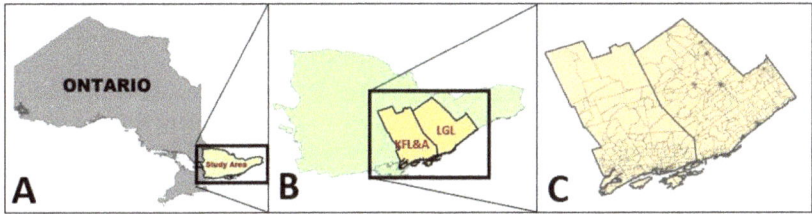

**Figure 2.** Extent of the tick validation study area. (**A**) The study area of the deer within the province of Ontario; (**B**) the boundaries of the two public health units with respect to the deer study area; and (**C**) the dissemination area (DA) boundaries of the two public health units. KFL&A, Kingston, Frontenac, Lexington and Addington Public Health; LGL, Leeds, Grenville and Lanark District Health Unit.

## 2.2. Data Collection

### 2.2.1. Tick Data

Tick data collected through passive surveillance from KFL&A and LGL range from January 2006, to December 2012. The parasitical specimens attached on humans and found by the public have been submitted to local health units. These tick data were then sent to provincial health laboratories of the Public Health Agency of Canada, where the identification of tick species (*I. scapularis, Amblyomma americanum, Dermacentor variabilis, Ixodes cookei, Ixodes marxi, Ixodes muris*) and of the bacteria, *B. burgdorferi*, was conducted. Tick data that were collected and that did not provide an acquisition location at the postal code level were eliminated from the study. Due to the confidentiality of the tick submitters' identities and personal health information, tick acquisition locations acquired at the six-digital postal code level were geocoded and converted into the DA unit level in our analysis for anonymous participation. There were a total of 3474 tick submissions collected from KFL&A and LGL via passive surveillance from 2006 to 2012, in which 1570 of them were acquired within our study area. The remainder was ticks acquired either outside the study area health unit boundaries or tick acquisitions without the complete 6-digit postal code information provided. Of the 1570 ticks, 1241 within the study area were identified as the *I. scapularis* species, in which 231 were positive for *B. burgdorferi*.

### 2.2.2. Deer Harvest Data

Deer harvest data for 2008 to 2011 were obtained from the Ontario Ministry of Natural Resources (MNR), which manages Ontario's ecosystems and biodiversity. Deer harvest numbers were obtained from deer surveys received from hunters following the hunting season. The harvest data used for this study were projected based on raw numbers of deer killed from a sample of hunters to more accurately represent the number of deer harvested. The extrapolation rate used to determine the projected deer harvest number is equal to the total number of hunters in the WMU divided by the total number of valid replies received from the hunter. Under the current management system, the hunting administration prescribes the annual harvest for each hunting unit through a deer validation tag program. MNR's harvest plan specifies the number of white-tailed deer to be harvested for each sex and outlines the specific time and dates during the year for hunting.

### 2.2.3. Deer Habitat Data

The data collected for the deer habitat suitability analysis ranged from GIS data files to remote sensing imagery. For deer habitat analysis, factors from two main categories are considered: (1) the need for food, shelter and water; and (2) the disturbances from human activities. The data on food, shelter and water were extracted from land cover, vegetation type, terrain slope and proximity to water bodies. The data on disturbances from human activities were extracted from the distance to the roads and urban areas, as well as landscape segmentation measured by land cover diversity. The regional land use/cover data generated by the Earth Observation for Sustainable Development (EOSD) project was downloaded from the GeoBase Portal [26] and mosaicked for the study area. This EOSD data was generated by classifying the multispectral Landsat 5 and 7 TM ortho-images acquired from 1999 to 2001 with 30-m spatial resolution. Based on a recent change detection analysis, the land use/cover in the study area is quite stable, with a change of less than 5% within the study area from 2001 to 2011. Therefore, we updated the land use/cover layer from the EOSD data in the detected change area. The final land use/cover map used for this study is shown in Figure 3. It can be seen that the main land use types are forest and agriculture, which use about 60% and 39% of the study area, respectively. The 2011 MODIS land cover classification data were also downloaded from NASA's Earth Observation System, and the Data and Information System was used as an index for food source variables, due to its information on vegetation types. The digital elevation model (DEM) used for the slope variable was obtained from the Government of Canada's Centre for Topographic Information. Obtained in raster format, the DEM has a spatial resolution of a minimum of 3 arc seconds to a maximum of 12 arc seconds. GIS data files collected for the urban areas, water regions and road network environmental variables were obtained from Statistics Canada based on the year 2011 in vector digital data format. The 2012 Ontario road network was acquired from DMTI Spatial Inc. in vector digital data format and was used for the distance to major roads and secondary road habitat variables.

### 2.3. Procedures

#### 2.3.1. Multi-Criterial Decision Making Model for Deer Habitat

Seven habitat variables generated from the data listed in Section 2.2.3 were imported as individual layers into ArcMap and converted into a raster dataset. Multi-criteria decision making (MCDM) was used to reclassify and assign numeric values to each of the eight factors or criteria, and a weighted sum analysis was applied to produce the final habitat suitability maps. Figure 4 illustrates the variables and steps used in MCDM.

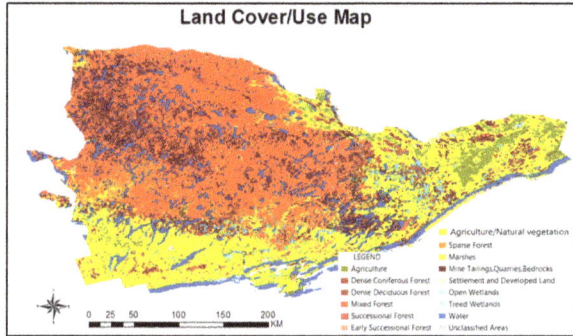

**Figure 3.** The land use/cover map of this study generated from 30-m TM imagery.

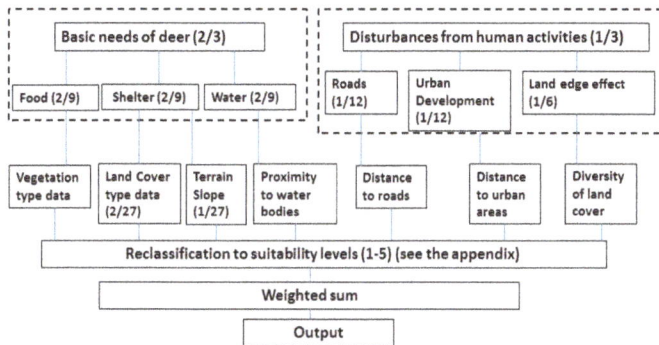

**Figure 4.** The factors, variables and processing steps used in multi-criteria decision making (MCDM) deer habitat suitability analysis (the numbers in () behind the factors are the weights assigned based on the rank sum method).

A classification scheme for each of the following 7 variables was developed (see the Appendix A). The data within each criterion's layer was then reclassified by assigning a numeric value within the range of 1 to 5 based on whether it would be favorable or unfavorable for white-tailed deer habitat. A reclass value of 5 represents conditions most suitable, while the lower end of the scale, with a value of 1, represents the least suitable conditions for deer habitat.

The white-tailed deer is a herbivore species that consumes a variety of grasses and plants. Studies have shown that the white-tailed deer generally confine themselves to woody riparian forest vegetation and shrub cover for foraging during the day time [27]. Examples of some of the vegetation species include shrubs, herbs, grass, fruits and fungi. The main diets are dominated by grasses in spring, flowering herbs in early summer, leaves of woody plants in late summer, acorns and other fruit in fall and evergreen woody shrubs and other woody twigs/buds in winter. Agricultural crops are also commonly consumed. As a result, these specific land cover types or "preferred feeding sites" were then assigned relatively high "new values" in our food source reclassification scheme. For instance, grasses/crops and shrubs were assigned "new values" of 5, whereas forested areas (e.g., evergreen and deciduous) were assigned lower values.

With respect to the shelter need, research studies suggest that the white-tailed deer tend to thrive in young forests and often benefit from recent disturbances in the forest, such as wildfire or forestry operations [28]. This species prefers conifer habitats, as these areas provide shelter during the winter when snow deepens and temperatures drop. Tree stands provide cover from falling snow and help moderate extreme temperatures [28]. Studies have also suggested that conifer canopies and flat bottomlands can benefit the deer with more radiation flux, little or no wind and slightly warmer temperatures, even with cold air drainage [29]. Therefore, the small slope terrain receives higher suitability values than the terrain with deep slopes.

Deer need to get access to water every day within their home range. The deer home range is small, although they can travel a long distance. Previous studies in New Jersey showed that 68% of deer had a home range of one mile or less, 27% ranged from 1 to 8 miles and only 5% of deer would move over 10 miles [30]. From a collar tracking study, it was found that deer usually remain within 1 to 1.5 miles of permanent water, and the maximum distance from a permanent water source was 2.4 miles [31]. Based on the distance to the permanent water bodies, distances beyond 8 miles were assigned a low score of 1, and distances within 1 mile were assigned a score of 5.

Human activities impact deer movement and habitat [27,31]. Deer reactions to human disturbances (such as sighting and observing humans) vary from a short run to a movement of 2 to 3.5 miles away. The human hunting activities have an obvious impact on deer movements and activities [31]. Therefore, the distance to the roads and urban areas is used to represent the potential human sighting and hunting impact on deer habitat. Distances farther away from roads and urban areas are considered more preferable to closer proximity. As a result, a suitability score of 5 would be assigned to all distances beyond a 5-km (3.1 miles) (average one-hour walking distance by humans (http://www.princeton.edu/~achaney/tmve/wiki100k/docs/Walking.html)) radius from roads and urban areas. In contrast, distances within a 0- to 1-km radius from major roads were assigned a low value score of 1. This is because closer proximity to roads and urban areas represents an increased probability of mortality and human anthropogenic disturbance.

Landscape edges caused by human activities, such as farming, timber cutting, road and urban construction, change the structure and the connectivity of vegetation patches, thus they can impact the deer habitat. White-tailed deer usually prefer habitats with more than one vegetation type and move between open canopy vegetation and forests [27,31]. Deer usually benefit from the "edge effect", since their main diet, grass and shrubs, are on the edges of forested areas [32]. An open edge also allows deer to move more easily. An area containing a diversity of plants is usually a better deer habitat than ones with a single vegetation type [33]. The edge length and diversity of land cover were calculated for each cell using a neighborhood of 1 mile (the home range for 68% of deer). The higher the diversity, the better suitability value that is assigned.

After all variables are recoded into a suitability value, a ranking method was used to assign a weighting scheme and to combine all variables together. With the rank comparison weighting, the importance of each factor with regards to one another is considered [34]. The relative importance of pairs of factors in two categories (deer needs and human disturbances) is assessed and ranked first. It is obvious that the food, shelter and water needs are much more important than human disturbances. Based on the rank sum method, the weight assigned to the factors of food, shelter and water is 2/3, and the factors of human disturbances take a total weight of 1/3. For human disturbance factors, studies have suggested that deer prefer habitats with more edges with a diversity of vegetation [32], so the edge/diversity variable was assigned a half weight of the total weight of the human disturbances. The distance to roads and urban areas is related to human sighting/hunting, and each takes half of the remaining weight for the human disturbance factors. For the factors in the category of basic needs of the deer, the needs of food, shelter and water were treated as equally important, due to the challenge of ranking them. For two variables under the shelter need, it is obvious that terrain slope is less important compared with land cover [31]. Therefore, the land cover type takes 2/3 of the weight assigned to shelter factors.

Since different weighting schemes would impact the final results of the suitability analysis, two other weight methods were also tested. One is equal weighting, in which all factors are ranked as equally important. The other one is weight assessment through the expected value method. Instead of assigning a 2/3 weight to the factors of deer needs, a 0.75 weight was assigned. The final weights for the factors from three weighting schemes are listed in Table 1 below.

**Table 1.** Weightings assigned to each variable in three tested weighting schemes.

| Variables | Equal Weights | Weights from Expected Value Method | Weights from Rank Sum Method |
|---|---|---|---|
| Vegetation (food) | 1/7 | 1/4 | 2/9 |
| Land cover (shelter) | 1/7 | 3/16 | 2/27 |
| Terrain slope (shelter) | 1/7 | 1/16 | 1/27 |
| Proximity to water | 1/7 | 1/4 | 2/9 |
| Distance to roads | 1/7 | 1/16 | 1/12 |
| Distance to urban areas | 1/7 | 1/16 | 1/12 |
| Diversity of land cover | 1/7 | 1/8 | 1/6 |

The suitability values from three different weighting schemes are significantly correlated. The correlation coefficient between the equal weights and the weighting from rank sum methods is 0.91, while the correlation coefficient between the results from weights obtained by the expected value method and rank sum method is 0.96. Therefore, only the results from the rank sum method are used in the following.

### 2.3.2. Deer Habitat Suitability and Deer Harvest Data Analysis

Once the deer habitat suitability map was developed, the results were compared with deer harvest data to evaluate whether it is appropriate to use the suitability values from the habitat model to represent the deer abundance for each WMU. By using the zonal statistics tool in ArcMap, a mean suitability value for each of the 35 WMU within the study area was generated. The zonal statistic tool calculates a statistic for each zone, based on the raster values from the habitat suitability map. In this case, the statistic is the average suitability value, and the zones are the WMUs. WMU 55 was eliminated from this analysis, as this WMU represents the Algonquin Provincial Park, an area where hunting is not permitted in Ontario.

Since the number of deer harvested is dependent on the accessibility of hunters within each WMU, the total deer harvest numbers for each WMU were adjusted based on road accessibility. In this study, the total length of major roads was used as the measurement of hunter accessibility. For example, WMU 65 (Ottawa region) had a projected deer harvest of 23,556 and a road length of 2862 km, while WMU 69B had a harvest of 477 and a road length of 171 km. By adjusting the deer harvest data, any biases associated with WMU size or the population of hunters would be eliminated. The correlation between the mean suitability value and the accessibility-adjusted number of deer was analyzed.

### 2.3.3. Deer Habitat Suitability and Tick Data Analysis

In order to test whether the tick abundance is related to the white deer abundance, a correlation analysis was expected to be conducted to analyze the relationship between the tick and the white-tailed deer abundance. However, we do not have direct information on deer and tick abundance in Ontario. Since the result from deer habitat modeling and adjusted deer harvest data suggests that there is a strong positive correlation between our deer habitat suitability model and the deer harvest data, the suitability values from the deer habitat model were used to predict the preferred location and potential density of white-tailed deer. The zonal statistics tool in ArcMap was used to generate a mean habitat suitability value for each dissemination area, the geographic level at which our ticks were analyzed and compared with the number of tick acquisitions and ticks positive for *B. burgdorferi*.

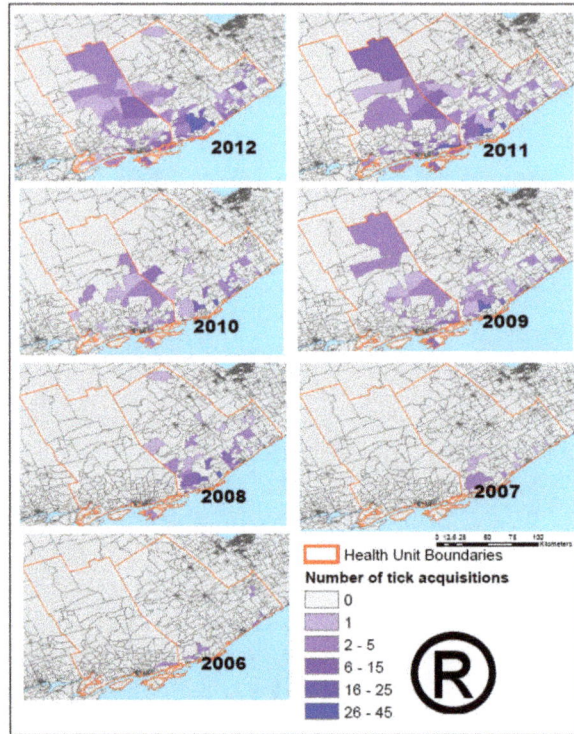

**Figure 5.** Distribution of Tick Acquisition by year from 2006 to 2012.

## 3. Results and Discussions

Figure 5 maps the distribution of the total tick number and the total numbers of positive tick acquisitions each year. As shown in the time series maps, it is clear that there has been an increase from 2006 with an expansion of tick acquisitions and positive ticks carrying the *B. burgdorferi* bacterium northwards from the U.S.-Canadian border. In 2006, the highest number of tick acquisitions for the year at a DA was two, increasing to 44 per year by 2012. A very similar trend was also identified in the number of positive tick acquisitions across the study area. The total number of tick acquisitions positive for *B. burgdorferi* per year almost quadrupled between 2010 and 2011 (Figure 5). From these results, we can infer that there has been an overall increase in the spatial distribution of the tick population in eastern Ontario.

In 2010, a passive surveillance study on Lyme disease risk in Quebec suggested that the number of ticks submited has increased each year from 2004 onward, with much of the increase submitted from regions near the U.S. border [35,36]. This is consistent with our findings in that the increase in the number of submitted ticks has occurred geographically coincident to the U.S.-Canadian border and spreading northwards over time in eastern Ontario.

Figure 6 shows the results of the habitat suitability from the weighted scheme, while Table 2 lists the results of the correlation coefficient among deer habitat suitability, adjusted deer harvest data, the number of tick submissions and ticks positive for *B. burgdorferi*. The deer habitat suitability map suggested that locations associated with high suitability scores were focused towards the central and northern regions of the study area. In contrast, the suitability map revealed that regions with low suitability scores are situated towards the southern border of the study area, where human activity and anthropogenic impacts are the highest.

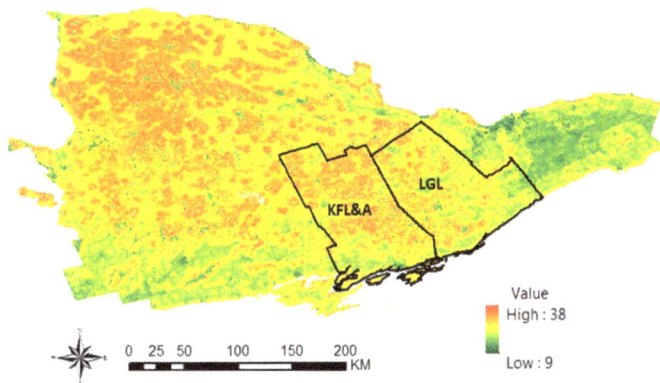

**Figure 6.** Map of habitat suitability analysis results from the rank sum weighting method.

**Table 2.** Correlation coefficients between variables tested in this study. WMU, wildlife management unit.

| Variables | Correlation Coefficient * |
|---|---|
| Mean habitat suitability and adjusted deer harvest for all WMUs | 0.645 |
| Mean habitat suitability and the number of positive ticks for all DAs in the KFL&A and LGL Health Units | 0.609 |
| The number of tick acquisition and the number of positive ticks for all Das in KFL&A and LGL Health Units | 0.892 |

\* Pearson correlation coefficient. All coefficients are over the 0.05 significance level.

Figure 7 suggests that there is a strong positive correlation between our deer habitat suitability model and the deer harvest data, which confirms that in the absence of deer density data, the deer habitat suitability values can be used to represent the deer abundance.

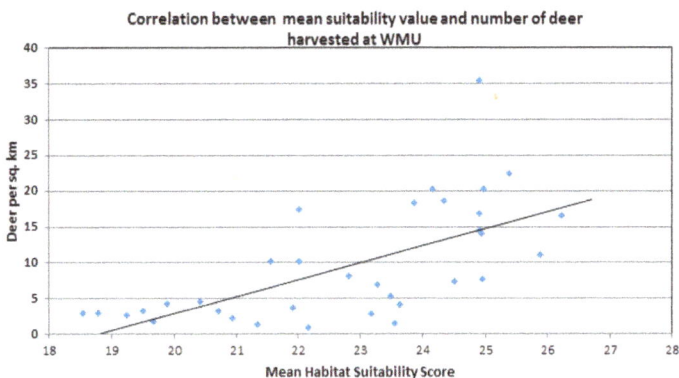

**Figure 7.** Correlation between the mean habitat suitability value and adjusted number of deer harvested in the WMU. The *p*-value is less than 0.05.

The positive correlation between the deer habitat suitability values and the positive tick number in our study demonstrates that as the primary host of the black-legged tick, the habitat and environmental conditions of the white-tailed deer may also impact the abundance of the tick. In many regions of eastern North America, research has been shown that white-tailed deer population density is positively correlated with black-legged tick abundance [19]. The results from this study seem consistent with this finding. As demonstrated by previous studies in the U.S., the high tick abundance usually corresponds to the high human Lyme cases [37].

With the growth in human population over the last decade, urbanization may have impacted the change of land use within the eastern Ontario region. It has been suggested that the transition of wooded forested regions into residential neighbourhoods may lead to greater interactions between humans and deer [35]. As deer lose their habitat to urban development, there may be a higher risk for endemic ticks to come in contact with humans and transfer the pathogen to newly-established neighbourhoods.

It is important to consider limitations and biases, as they play roles in the results of our study. First, the tick acquisition locations collected incorporated only the areas that the general public is able to access; we do not have data of endemic ticks located in isolated or inaccessible areas. For instance, regions with limited access to humans, such as islands or highly elevated areas, may also be areas that contain positive ticks. Due to the nature of this study, only ticks acquired by humans are included in the analysis. Therefore, our findings may not provide an accurate representation of all locations with endemic tick populations within eastern Ontario.

Our sources of data were derived from only two public health units. Any tick acquisitions submitted to other public health units with ticks acquired within our study area were not included in our analysis. In addition, our tick data do not include ticks that have fed on animals (e.g., dogs and cats), which may contribute to a large proportion of positive ticks. Since our data do not incorporate data on ticks submitted to local hospitals, physicians and veterinarian clinic, our results may not provide a true representation of the severity and extent of the endemic tick population in eastern Ontario. It should also be noticed that not all data used to extract different factors are in the same period that tick data were collected. There is a time lag among the different data used in this study. The tick data were collected between 2006 and 2012, while the deer harvest data were from 2008 to 2011. The time inconsistency may cause a bias in the result. However, considering the land use/cover stability and slow population changes in the study area, the bias caused by time lag may be very small.

Our deer harvest data were analyzed using the WMU scale (the finest scale available from the Ontario Ministry of Natural Resources). However, this may be considered a fairly coarse scale relative to our tick data analysis, which was analyzed at the DA level. Since deer hunting is controlled by the number of tags or permits issued to hunters in Ontario, the change in the number of deer harvested over the years may not truly and accurately reflect deer population change. Rather, the number of deer harvested at each WMU may be associated with the hunting and deer control objectives of the Ministry of Natural Resources. For instance, if the objective of the government were to stabilize deer populations at each WMU ever year, deer harvest numbers would not show a large fluctuation or variation, as a fixed number of tags would be allocated per year. As a result, our deer harvest data may not allow us to accurately estimate deer density within each WMU. Nevertheless, our deer harvest data can be useful in providing some insight into relative deer population numbers.

Previous studies have shown that increasing temperatures as a result of climate change may be one of the most significant factors contributing to the northward dispersal of Lyme endemic areas in the U.S. and Canada [19,37,38]. Researchers have suggested that climate impacts tick survival rates, the densities of tick populations and the threshold number of immigrating ticks required to establish a new population [20,21,38]. Beyond certain ranges of temperature, humidity and rainfall thresholds, ticks will not be able to survive, as these conditions will kill the tick. With the warming of temperatures over recent years, tick populations may spread from the U.S. into Ontario, where temperatures have been historically lower. Optimal survival conditions may allow ticks to become more robust in migrating greater distances in shorter time periods, while arid and hot temperatures may potentially hinder tick activity. The temperature preferences of the deer populations may also play a role in the dispersal of ticks. The role of climate change should be studied in more detail for this region in the future.

## 4. Conclusions

In this paper, the spatial relation between the black-legged Lyme tick and its primary host, the white-tailed deer, in eastern Ontario was examined. By developing a habitat suitability model through a Geographic Information System (GIS)-based multi-criteria decision making (MCDM) analysis, a deer habitat suitability map was generated, and the results were compared with deer harvest data. Tick submission data collected from two public health units between 2006 and 2012 were used to explore the relationship between endemic ticks and deer habitat suitability in eastern Ontario. A northwards expansion of ticks into eastern Ontario was observed.

While the dynamic relationship between the deer and the tick is not entirely clear, a deer habitat suitability model may provide some insight into the spread and distribution of Lyme disease in eastern Ontario. Since the deer is the primary host of the disease vector, a spatial analysis of the deer is crucial to understanding the spatial movement of the tick and the disease. The results suggest that a positive relationship exists between deer harvest data and deer habitat suitability. It was also found that the high tick population corresponds with the high suitability. The results suggest that locations associated with high suitability scores are focused towards the central and northern regions of the study area. In contrast, the suitability map revealed that regions with low suitability scores are situated towards the southern border of the study area, where human activity and anthropogenic impacts are highest. Nevertheless, the results suggest that a positive relationship could exist between our deer suitability map and endemic ticks in eastern Ontario.

These results are useful for developing management strategies that aim at preventing Lyme from becoming a threat to public health in Canada. A stronger understanding of deer habitat in relation to the distribution of black-legged ticks at a local scale can better equip health units with the prevention and occurrence of Lyme disease infection in eastern Ontario. As demonstrated by previous studies in the U.S., a high tick abundance usually corresponds to high human Lyme cases [33]. The potential to locate high and low risk areas within neighbourhoods may allow public health professionals to proactively formulate Lyme prevention methods, generate infectious disease awareness and develop strategies aimed exclusively at their respective local communities. The higher-scale geographic information, such as at the dissemination area level, provided for public health allows for greater precision in locating endemic tick areas for residents. As a result, public health units now have the ability and confidence to provide residents with site-specific information of black-legged tick risk areas, such as particular parks or neighbourhood areas where tick abundance had previously been noted. Since previous studies have focused on analyzing Lyme disease at a provincial and national level, this study demonstrates the ability for local public health units to formulate site-specific awareness and prevention strategies.

It should be noted that this study did not consider the impact of climate factors and forest management on the deer habitat change and tick expansion. Further studies are required to investigate how tick survival, behaviour and seasonal activity may change with past and projected climate change. A strong understanding of deer population dynamics and habitat suitability is essential to realizing the spatial trends and patterns of black-legged ticks and Lyme disease. The effect (or relationship) of land use edges and landscape fragmentation of the deer habitats and changes in both deer and tick populations should also be further studied in the future. As the number of endemic tick populations continues to expand and colonize new regions in Ontario, Lyme will become an emergent challenge for public health.

**Acknowledgments:** This research was initially supported by a National Science and Engineering Research Council (NSERC) Undergraduate Student Research Award and a grant from the Geomatics for Information Decision (GEOIDE), National Centre of Excellence, Canada. The authors would like to thank Michael Gatt, the wildlife biologist at Ontario Ministry of Natural Resources (MNR), for helping us access, organize and interpret the deer harvest data. We would also like to thank three anonymous reviewers for their constructive comments and suggestions, which have greatly helped to improve the quality of the paper.

**Author Contributions:** D. Chen designed the study and wrote detailed methodology on MCDM, as well as run the MCDM procedure and produced the final Figures 3, 4, 6 and 7 and analysis. H. Wang conducted initial literature review and writing, as well as initial data and result analysis and generated maps in Figure 1, Figure 2, and Figure 5. D. P. Belanger, K. Moore, M. Peterson, and J. Cunningham collected original Lyme data from KFL & A and LGL. They also contributed to interpret the Lyme data and provided comments on result analysis.

**Conflicts of Interest:** The authors declare no conflict of interest.

## Appendix A

**Table A1.** The recoding schemes and weights used for different layers in the MCDM analysis.

| Layers (Weights) | Original Values | Recoding Values * |
|---|---|---|
| | Water | 1 |
| Food sources (2/9) | Barren or sparsely vegetated Build-up, wetland | 2 |
| | Deciduous needleleaf forest Deciduous broadleaf forest Evergreen broadleaf forest | 3 |
| | Evergreen needleleaf forest Wood savannah | 4 |
| | Crops, shrubs, grass, mixed vegetation | 5 |
| Land cover shelter (2/27) | Water | 1 |
| | Settlement and developed land Mine tailings, quarries, redrocks Marshes, open wetlands Treed wetlands | 2 |
| | Agriculture Agriculture/natural vegetation Early successional forest | 3 |
| | Successional forest Sparse forest Dense coniferous forest | 4 |
| | Dense deciduous forest Mixed forest | 5 |
| Terrain slope (1/27) | Water | 0 |
| | 13.41°–20.89° | 1 |
| | 8.44°–13.41° | 2 |
| | 5.02°–8.44° | 3 |
| | 2.36 °–5.02° | 4 |
| | 0°–2.36° | 5 |
| Proximity to water bodies (2/9) | >8 miles | 1 |
| | 2.4 to 8 miles | 2 |
| | 1.5 to 2.4 miles | 3 |
| | 1 to 1.5 miles | 4 |
| | 0–1 miles | 5 |
| Distance to roads (1/12) | 0–0.8 miles | 1 |
| | 0.8 to 1.6 miles | 2 |
| | 1.6 to 2.4 miles | 3 |
| | 2.4 to 3.1 miles | 4 |
| | >3.1 miles | 5 |
| Distance to urban areas (1/12) | 0–0.8 miles | 1 |
| | 0.8 to 1.6 miles | 2 |
| | 1.6 to 2.4 miles | 3 |
| | 2.4 to 3.1 miles | 4 |
| | 3.1 miles | 5 |
| Diversity of land cover (1/6) | 1 | 1 |
| | 2 | 2 |
| | 3 | 3 |
| | 4 | 4 |
| | >5 | 5 |

* all suitable values are recoded as 1 to 5, with 1 indicating not suitable at all and 5 as the most suitable.

*ISPRS Int. J. Geo-Inf.* **2015**, *4*, 105–123

## References

1. Ogden, N.H.; Maarouf, A.; Barker, I.K.; Poulin, M.B.; Lindsay, L.R.; Morshed, M.G.; O'Callaghan, C.J.; Ramay, F.; Waltner-Toews, D.; Charron, D.F. Climate change and the potential for range expansion of the lyme disease vector *Ixodes scapularis* in Canada. *Int. J. Parasitol.* **2005**, *36*, 63–70. [CrossRef] [PubMed]
2. Public Health Agency of Canada. *Lyme Disease Fact Sheet*; Government of Canada: Ottawa, ON, Canada, 2012.
3. Ogden, N.H.; Bigras-Poulin, M.; O'Callaghan, C.J.; Barker, I.K.; Lindsay, L.R.; Maarouf, A.; Smoyer-Tomic, K.E.; Waltner-Toews, D.; Charron, D. A dynamic population model to investigate effects of climate on geographic range and seasonality of the tick *Ixodes scapularis*. *Int. J. Parasitol.* **2005**, *35*, 375–389. [CrossRef] [PubMed]
4. Watson, T.G.; Anderson, R.C. Ixodes scapularis say on white-tailed deer (*Odocoileus virginianus*) from Long Point, Ontario. *J. Wildl. Dis.* **1976**, *12*, 66–71. [CrossRef] [PubMed]
5. Warden, L. Factors Affecting the Abundance of Blacklegged Ticks (*Ixodes scapularis*) and the PREVALENCE of *Borrelia burgdorferi* in Ticks and Small Mammals in the Thousand Islands Region. Master's Thesis, The University of Guelph, Guelph, ON, Canada, 2012.
6. Spielman, A. The emergence of lyme disease and human babesiosis in a changing environment. *Ann. N. Y. Acad. Sci.* **1994**, *740*, 146–156. [CrossRef] [PubMed]
7. Odgen, N.H.; St-Onge, L.; Barker, I.K.; Brazeau, S.; Bigras-Poulin, M.; Charron, D.F.; Francis, C.M.; Heagy, A.; Lindsay, L.R.; Maarouf, A.; *et al.* Risk maps for range expansion of the Lyme disease vector, *Ixodes scapularis*, in Canada now and with climate change. *Int. J. Health Geogr.* **2008**, *7*. [CrossRef]
8. Scott, J.D.; Fernado, K.; Durden, L.A.; Morshed, M.G. Lyme disease spirochete, *Borrelia burgdorferi*, endemic in epicentre at Turkey Point, Ontario. *J. Med. Entomol.* **2004**, *41*, 226–230. [CrossRef] [PubMed]
9. Rand, P.W.; Lubelczyk, C.; Lavigne, G.R.; Elias, S.; Holman, M.S.; Lacombe, E.H.; Smith, R.P. Deer density and the abundance of *Ixodes scapularis* (Acari: Ixodidae). *J. Med. Entomol.* **2003**, *40*, 179–184. [CrossRef] [PubMed]
10. Jones, C.G.; Ostfeld, R.S.; Richard, M.P.; Schauber, E.M.; Wolff, J.O. Chain reactions linking acorns to gypsy moth outbreaks and Lyme disease risk. *Science* **1998**, *279*, 1023–1026. [CrossRef] [PubMed]
11. Bosler, E.M.; Ormiston, B.G.; Coleman, J.L.; Hanrahan, J.P.; Benach, J.L. Prevalance of the Lyme disease spirochete in populations of white-tailed deer and white-footed mice. *Yale J. Biol. Med.* **1984**, *57*, 651–659. [PubMed]
12. Rand, P.W.; Lubelczyk, C.; Holman, M.S.; Lacombe, E.H.; Smith, R.P. Abundance of *Ixodes scapularis* (Acari: Ixodidae) after complete removal of deer from an isolated offshore island, endemic for Lyme disease. *J. Med. Entomol.* **2004**, *41*, 779–784. [CrossRef] [PubMed]
13. Thompson, C.; Spielman, A.; Krause, P.J. Coninfecting deer-associated zoonoses: Lyme disease, babesiosis, and ehrlichiosis. *CID* **2001**, *33*, 676–685. [CrossRef]
14. Garnett, J.M.; Connally, N.P.; Stafford, K.C.; Cartter, M.L. Evaluation of deer-targeted interventions on Lyme disease incidence in Connecticut. *Public Health Rep.* **2011**, *126*, 446–454. [PubMed]
15. Fish, D.; Childs, J.E. Community-based prevention of Lyme disease and other tick-borne diseases through topical application of acaricide to white-tailed deer: Background and rationale. *Vector-Borne Zoonotic Dis.* **2009**, *9*, 357–364. [CrossRef] [PubMed]
16. Piesman, J. Ecology of *Borrelia burgdorferi* sensu lato in North America. In *Lyme Borreliosis: Biology, Epidemiology and Control*; Gray, J.S., Kahl, O., Lane, R.S., Stanek, G., Eds.; CABI Publishing: New York, NY, USA, 2002; pp. 223–249.
17. Shope, R. Global climate change and infectious diseases. *Environ. Health Perspect.* **1991**, *96*, 171–174. [CrossRef] [PubMed]
18. Bunnell, J.E.; Price, S.D.; Das, A.; Shields, T.; Glass, G.E. Geographic information system and spatial analysis of adult *Ixodes scapularis* (Acari: Ixodidae) in the Middle Atlantic region of the U.S.A. *J. Med. Entomol.* **2003**, *40*, 570–576. [CrossRef] [PubMed]
19. Ogden, N.H.; Lindsay, L.R.; Beauchamp, G.; Charron, D.; Maarout, A.; O'Callaghan, C.J.; Walternet-Toews, D.; Barker, I.K. Investigation of the relationships between temperature and development rates of the tick *Ixodes scapularis* (Acari: Ixodidae) in the laboratory and field. *J. Med. Entomol.* **2004**, *41*, 622–633. [CrossRef] [PubMed]

20. Brownstein, J.S.; Holford, T.R.; Fish, D. A climate-based model predicts the spatial distribution of the Lyme disease vector *Ixodes scapularis* in the United States. *Environ. Health Perspect.* **2003**, *111*, 1152–1157. [CrossRef] [PubMed]

21. Brownstein, J.S.; Holdford, T.R.; Fish, D. Effect of climate change on Lyme disease risk in North America. *EcoHealth* **2005**, *2*, 38–46. [CrossRef] [PubMed]

22. Ogden, N.H.; Trudel, L.; Artsob, H.; Barker, I.K.; Beauchamp, G.; Charron, D.F.; Drebot, M.A.; Galloway, T.D.; O'Handley, R.; Thompson, R.A.; *et al.* *Ixodes scapularis* ticks collected by passive surveillance in Canada: Analysis of geographic distribution and infection with *Lyme borreliosis* agent *Borrelia burgdorferi*. *J. Med. Entomol.* **2006**, *43*, 600–609. [CrossRef] [PubMed]

23. Phua, M.H.; Minowa, M. A GIS-based multi-criteria decision making approach to forest conservation planning at a landscape scale: A case study in the Kinabalu Area, Sabah, Malaysia. *Landsc. Urban Plan.* **2004**, *71*, 207–222. [CrossRef]

24. Keisler, J.M.; Sundell, R.C. Combining multi-attribute utility and geographic information for boundary decision: An application to park planning. *J. Geogr. Inf. Decis. Anal.* **1997**, *1*, 101–118.

25. Eastman, J.R.; Jin, W.; Kyem, P.A.K.; Toledano, J. Raster procedures for multi-criteria/multi-objective decisions. *Photogramm. Eng. Remote Sens.* **1995**, *61*, 539–547.

26. Land Cover, Circa 2000—Vector. Available online: http://www.geobase.ca/geobase/en/data/landcover/csc2000v/description.html (accessed on 10 October 2014).

27. Compton, B.; Mackie, R.; Dusek, G.L. Factors influencing distribution of white-tailed deer in Riparian Habitats. *J. Wildl. Manag.* **1988**, *52*, 542–548. [CrossRef]

28. Ministry of Natural Resources. White-Tailed Deer Biology. Ontario Ministry of Natural Resources. 2012. Available online: http://www.mnr.gov.on.ca/en/Business/FW/2ColumnSubPage/STDPROD_097096.html (accessed on 3 April 2014).

29. Moen, A.N. Energy conservation by white-tailed deer in the winter. *Ecology* **1976**, *57*, 192–198. [CrossRef]

30. Clef, M.V. Review of the Ecological Effects and Management of White-Tailed Deer in New Jersey. The Nature Conservancy, New Jersey Chapter, Oct. 2004. Avilable online: http://deerinbalance.files.wordpress.com/2010/01/review-of-the-ecological-effects-and-management-of.pdf (accessed on 20 November 2014).

31. Rodgers, K.J.; Ffolliott, P.F.; Patton, D.R. Home range and movement of five mule deer in a semidesert grass-shrub community. In *Rocky Mountain Forest and Range Experiment Station*; Forest Service US, Department of Agriculture: Fort Collins, CO, USA, 1978; Volume 355, pp. 1–6.

32. Beier, P.; McCullough, D.R. Factors influencing white-tailed deer activity patterns and habitat use. *Wildl. Monogr.* **1990**, *109*, 3–51.

33. Richardson, C.L. Brush Management Effects on Deer Habitat. Texas A & M AgriLife Extension. 1914. E-129. Available online: http://gillespie.agrilife.org/files/2013/02/Brush-Management-Effects-on-Deer-Habitats.pdf (accessed on 20 November 2014).

34. Janssen, R.; van Herwijnen, M. *Multiobjective Decision Support for Environmental Management + DEFINITE DEcisions on an FINITE Set of Alternatives: Demonstration Disks and Instruction*; Kluwer Academic Publishers: Dordrecht, The Netherlands, 1994; p. 232.

35. Ogden, N.H.; Bouchard, C.; Kurtenbach, K.; Margos, G.; Lindsay, L.R.; Trudel, L.; Nguon, S.; Milord, F. Active and passive surveillance and phylogenetic analysis of *Borrelia burgdorferi* elucidate the process of Lyme disease risk emergence in Canada. *Environ. Health Perspect.* **2010**, *118*, 909–914. [CrossRef] [PubMed]

36. Estrada-Pena, A. Increasing habitat suitability in the United States for the tick that transmits Lyme disease: A remote sensing approach. *Environ. Health Perspect.* **2002**, *110*, 635–640. [CrossRef] [PubMed]

37. Kitron, U.; Kazmierczak, J.J. Spatial analysis of the distribution of Lyme disease in Wisconsin. *Am. J. Epidemiol.* **1997**, *145*, 558–566. [CrossRef] [PubMed]

38. Gubler, D.J.; Reiter, P.; Ebi, K.L.; Yap, W.; Nasci, R.; Patz, J.A. Climate variability and change in the United States: Potential impacts on vector and rodent-borne diseases. *Environ. Health Perspect.* **2001**, *109*, 223–233. [CrossRef] [PubMed]

International Journal of
**Geo-Information**

isprs

MDPI

*Article*

# Geospatial Technology: A Tool to Aid in the Elimination of Malaria in Bangladesh

**Karen E. Kirk** [1], **M. Zahirul Haq** [2], **Mohammad Shafiul Alam** [2] and **Ubydul Haque** [3,4,*]

[1]    Johns Hopkins Bloomberg School of Public Health, Baltimore, MD 21205, USA; kkirk8@jhu.edu
[2]    International Center for Diarrheal Disease Research Bangladesh, Dhaka 1212, Bangladesh;
       mzhaq@icddrb.org (M.Z.H.); shafiul@icddrb.org (M.S.A.)
[3]    Emerging Pathogens Institute, University of Florida, Gainesville, FL 32610, USA
[4]    Department of Geography, University of Florida, Gainesville, FL 32611, USA
*      Author to whom correspondence should be addressed; ubydul.kth@gmail.com;
       Tel.: +1-443-839-6119; Fax: +1-352-273-6890.

Academic Editors: Fazlay S. Faruque and Wolfgang Kainz
Received: 29 May 2014; Accepted: 11 December 2014; Published: 31 December 2014

**Abstract:** Bangladesh is a malaria endemic country. There are 13 districts in the country bordering India and Myanmar that are at risk of malaria. The majority of malaria morbidity and mortality cases are in the Chittagong Hill Tracts, the mountainous southeastern region of Bangladesh. In recent years, malaria burden has declined in the country. In this study, we reviewed and summarized published data (through 2014) on the use of geospatial technologies on malaria epidemiology in Bangladesh and outlined potential contributions of geospatial technologies for eliminating malaria in the country. We completed a literature review using "malaria, Bangladesh" search terms and found 218 articles published in peer-reviewed journals listed in PubMed. After a detailed review, 201 articles were excluded because they did not meet our inclusion criteria, 17 articles were selected for final evaluation. Published studies indicated geospatial technologies tools (Geographic Information System, Global Positioning System, and Remote Sensing) were used to determine vector-breeding sites, land cover classification, accessibility to health facility, treatment seeking behaviors, and risk mapping at the household, regional, and national levels in Bangladesh. To achieve the goal of malaria elimination in Bangladesh, we concluded that further research using geospatial technologies should be integrated into the country's ongoing surveillance system to identify and better assess progress towards malaria elimination.

**Keywords:** malaria; Bangladesh; GIS; GPS; remote sensing

## 1. Introduction

Malaria is a major public health problem in many developing countries, including Bangladesh. In 2012, the official number of laboratory confirmed malaria cases in the country was 29,522 with 11 confirmed malaria deaths [1]. Thirteen million people are living in malaria risk areas in 13 of the country's 64 administrative districts [2]. The Bangladesh National Malaria Control Program (NMCP), which is responsible for overseeing malaria control activities at the national level, has had some success in controlling malaria with a reduction in prevalence of all malaria in the country by 65% (95% CI: 65-66) between 2008 and 2012 [1,3]. Between 2008 and 2012, the NMCP, through the support of the Global Fund, implemented in the 13 endemic districts, malaria control programs, including test and treat and the distribution of long lasting insecticide-treated nets (LLINs) [1]. During this timeframe, the prevalence rate in these 13 endemic districts decreased from 6.2 cases per 1000 population in 2008 to 2.1 cases per 1000 population in 2012 [1]. This steep decline in malaria prevalence has been attributed to the increase distribution of LLINs [1]. After successful implementation of its malaria control program in certain regions of the country, Bangladesh has started to eliminate malaria in eight of the 13 malaria endemic districts (Figure 1).

Bangladesh's malaria control activities have been integrated into the government's general health services system; however, this system relies mostly on passive case detection in health facilities and with community health workers at the community level. This delivery system still faces many challenges, including inadequate accessibility to proper treatment; lack of trained health workers; and marginalized, at-risk populations with limited education [4]. In addition, there are a limited number of health facilities in the country that are equipped to manage severe malaria cases. The country's overall surveillance and vector control programs have been insufficient to eliminate malaria from these at-risk populations particularly in the endemic regions of the country.

Geospatial technology, which has been used successfully in other malaria control programs in developing countries, includes Geographic Information System (GIS), Global Positioning System (GPS), and remote sensing (RS) [5]. GIS is defined as an organized collection of computer hardware and software, and geographic data to efficiently capture, store, update, manipulate, analyze, and display all forms of geographically referenced information [5]. With GIS it is possible to analyze differences in multiple spatial data layers related to the geographic position of a phenomenon, its attributes, and spatial relationships and to create new spatial information not available by studying the data layers separately. GPS provides users with navigation, position and timing services captured through satellite transmission [6]. Used in conjunction with GIS, this technology provides real-time data collection with accurate position information that can be used to analyze geospatial information. RS, an earth-observing instrument on satellite platforms, provides information on landscape features and climatic factors, and can be used to associate these factors with the risk of vector-borne diseases. Geospatial technologies have been used extensively in malaria risk mapping and malaria control throughout the world [7]. Providing accurate malaria risk maps can effectively guide the allocation of malaria resources and interventions in developing countries [8].

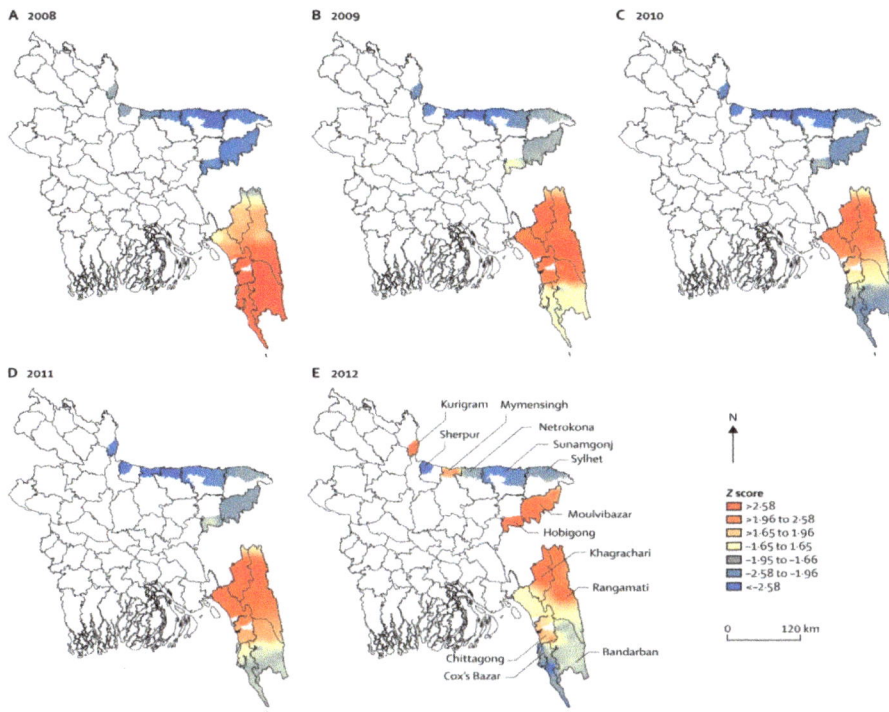

**Figure 1.** Malaria prevalence in Bangladesh in 2008, and progress in malaria control from 2009 to 2012
\*. \* For 2009–2012, red areas show the regions with the largest decreases in malaria prevalence. (Z-score: A statistically positive Z-score indicates high rates and a negative Z-score indicates of low rates).

Geospatial technology has been previously implemented in Bangladesh to develop risk mapping in parts of the country. In this study, we investigated the recent progress of malaria mapping in Bangladesh with GIS, GPS, and RS, and identified potential future applications and contributions of geospatial technologies to eliminate malaria in the country.

## 2. Materials and Methods

### 2.1. Study Area

Approximately 98% of all malaria morbidity and mortality cases reported in Bangladesh are located in 13 malaria endemic districts along the India and Myanmar borders. These districts are situated in densely forested, hilly areas with an average altitude of 500 meters above sea level. Eight of the districts, which are considered hyperendemic, are located in the country's northern region. The three districts with the highest malaria prevalence rates are Bandarban, Khagrachari, and Rangamati districts, located in the southwestern area of the country, in the Chittagong Hill Tracts (CHT) region, home to 1.3 million people [1–3]. CHT topographically encompasses hilly forests, lakes, rivers, canals, and waterfalls that provide an excellent breeding ground for *Anopheline* species, the vector responsible for carrying malaria parasites. Of the 35 A*nopheline* species found in the country, at least 26 were reported to be present in the CHT region in recent times. More than 10 of these vectors have been implicated, through laboratory testing, with the malaria parasite [9]. Historically, *Anopheles minimus* and *An. Baimaii* [10] were the most abundant and efficient vectors in the area. Due to habitat destruction (deforestation) for the expansion of human settlements and the introduction of irrigation

*ISPRS Int. J. Geo-Inf.* **2015**, *4*, 47–58

systems among the farmlands, these two species have declined in abundance and have been replaced by a number of plain land vector species (*Anopheles philippinensis, An. vagus, An. anularis, An. aconitus, An. sundaicus*) [11,12].

*2.2. Data Search*

We researched the PubMed database for studies on the use of geospatial technology to study malaria in Bangladesh through to June 2014 (Box 1). The following search terms were used: "Malaria, Bangladesh". The results in PubMed showed there were 218 journal articles listed; however, only 17 of 218 articles included GIS, GPS, and RS in their studies on malaria epidemiology and risk mapping in Bangladesh (Table 1). Five studies included all 13 endemic districts of the country while other studies focused specifically on the districts with the highest malaria endemic rates with five studies in Bandarban, six in Rangamati, and two in the Khagrachari district.

**Box 1**

Box 1. Literature search for data on the use of geospatial technology to study malaria in Bangladesh.

- Period searched: through to June 2014
- Source: PubMed databases
- Search terms: ("Bangladesh") AND ("malaria")
- Articles found: 218
- Inclusion criterion: Referral to any article using GIS/RS/GPS to study malaria in Bangladesh
- Articles retained for evaluation after detailed review: 17 [1,2,4,8,13–25]

**Table 1.** Malaria risk mapping studies in Bangladesh, 1992–2012.

| Study Area | Ref. # | Study Year | Sample Size | Geospatial Application | | | Findings of the Study |
|---|---|---|---|---|---|---|---|
| | | | | GIS | GPS | RS | |
| 13 districts | [1] | 2008–2012 | - | ✓ | | | Malaria mapping of Bangladesh from 2008 to 2012 showing a decrease in the country's prevalence rates (65%). |
| 13 districts | [2] | 2007 | - | ✓ | | | Findings reported statistically significant positive associations between the incidence of reported P. vivax and P. falciparum cases and rainfall and maximum temperature. |
| Rangamati | [4] | 2009 | 5322 | ✓ | ✓ | | Housing materials, household densities, education levels, and proximity to the regional urban center, were found to be effective predictors of treatment-seeking preferences for malaria. |
| 13 districts | [8] | 2007 | 9750 | ✓ | ✓ | ✓ | Bayesian modeling found statistically significant correlation between malaria prevalence and rainfall, temperature, and elevation as major factors influencing spatiotemporal patterns. |
| Khagrachari | [13] | 2007 | 750 | ✓ | ✓ | ✓ | Risk mapping used to predict areas of high and low malaria prevalence based on risk factors that included age and location of fragmented forests. |
| Khagrachari | [14] | 2007 | 750 | ✓ | ✓ | ✓ | Proximity (3 km) to water proved significant as a risk factor for malaria. |
| Rangamati | [15] | 2009 | 1400 | ✓ | ✓ | ✓ | Hot-spot clustering of cases with statistically significant risk factors between malaria positivity and ethnicity, forest cover, altitude, treatment preference, floor construction, and household density. |
| Rangamati | [16] | 2009–2010 | 1634 | ✓ | ✓ | ✓ | Identified malaria hotspots and with risk factors of low bed net ratio, home construction material, and high density of homes. |
| Rangamati | [17] | 2009 | 5322 | ✓ | ✓ | | Mapping of treatment seeking behaviors showed place of preference for malaria treatment were government health facilities if it was located 2 km from government health facilities preferred drug vendors. |
| Bandarban | [18] | 2010–2012 | 24,074 | ✓ | ✓ | | Risk factors for malaria were higher among jhum cultivators than non-cultivators living in the same household. |
| All Bangladesh | [19] | 1992–2001 | - | | | ✓ | VCI and TCI were strong predictors of malaria risk in Bangladesh. |
| Bandarban | [20] | 1992–2004 | - | | | ✓ | Estimated epidemic risks can be achieved using VHI and high summer TCI. |
| Bandarban | [21] | 2009–2010 | 20,563 | ✓ | ✓ | | Mapping of symptomatic & asymptomatic cases with high clustering within CHT; 80% of cases occurred during the rainy season. |
| Rangamati | [22] | 2009 | 1400 | ✓ | ✓ | | Age, ethnicity, proximity to forest, household density, and elevation were significant risk factors for malaria with 44.12% households living in areas with ≥10% prevalence rates. |
| 13 districts | [23] | 2007 | 9750 | ✓ | ✓ | | Malaria risk mapping of CHT with unequal distribution of prevalence rates -Khagrachari (15.25%), Bandarban (10.97%), and Rangamati (7.42%). |
| Bandarban | [24] | 2010–2013 | 1753 | ✓ | ✓ | | Higher risk of P. falciparum infection in pregnant women than other adults with asymptomatic infections. |
| Bandarban | [25] | 2009–2012 | 4782 | ✓ | ✓ | | Cases were geographically limited to hotspots with 80% infections occurring in one third of the population; incidence rates were highly seasonal with 85.8% of cases during rainy season (May–October). |

VCI: Vegetation Condition Indices; TCI: Temperature Condition Index

45

## 3. Results and Discussion

### 3.1. Results

Through June 2014, there were 17 published studies that have used geospatial technologies (GIS, GPS and/or RS) to understand and predict malaria risks [1,2,8,13–21], develop malaria risk maps, [21,22] and provide findings for targeted interventions strategies in Bangladesh. To supply feedback to the NMCP, national malaria risk maps were produced [2,23]. Long-term average monthly rainfalls and minimum/maximum temperatures ranges were interpolated, and elevation and vegetation coverage data were obtained using satellite imagery. GIS was also employed to develop maps of posterior distributions of predicted prevalence [13]. In one study, environmental variables were extracted from RS data to predict malaria risk at the national level [8]. Since 2008, the NMCP has confirmed a reduction of malaria in all endemic districts including CHT, which reported prevalence rate above 10%, the highest in the country [1,23].

There were four published studies based in CHT that used satellite images for land cover classification, risk mapping, and GIS mapping to investigate spatial relationships between malaria and risk factors. In addition, GIS and GPS were used to understand malaria incidence and prevalence rates with both active and passive surveillance at the individual and household levels. In another study, in the Bandarban District, a cohort framework of over 4000 households was established and mapped for a longitudinal study [21]. Four studies prepared risk maps in the same district focusing on pregnant women and jhum cultivators—subsistent farmers that use slash and burn methods of farming [18,24]. There were six malaria risk mapping studies conducted in Rajasthali, a sub-district of Rangamati district, which in 2007 had the highest (36%) malaria prevalence rates in the country [23]. All households ($n$ = 5322) studied in this sub-district were mapped using GPS and a household census was conducted collecting socio-demographic information, linking this data to geocoded household locations [4]. In addition, all health facilities and providers were mapped including the locations of community health workers, satellite clinics, hospitals, and drug stores [4]. Spatial models were explored in these different studies using ArcGIS, SaTScan, SAGA, and WinBugs mapping software. SAGA was used to extract topographical wetness (a measure of the extent of flow accumulation at the given point of the topographic surface) and elevation for all households surveyed in the Rangamati district. Satellite images using Advanced Spaceborne Thermal Emission and Reflection Radiometer digital elevation models was used to create water flow direction, accumulation, watershed, stream network, and stream link layers. Using cross-sectional and longitudinal data from these studies, spatial modeling of treatment seeking behaviors [17] and risk maps [8,13,16] were established and used to determine the progress and challenges to the NMCP.

### 3.2. Discussion

Geospatial technologies have been used for malaria risk mapping in Bangladesh since 2006 with the updating of these maps in 2008 and 2012 by the NMCP in collaboration with other research institutions [1,23]. Bangladesh has also developed spatial models for treatment seeking behavior, hydrological and prediction analysis using geostatistical modeling. Findings have confirmed that Bangladesh is closer to malaria elimination than ever before. The implementation of risk mapping with the NMCP, focusing particularly on targeted interventions in high risk areas in CHT, could provide decision makers crucial information for their elimination strategies [16].

While GIS, GPS and RS have all shown to be beneficial in providing data for malaria elimination strategies in both Bangladesh and other developing countries, improvements are still needed in the data collection process and dissemination of findings from the different malaria studies. Using household surveys, malaria hot-spots (the clustering of high malaria incidence cases) data, and more readily available spatial malaria databases along with a better understanding by decision makers on the usages and capabilities of these spatial analytical methods, can lead to more successful malaria elimination strategies [26]. The use of GIS and GPS systems connected to mobile phones [27] can also aid in malaria case detection and delivery of health services particularly in remote areas of CHT where tracking and analyzing malaria prevalence data can often be difficult [28]. With changing malaria epidemiology and reduced burden of malaria throughout the country, Bangladesh needs to focus on CHT region elimination strategies to address malaria hot-spots efficiently and effectively as well as reduce malaria importation, insecticide resistance, drug resistance [29], and the mapping of asymptomatic carrier [24]. Geospatial technology can provide the necessary data to assist in these elimination strategies.

Bangladesh NMCP has an extensive network of community health workers in all endemic districts but travel is often challenging in certain areas because of the difficult terrain and dense forest cover with few roads and limited waterways, which is often the only way to travel in these remote regions during the wet season. Thus, people living in these isolated areas are often deprived of many essential health services including proper malaria prevention and treatment methods. Careful planning using geospatial technologies to locate hot-spots particularly in the endemic districts of Bangladesh will be crucial in achieving malaria elimination goals in a cost effective way by focusing resourcing in areas with high prevalence rates (Table 2).

**Table 2.** Recommended Geospatial Technologies to use toward Malaria Elimination in Bangladesh.

| Geospatial Technology | Recommended Applications |
|---|---|
| GIS | Create malaria risk map at the lowest administrative level and update malaria maps each month. |
| GPS | Locate hospitals, health facilities, clinics, and households to create malaria information systems for improved mapping of risk areas. |
| Mobile Telecommunication Systems | Target interventions using surveillance data, satellite imagery, and mobile phone call records to improve coordination of services. |
| Spatiotemporal Cluster Detection | Analyze malaria cases in each month, detect spatio-temporal clustering, and locate hotspots. |
| Geostatistics | Detect spatial auto-correlation, prediction, and modeling for cost effective interventions. |
| Ecological Niche Models | Prepare malaria vector distribution maps, ecological suitability, predict vector distribution maps to locate vector-breeding areas and determine the indoor residual spraying strategies. |
| Bayesian methods | Interpolate, predict, and develop models using multiple malaria risk factors. |
| Spatially Explicit Mathematical Models | Create hierarchical models, hierarchical linear regression models, and mixed linear regressions to predict current and future malaria risk scenarios. |

With the abundance of malaria vector breeding sites in the CHT region of the country, RS can be used to look at vector's oviposition site/breeding ground and thereby predict their abundance. There has been a proliferation of some vector species in this region with an increase in standing water, particularly during the rainy seasons, from deforestation for agricultural purposes, providing more breeding sites for the mosquitos [30]. It has been noted that during the dry and malaria off-seasons, cases were found clustered around natural streams and canals in the CHT region [25]. Locating land use patterns across the vector breeding sites and using remotely sensed climatic environmental factors, particularly temperature, humidity, and rainfall as they correlate to malaria outbreaks, could provide useful information on the planning of vector control in this region. This same approach was applied in Indonesia and revealed malaria vector abundance in certain areas where malaria control programs were then implemented with much success [31]. The topographic variable can also be remotely-sensed to predict malaria vector's breeding site in malaria endemic areas [32].

It should be noted that Bangladesh performed much better in malaria mapping of its endemic areas compared to its neighboring malaria endemic countries, India and Myanmar. Countries engage in malaria control [33–36] and elimination [37] may learn from Bangladeshi experiences. Spatial analytical studies have shown a higher incidence of malaria in areas adjacent to these international borders [26]. To sustain recent gains in malaria elimination, the Bangladesh NMCP should look at cross-border movement between Bangladesh and India and Myanmar. The mapping of malaria prevalence as well as surveying for household movement along the borders can also be used to identify hotspots in these areas.

Based on the findings of these studies, the following are recommendations for the NMCP to implement to further the malaria elimination program in Bangladesh: identify high-risk malaria zones to determine spatial distribution; combine surveillance with GIS and weather pattern database to determine geographic and climate risks [21]; conduct further studies to understand malaria transmission among impacted populations during high transmission season; develop strategies to interrupt transmission in low transmission settings [18]; use mass screening in stable hot-spot areas to identify asymptomatic reservoirs and provide treatment to reduce the malaria burden [16]; promote malaria test and treatment services particularly among vulnerable populations, such as children under five years and jhum cultivators [18]; monitor asymptomatic malaria particularly among pregnant women living in low-intensity malaria transmission areas [24]; target interventions at fine spatial scales for malaria control programs; improve equity of access to interventions [15]; improve collaboration between researchers using geostatistical techniques and malaria control managers; target high malaria transmission areas [8] to provide more GPS, GIS, and RS strategies to characterize spatial heterogeneity with malaria risk at a fine scale and identify high risk areas that have not been studied [19]; and lastly focus on interventions to be targeted and timed according to risk profiles of endemic areas [38].

## 4. Conclusions

Bangladesh has made significant advancements in malaria reduction and has successfully used geospatial technologies at various levels for risk mapping and targeted interventions in the 13 malaria endemic districts. However, with the continued reduction in malaria burden along with the country's goal of elimination, this technology should be prioritized by the NMCP in their elimination strategy. More effort should be made by the NMCP to incorporate these strategies with improved surveillance systems, particularly in the remote areas of the CHT region, to detect and address malaria hotspots. Incorporating malaria ecology with spatial data along with malaria incidence maps, land usage maps, and population distributions can be helpful for decision makers to establish elimination strategies in these malaria endemic areas. Rapid advances in technology and analytical methods have allowed the development of spatial decision support systems, which can improve the elimination programs by enabling more accurate and timely resource allocation in high-risk areas. Geospatial technologies can play an important role in identifying inequities in health services in endemic areas of the country. GIS, GPS and RS are proven powerful tools that can provide important data and should be integrated with active and passive surveillance systems to achieve malaria elimination in Bangladesh.

**Acknowledgments:** This work was funded in part by the Emerging Pathogens Institute at the University of Florida and the College of Liberal Arts and Sciences, as part of the University of Florida Preeminence Initiative.

**Author Contributions:** K.E.K. and U.H. conceived the study design, analyzed data and drafted the manuscript. M.Z.H. and M.S.A. contributed in writing and critically reviewed the manuscript.

**Conflicts of Interest:** The authors declare no conflict of interest.

## References

1.  Haque, U.; Overgaard, H.J.; Clements, A.C.; Norris, D.E.; Islam, N.; Karim, J.; Roy, S.; Haque, W.; Kabir, M.; Smith, D.L.; *et al.* Malaria burden and control in Bangladesh and prospects for elimination: An epidemiological and economic assessment. *Lancet Glob. Health* **2014**, *2*, e98–e105. [CrossRef] [PubMed]
2.  Reid, H.L.; Haque, U.; Roy, S.; Islam, N.; Clements, A.C. Characterizing the spatial and temporal variation of malaria incidence in Bangladesh, 2007. *Malar. J.* **2012**, *11*. [CrossRef] [PubMed]

3.  National Malaria Control Program About NMCP in Bangladesh. Available online: http://www.Nmcp.Info/nmcp.Aspx (accessed on 9 June 2014).

4.  Haque, U.; Hashizume, M.; Sunahara, T.; Hossain, S.; Ahmed, S.M.; Haque, R.; Yamamoto, T.; Glass, G.E. Progress and challenges to control malaria in a remote area of Chittagong hill tracts, Bangladesh. *Malar. J.* **2010**, *9*. [CrossRef] [PubMed]

5.  What is a GIS? Available online: http://www.Volusia.Org/gis/whatsgis.Htm (accessed on 11 September 2014).

6.  GPS. Available online: http://www.Gps.Gov/systems/gps/ (accessed on 28 May 2014).

7.  Bautista, C.T.; Chan, A.S.; Ryan, J.R.; Calampa, C.; Roper, M.H.; Hightower, A.W.; Magill, A.J. Epidemiology and spatial analysis of malaria in the northern Peruvian Amazon. *Am. J. Trop. Med. Hyg.* **2006**, *75*, 1216–1222. [PubMed]

8.  Reid, H.; Haque, U.; Clements, A.C.A.; Tatem, A.J.; Vallely, A.; Ahmed, S.M.; Islam, A.; Haque, R. Mapping malaria risk in Bangladesh using bayesian geostatistical models. *Am. J. Trop. Med. Hyg.* **2010**, *83*, 861–867. [CrossRef] [PubMed]

9.  Alam, M.S.; Chakma, S.; Khan, W.A.; Glass, G.E.; Mohon, A.N.; Elahi, R.; Norris, L.C.; Podder, M.P.; Ahmed, S.; Haque, R.; *et al.* Diversity of anopheline species and their plasmodium infection status in rural Bandarban, Bangladesh. *Parasit. Vector* **2012**, *5*. [CrossRef]

10. Khan, A.Q.; Talibi, S.A. Epidemiological assessment of malaria transmission in an endemic area of East Pakistan and the significance of congenital immunity. *Bull. World Health Organ.* **1972**, *46*, 783–792. [PubMed]

11. Elias, M.; Dewan, R.; Ahmed, R. Vectors of malaria in Bangladesh. *J. Prev. Soc. Med.* **1982**, *1*, 20–28.

12. Alam, M.; Khan, M.; Chaudhury, N.; Deloer, S.; Nazib, F.; Bangali, A.; Haque, R. Prevalence of anopheline species and their plasmodium infection status in epidemic-prone border areas of Bangladesh. *Malar. J.* **2010**, *9*. [CrossRef] [PubMed]

13. Haque, U.; Magalhaes, R.J.S.; Reid, H.L.; Clements, A.C.A.; Ahmed, S.M.; Islam, A.; Yamamoto, T.; Haque, R.; Glass, G.E. Spatial prediction of malaria prevalence in an endemic area of Bangladesh. *Malar. J.* **2010**, *9*. [CrossRef] [PubMed]

14. Haque, U.; Huda, M.; Hossain, A.; Ahmed, S.M.; Moniruzzaman, M.; Haque, R. Spatial malaria epidemiology in Bangladeshi highlands. *Malar. J.* **2009**, *8*. [CrossRef] [PubMed]

15. Haque, U.; Sunahara, T.; Hashizume, M.; Shields, T.; Yamamoto, T.; Haque, R.; Glass, G.E. Malaria prevalence, risk factors and spatial distribution in a hilly forest area of Bangladesh. *PLoS One* **2011**, *6*. [CrossRef] [PubMed]

16. Haque, U.; Glass, G.E.; Bomblies, A.; Hashizume, M.; Mitra, D.; Noman, N.; Haque, W.; Kabir, M.M.; Yamamoto, T.; Overgaard, H.J. Risk factors associated with clinical malaria episodes in Bangladesh: A longitudinal study. *Am. J. Trop. Med. Hyg.* **2013**, *88*, 727–732. [CrossRef] [PubMed]

17. Haque, U.; Scott, L.M.; Hashizume, M.; Fisher, E.; Haque, R.; Yamamoto, T.; Glass, G.E. Modelling malaria treatment practices in Bangladesh using spatial statistics. *Malar. J.* **2012**, *11*. [CrossRef] [PubMed]

18. Galagan, S.R.; Prue, C.S.; Khyang, J.; Khan, W.A.; Ahmed, S.; Ram, M.; Alam, M.S.; Haq, M.Z.; Akter, J.; Streatfield, P.K.; *et al.* The practice of jhum cultivation and its relationship to plasmodium falciparum infection in the Chittagong hill districts of Bangladesh. *Am. J. Trop. Med. Hyg.* **2014**, *91*, 374–383. [CrossRef] [PubMed]

19. Rahman, A.; Kogan, F.; Roytman, L. Short report: Analysis of malaria cases in Bangladesh with remote sensing data. *Am. J. Trop. Med. Hyg.* **2006**, *74*, 17–19. [PubMed]

20. Rahman, A.; Krakauer, N.; Roytman, L.; Goldberg, M.; Kogan, F. Application of advanced very high resolution radiometer (AVHRR)-based vegetation health indices for estimation of malaria cases. *Am. J. Trop. Med. Hyg.* **2010**, *82*, 1004–1009. [CrossRef] [PubMed]

21. Khan, W.A.; Sack, D.A.; Ahmed, S.; Prue, C.S.; Alam, M.S.; Haque, R.; Khyang, J.; Ram, M.; Akter, J.; Nyunt, M.M.; *et al.* Mapping hypoendemic, seasonal malaria in rural Bandarban, Bangladesh: A prospective surveillance. *Malar. J.* **2011**, *10*. [CrossRef] [PubMed]

22. Haque, U.; Soares Magalhaes, R.J.; Mitra, D.; Kolivras, K.N.; Schmidt, W.P.; Haque, R.; Glass, G.E. The role of age, ethnicity and environmental factors in modulating malaria risk in Rajasthali, Bangladesh. *Malar. J.* **2011**, *10*. [CrossRef] [PubMed]

23. Haque, U.; Ahmed, S.M.; Hossain, S.; Huda, M.; Hossain, A.; Alam, M.S.; Mondal, D.; Khan, W.A.; Khalequzzaman, M.; Haque, R. Malaria prevalence in endemic districts of Bangladesh. *PLoS One* **2009**, *4*, e6737. [CrossRef] [PubMed]

24. Khan, W.A.; Galagan, S.R.; Prue, C.S.; Khyang, J.; Ahmed, S.; Ram, M.; Alam, M.S.; Haq, M.Z.; Akter, J.; Glass, G.; *et al.* Asymptomatic plasmodium falciparum malaria in pregnant women in the Chittagong hill districts of Bangladesh. *PLoS One* **2014**, *9*. [CrossRef] [PubMed]

25. Glass, G.; Alam, M.S.; Khan, W.A.; Sack, D.A.; Sullivan, D.J. Spatial clustering of malaria cases during low-transmission season in Kuhalong, Bangladesh. In Proceedings of the 13th Ascon Conference, Dhaka, Bangladesh, 28 March 2011.

26. Clement, A.C.; Reid, H.; Kelly, G.; Hay, S. Further shrinking the malaria map: How can geospatial science help to achieve malaria elimination? *Lancet Infect. Dis.* **2013**, *13*, 709–718. [CrossRef] [PubMed]

27. Tatem, A.J.; Huang, Z.; Narib, C.; Kumar, U.; Kandula, D.; Pindolia, D.K.; Smith, D.L.; Cohen, J.M.; Graupe, B.; Uusiku, P.; *et al.* Integrating rapid risk mapping and mobile phone call record data for strategic malaria elimination planning. *Malar. J.* **2014**, *13*. [CrossRef] [PubMed]

28. Prue, C.S.; Shannon, K.L.; Khyang, J.; Edwards, L.J.; Ahmed, S.; Ram, M.; Shields, T.; Hossain, M.S.; Glass, G.E.; Nyunt, M.M.; *et al.* Mobile phones improve case detection and management of malaria in rural Bangladesh. *Malar. J.* **2013**, *12*. [CrossRef] [PubMed]

29. Haque, U.; Glass, G.E.; Haque, W.; Islam, N.; Roy, S.; Karim, J.; Noedl, H. Antimalarial drug resistance in Bangladesh, 1996–2012. *Trans. R. Soc. Trop. Med. Hyg.* **2013**, *107*, 745–752. [CrossRef] [PubMed]

30. Alam, M.S.; Chakma, S.; Al-Amin, H.M.; Elahi, R.; Mohon, A.N.; Khan, W.A.; Haque, R.; Glass, G.E.; Sack, D.A.; Sullivan, D.J.; *et al.* Role of artificial containers as breeding sites for anopheline mosquitoes in Malaria hypo endemic areas of rural Bandarban, Bangladesh: Evidence form a baseline survey. In Proceedings of the Astmh Conference 2012, Atlanta, GA, USA, 11–15 November 2012.

31. Stoops, C.A.; Gionar, Y.R.; Shinta; Sismadi, P.; Rachmat, A.; Elyazar, I.F.; Sukowati, S. Remotely-sensed land use patterns and the presence of anopheles larvae (diptera: Culicidae) in Sukabumi, West Java, Indonesia. *J. Vector. Ecol.* **2008**, *33*, 30–39. [CrossRef] [PubMed]

32. Nmor, J.C.; Sunahara, T.; Goto, K.; Futami, K.; Sonye, G.; Akweywa, P.; Dida, G.; Minakawa, N. Topographic models for predicting malaria vector breeding habitats: Potential tools for vector control managers. *Parasit. Vector* **2013**, *6*. [CrossRef]

33. Kamuliwo, M.; Chanda, E.; Haque, U.; Mwanza-Ingwe, M.; Sikaala, C.; Katebe-Sakala, C.; Mukonka, V.M.; Norris, D.E.; Smith, D.L.; Glass, G.E.; *et al.* The changing burden of malaria and association with vector control interventions in Zambia using district-level surveillance data, 2006–2011. *Malar. J.* **2013**, *12*. [CrossRef] [PubMed]

34. Mukonka, V.M.; Chanda, E.; Haque, U.; Kamuliwo, M.; Mushinge, G.; Chileshe, J.; Chibwe, K.A.; Norris, D.E.; Mulenga, M.; Chaponda, M.; *et al.* High burden of malaria following scale-up of control interventions in Nchelenge district, Luapula province, Zambia. *Malar. J.* **2014**, *13*. [CrossRef] [PubMed]

35. Chanda, E.; Mukonka, V.M.; Kamuliwo, M.; Macdonald, M.B.; Haque, U. Operational scale entomological intervention for malaria control: Strategies, achievements and challenges in Zambia. *Malar. J.* **2013**, *12*. [CrossRef] [PubMed]

36. Chanda, E.; Govere, J.M.; Macdonald, M.B.; Lako, R.L.; Haque, U.; Baba, S.P.; Mnzava, A. Integrated vector management: A critical strategy for combating vector-borne diseases in South Sudan. *Malar. J.* **2013**, *12*. [CrossRef] [PubMed]

37. Simon, C.; Moakofhi, K.; Mosweunyane, T.; Jibril, H.B.; Nkomo, B.; Motlaleng, M.; Ntebela, D.S.; Chanda, E.; Haque, U. Malaria control in Botswana, 2008–2012: The path towards elimination. *Malar. J.* **2013**, *12*. [CrossRef] [PubMed]

38. Ahmed, S.; Galagan, S.; Scobie, H.; Khyang, J.; Prue, C.S.; Khan, W.A.; Ram, M.; Alam, M.S.; Haq, M.Z.; Akter, J.; *et al.* Malaria hotspots drive hypoendemic transmission in the Chittagong hill districts of Bangladesh. *PLoS One* **2013**, *8*. [CrossRef] [PubMed]

International Journal of
*Geo-Information*

MDPI

*Article*

# Examining Personal Air Pollution Exposure, Intake, and Health Danger Zone Using Time Geography and 3D Geovisualization

**Yongmei Lu [1],\* and Tianfang Bernie Fang [2]**

1   Department of Geography, Texas State University, 601 University Drive, San Marcos, TX 78666, USA
2   Blanton & Associates, Inc., 5 Lakeway Centre Court, Suite 200, Austin, TX 78734, USA;
    bfang@blantonassociates.com
\*   Author to whom correspondence should be addressed; YL10@txstate.edu; Tel.: +1-512-245-1337;
    Fax: +1-512-245-8353.

Academic Editors: Fazlay S. Faruque and Wolfgang Kainz
Received: 25 August 2014; Accepted: 22 December 2014; Published: 30 December 2014

**Abstract:** Expanding traditional time geography, this study examines personal exposure to air pollution and personal pollutant intake, and defines personal health danger zones by accounting for individual level space-time behavior. A 3D personal air pollution and health risk map is constructed to visualize individual space-time path, personal Air Quality Indexes (AQIs), and personal health danger zones. Personal air pollution exposure level and its variation through space and time is measured by a portable air pollutant sensor coupled with a portable GPS unit. Personal pollutant intake is estimated by accounting for air pollutant concentration in immediate surroundings, individual's biophysical characteristics, and individual's space-time activities. Personal air pollution danger zones are defined by comparing personal pollutant intake with air quality standard; these zones are particular space-time-activity segments along an individual's space-time path. Being able to identify personal air pollution danger zones can help plan for proper actions aiming at controlling health impacts from air pollution. As a case study, this paper reports on an examination and visualization of an individual's two-day ozone exposure, intake and danger zones in Houston, Texas.

**Keywords:** air pollution exposure; air pollutant intake; space-time path; time geography; personal health danger zone

---

## 1. Introduction

Air pollution refers to the contamination of the atmosphere that may lead to adverse health effects to human beings, animals, plants, and environments [1]. The U.S. Environmental Protection Agency (EPA) has set National Ambient Air Quality Standards (NAAQS) for six common air pollutants (*i.e.*, criteria pollutants), including particulate matter (PM), ground-level ozone ($O_3$), carbon monoxide (CO), sulfur oxide ($SO_x$), nitrogen oxide ($NO_x$), and lead. If the levels of one or more pollutants are higher than the EPA standards, the air quality is considered bad and may cause severe health effects. PM and ground-level $O_3$ are the most widespread health threats to human beings. High air pollution exposure can cause an increase in morbidity and mortality rates; recent evidences revealed that accumulated exposure to low air pollution can also produce severe health effects, including death, disability, and illness [2].

This study focuses on examining the spatial-temporal dynamic patterns of personal air pollution exposure and intake by using an extended time geography approach and 3D geovisualization. This paper is developed to address three inter-mingled questions—What is the air pollution level in the ambient air? What is an individual's exposure to polluted air when personal spatiotemporal trajectory and activities are considered? And what are an individual's personal health danger zones due to air pollution?

### 1.1. AQI for Ambient Air Quality and Individual Health Impact

Air quality index (AQI) indicates the degree of air pollution and the potential health effects from air pollution. It is a tool designed to help the public understand the local air quality and the adverse health effects of ambient air [3]. AQI is reported as a positive number, and its standard varies across different nations. The U.S. EPA calculates AQI based the concentration of major air pollutants, *i.e.*, ground-level $O_3$, $PM_{2.5}$, $PM_{10}$, carbon monoxide (CO), nitrogen dioxide ($NO_2$), and nitrogen oxide ($NO_x$) [4]. AQI is reported following a six-color scheme from green to maroon, corresponding to good air quality to hazardous air quality, respectively (Table 1) [5].

**Table 1.** The U.S. AQI standard (Source: EPA 2009).

| AQI | Health Concern | Color | Explanation |
|---|---|---|---|
| 0–50 | Good | Green | Clean air, no health risk |
| 51–100 | Moderate | Yellow | Light air pollution, little health risk |
| 101–150 | Unhealthy for sensitive groups (USG) | Orange | Only sensitive groups are affected |
| 151–200 | Unhealthy | Red | Unhealthy air for everyone |
| 201–300 | Very Unhealthy | Purple | Serious health effects for everyone |
| 301–500 | Hazardous | Maroon | Severe adverse health effects, even death |

Air pollutants concentrations are measured at many locations, local and nation-wide. Separate AQI is calculated for each pollutant using the standard EPA formula below:

$$I = \frac{I_h - I_l}{B_h - B_l}(C - B_l) + I_l \tag{1}$$

where $I$ is the AQI value for a pollutant of concern, $C$ is the air pollutant concentration, $B_h$ is the high break point ($\geq C$) for the concentration of the pollutant, $B_l$ is the low break point ($\leq C$) for the concentration of the same pollutant, $I_h$ is the high AQI limit corresponding to $B_h$, $I_l$ is the low AQI limit corresponding to $B_l$. Note that, given an air pollutant, EPA has defined the threshold concentration values of $B_h$, $B_l$, and the corresponding AQI values of $I_h$ and $I_l$ to reflect the health impacts of the pollutant [5]. The highest AQI value during a day is recorded as the AQI of that day. The hourly reports for local and national AQI (such as $PM_{2.5}$, $O_3$, and $PM_{2.5}$-$O_3$ combined) are updated and published through multiple channels [6,7].

AQI maps show air quality across a mapping area by using the AQI six-color scheme (as Table 1). These maps are usually used for AQI reporting and forecasting. For example, a public web site—WWW. AIRNOW.GOV—provides near real-time hourly AQI maps for the U.S. and AQI readings for major U.S. cities. Figure 1 shows one such map.

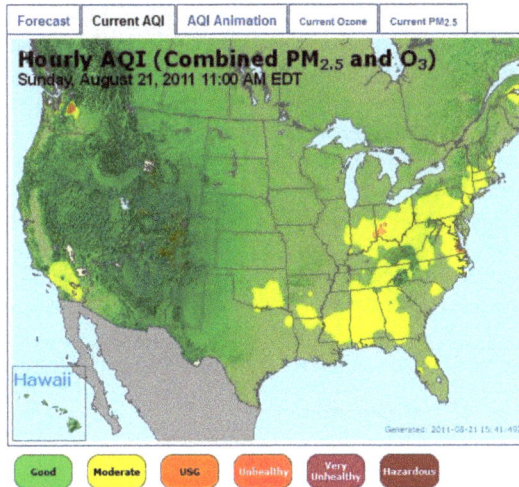

**Figure 1.** A U.S. national PM$_{2.5}$-O$_3$ combined AQI map (Source: AIRNow 2011).

AQI maps provide a good visualization of air quality and its variation across the mapping area. However, it is very limited for assessing air quality and its adverse health effects for individual human beings. The limitation is related to both the spatial and the temporal resolution of the AQI values. First, most AQI maps show AQIs on a city, township, or county level. To derive directly from these maps the health implication for individuals is subject to ecological fallacy. As Kwan [8] pointed out, there is a clear rising need for the assessment of health effects to gear away from deriving environmental effects on the individual level from an aggregated neighborhood level and to move towards assessing the health effects on a personal level. There has emerged in the past few years a number of studies and research projects that piloted the exploration of assessing individual level health effects of air pollution, although most of them are limited in both space and time scales partially due to the technical challenges related to sensors and data collection [9]. Second, AQI maps are 2D maps that reflect the spatial variation of air quality and its health effects at one specific time point or as an average over a period of time. However, air quality and its health effects are present continuously through time. The traditional AQI values and maps lack the capability to assist continuous assessment of air quality and health effects.

The traditional AQI is further limited for personal level health effects assessment due to its negligence of the individual characteristics, including individual's activities and biophysical characteristics. The health implication of air pollution is as much an individual level impact as it is for the general population. While an elevated concentration of a certain pollutant may impact human beings' health condition in a similar way, the adverse health effect on an individual is more a function of the type and patterns of the activities an individual conducts and his/her physical and biological characteristics. Therefore, as dynamic as the spatiotemporal patterns of air pollution and thus AQI value, individuals can benefit greatly from an individualized health effects assessment that specifically reflects the patterns and sequences of individual level space-time trajectory and activities.

*1.2. Personal Exposure to and Intake of Polluted Air*

Human exposure to air pollution occurs when contacting with air contaminants in a place and at a time [10,11]. Personal exposure can be measured either directly (e.g., personal sampling and biological marker measurement) or indirectly (e.g., ambient measurement/modeling and survey) [12]. Among the different measurement methods, personal sampling has a high accuracy. It is often used to collect data on air pollutant concentration in an individual's immediate surroundings, personal exposure frequency, and exposure duration. Personal exposure to air pollution is an accumulated process that is related to not only air pollutant concentration but also the periods of time and sequence of locations of exposure.

Personal air pollutant intake directly contributes to the health effects at individual level. It is related to a series of environment-human interaction processes, including human contacting with the air pollutants, the concentration of the pollutants over space and time, and the absorption of the pollutants by human body. Among the different absorption ways, inhalation is the major means for air pollutants to enter human body [13]. Inhalation rate varies across individuals; it changes for the same individual across different situations. Besides, health effects of air pollution are related to both air pollutant exposure and individual level biophysical characteristics [14]. Table 2 was adapted from Holmes [15]; it reports on the air inhalation rate when the different types of physical activities and the different population groups are considered. Equation 2 explains how air pollutant intake can be estimated by considering pollutant concentration, individual inhalation rate, and individual exposure time and place:

$$AI_p = \int_{l1}^{l2} \int_{t_1}^{t_2} C(t,l)R(t,l)\, dt\, dl \tag{2}$$

where $AI_p$ indicates personal air pollutant intake (*inhalation dose*), $C(t,l)$ is air pollutant concentration at time $t$ and location $l$, $R(t,l)$ is the real-time inhalation rate, $dt$ is the time span ($t_1$ to $t_2$) of exposure, and $dl$ is the location unit that collectively make the whole spatial trajectory ($l_1$, $l_2$).

**Table 2.** Individual average air intake volume per minute (adapted from Holmes 1994 [15]).

| Group | Staying/Sleeping/In Car | | Walking | | Running/Cycling | | Playing/Light Physical Labor | |
|---|---|---|---|---|---|---|---|---|
| | Speed | Air Volume | Speed | Air Volume | Speed | Air Volume | Speed | Air Volume |
| Children | 24 | 5–10 | 1–5 | 12.5–17.5 | 5–24 | 30–35 | <2 | 15–20 |
| Adult females | 24 | 5–10 | 1–5 | 17.5–22.5 | 5–24 | 45–50 | <2 | 15–20 |
| Adult males | 24 | 7.5–12.5 | 1–7 | 25–35 | 7–24 | 55–60 | <2 | 20–30 |

Unit: Speed (km/h), Air volume (liter)

## 2. Time Geography and Individual Space-Time Behavior Analysis

Individual space-time behavior refers to an individual's movements and activities through time and across space. There are several aspects for describing individual space-time behavior, such as travel type, space-time trajectory, stops, duration, speed, and sequence [16]. Two groups of quantitative methods are commonly employed for individual space-time behavior analysis. One group takes a holistic view to analyze the trajectories or activity types, such as the approach by location sequence alignment [17]. The other group focuses on a certain aspect of individual behavior, such as the travel distance from major employment centers [18]. As pointed out in literature [16], many quantitative analyses of individual space-time behavior are limited because they rely on a series of isolated events to describe continuous space-time behavior. In other words, discrete spatial variables (e.g., dividing space into several homogeneous zones) and temporal variables (e.g., using hourly average activity instead of real-time activity) are used, and these discrete variables may lead to errors and are subject to issues such as the Modifiable Areal Unit Problem (MAUP) [8]. Therefore, most studies failed to reflect the full picture of individual space-time behavior (and the environment effects of these behavior) due to the limitation of the spatial and temporal analysis scheme.

Swedish geographer Torsten Hägerstrand developed the conceptual framework of time geography in the late 1960s. According to Hägerstrand [19], individual space-time behavior is always limited by a series of constraints of space and time, including authority constraints, capability constraints, and coupling constraints. Under these constraints, an individual travels through a 3D space-time cube, *i.e.*, a two-dimensional space and a one-dimensional time. The 3D travel trajectory in this cube is called a space-time path (Figure 2). Space-time prism [19] was developed to define an individual's travel possibility across space and through time. The concept of space-time prism was introduced into Geographic Information System (GIS) by Miller [20] more than two decades later to examine the spatiotemporal accessibility of individuals.

According to time geography, spatiotemporal events are presented through three components—theme, location, and time; these components are measured, controlled, and fixed respectively, to form different combinations [21]. For example, in a regional air pollution map, location is fixed, time is controlled, and theme (*i.e.*, pollution) is measured. However, due to the limitation in information and visualization technologies as well as that in computation power, Hägerstrand's time geography has been mostly a framework for a long time. Individual space-time behavior has been studied at relatively coarse spatial scales and discrete time slices until recently. The visualization has been mostly 2D-maps showing distribution across space at selected time points.

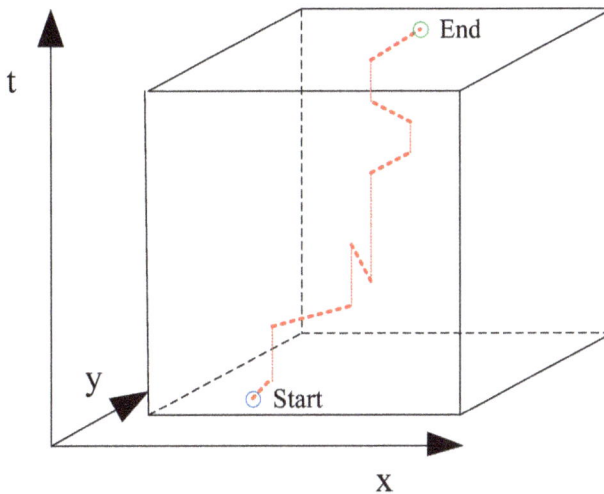

**Figure 2.** A space-time path in a time geography space-time cube.

With the development in GIS and visualization technologies in the 21st century, important progresses were made to profile individual space-time behavior using GIS-based 3D modeling and visualization. For example, Kwan [22] explored the activity-travel behavior of more than 10,000 individuals in Portland, Oregon. The space-time behaviors of these individuals were mapped using a GIS-based 3D geovisualization where individual space-time paths was integrated with the base map. Shaw and Yu [23] believe that individual's daily activity and travel behavior are highly influenced by modern Information and Communication Technologies (ICT), and therefore the virtual space makes a significant component for an individual's living space. They developed individual space-time path maps that integrate individual's virtual activities (such as teleconference and mobile phone calls) with his/her physical activities (such as working, commuting trips, and meeting). Chen *et al.* [24] developed an ArcGIS space-time extension—Activity Pattern Analyst (APA) to visualize and analyze individual space-time behavior aiming at exploring the hidden aggregate patterns in large spatiotemporal datasets.

In the last ten years or so, development in hand-held Global Position System (GPS) and that in ICT (especially mobile and wireless communication technologies) have greatly enhanced the advancement in pervasive location acquisition [25]. This made the data collection for individual level space-time trajectory more practical and benefited the space-time analysis of individual behavior. Hägerstrand's space-time cube was implemented by recent studies for both visualization (e.g., [26]) and analysis (e.g., [27,28]) of individual's space-time trajectories. Rossmo *et al.* [29] analyzed the travel patterns of 19 parolees and mapped their 3D space-time paths using GeoTime software; their individual space-time trajectory data were collected by portable GPS units. However, these studies examined individual-level space-time behavior by focusing mostly on an individual's space-time trajectory while considering away the dynamics of individual-environment interaction and environmental exposure. A trajectory is described as a collection of tri-tuples, $I (x_i, y_i, t_i)$. The environment impacts along the trajectory are either overlooked or treated as constant. This approach is limited for modeling the dynamic nature of the interaction between an individual and the environment, which is critical for understanding individual-level environment exposure and the related health effects. The dynamics of an individual's exposure to air pollution as one moves along a spatiotemporal trajectory must be properly accounted for in order to accurately assess the health effects of air pollution on an individual.

There is a lack in literature to connect an individual's spatiotemporal trajectory (*i.e.*, space-time path in Hägerstrand's space-time cube) with the changing nature of his/her environment exposure in general and air pollution exposure in particular. To the authors' knowledge, the only study that has explicitly developed the traditional time geography space-time cube to include a description of the environment dynamics was Fang and Lu [30]. Hägerstrand's space-time cube was extended to an air pollution cube of space-time in this study where the base 2D map shows the air pollution concentration and variation across space and the third-dimension shows the progress of time along which the spatial distribution of air pollution concentration changes. Following Fang and Lu [30], we propose that an individual's interaction with environment along a particular space-time path through a space-time cube can be modeled as a collection of qua-tuples, $I (x_i, y_i, t_i, E_i)$, where $E_i$ represents the environmental exposure of an individual at $i$, a particular space-time position. For air pollution and personal health effect study, $E_i$ is defined collectively by the air pollutant concentration, personal pollutant intake, and personal physical activity and biophysical characteristics. The dynamics of $E_i$ determines the continuous change of the health effects of air pollution for a person given his/her space-time behavior in the study area and during the study period.

The study reported below seeks to expand the traditional space-time path in time geography to account for the dynamics of environment exposure of an individual through space and time. Using data from portable GPS and air pollution sensors, this study presents an approach to measure individual-level air pollution exposure by considering an individual's space-time behavior. The study further seeks to assess individual-level health effects by defining individual-level air pollution danger zones to reflect the violation of EPA's air quality standards.

### 3. An Experiment in Houston

On 27 and 28 December 2010, an adult male volunteer traveled in Houston, Texas and collected two essential data sets and some supplementary data. The first data set was real-time air pollution level data, particularly ozone ($O_3$) concentration level. A BW GasAlert Extreme single gas detector ($O_3$) with 10 ppb increments was used as the air pollutant sampler. This sampler was carried by the data collector to record pollutant concentration data at his immediate surroundings. The second data set was individual space-time trajectory data. A Garmin eTrex Vista H handheld GPS navigator was used to record travel data (*i.e.*, travel trajectories, travel speeds, and stops). Both instruments were set to collect data at a 10-s interval. The volunteer also maintained a travel diary describing the location and nature of activities during the data collection period.

The volunteer's space-time travel data was processed using ArcScene in ArcGIS software package. Figure 3 is an ArcScene generated 3D map showing the individual's space-time path as laid on top of a Houston base map. The vertical dimension (z) represents the two-day travel time; the horizontal dimensions (x, y) represent the space of Houston, Texas. The real-time $O_3$ concentration data for the individual's immediate surroundings was downloaded from the portable sensors. The real-time AQI values along the individual's space-time path were calculated following Equation (1). The real-time personal $O_3$ intake rates were computed using Equation (2) by considering the real-time ambient $O_3$ concentration data, individual physical activities (as recorded in travel diary), and physical activity-related air intake rate (as listed in Table 1). When the real-time personal $O_3$ intake volume during a unit time was higher than EPA air quality standards, the person was determined to be in an air pollution danger zone.

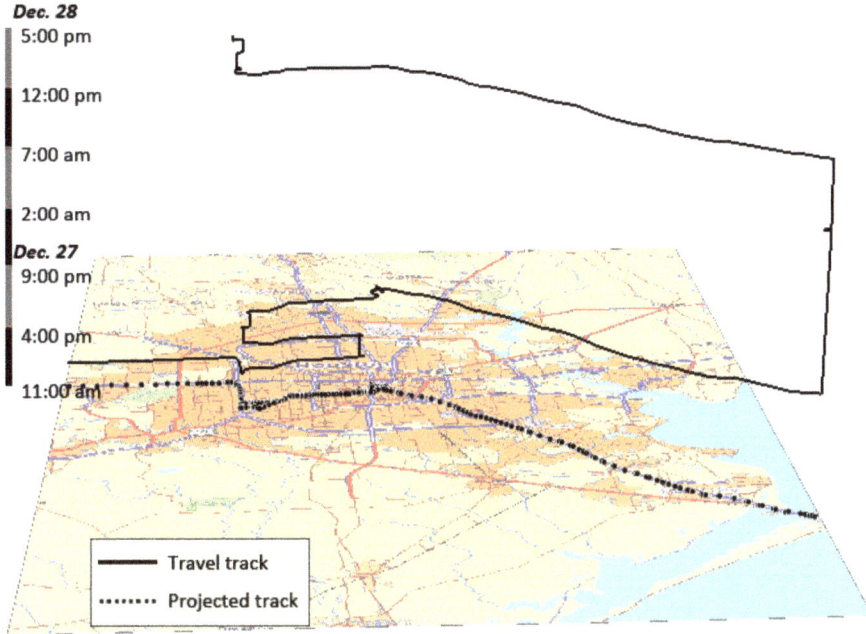

**Figure 3.** The space-time path of the volunteer in Houston, Texas on 27 and 28 December 2010.

The recorded $O_3$ concentration values ranged from 20 ppb to 80 ppb on 27 and 28 December 2010. During the afternoon of 27 December, the measured $O_3$ concentration values varied a lot. By connecting to the GPS data and the travel diary, it was found that $O_3$ concentration was mostly at medium level when the volunteer traveled on major roads (e.g., Interstate Highway 10, Interstate Highway 45, and W Sam Houston Parkway S); $O_3$ concentration was low when the volunteer traveled on minor roads or stayed indoors (e.g., shopping and rest). This indicates that $O_3$ concentration level is closely related to traffic emissions in urban Houston. During the night of 27 December and the morning of 28 December, the $O_3$ concentration level remained low and stable. The volunteer stayed inside mostly during this time period except for a 40-min outdoor morning exercise, which included jogging and walking. When the volunteer traveled through the city of Houston along highways during midday on 28 December, the $O_3$ concentration was recorded as being relatively high. It was low in the afternoon of 28 December, during which time the volunteer traveled short distances while conducting a variety of activities, including staying, walking, and running.

Using the recorded $O_3$ concentration data, real-time AQI values were calculated for the volunteer throughout the study period. Following the U.S. AQI standard (see Table 1), different colors are used to visualize the volunteer's real-time personal AQIs. In Figure 4, green indicates good air quality (AQI range: 0–50), yellow indicates moderate air quality (AQI range: 51–100), and orange indicates that the air quality is unhealthy for sensitive groups (USG) (AQI range: 101–150). The continuous AQI values and its color scheme is an important individual property that is uniquely associated with the volunteer's space-time trajectory. As discussed in the previous section, this aspect can be represented as the $E_i$ for the individual's space-time path that is made up of instances $I$ $(x_i, y_i, t_i, E_i)$. Figure 4 is a visualization of the dynamic individual-environment interaction along the space-time path of the volunteer; particularly, his real-time AQI along the space-time path is visualized together with his position in the space-time cube.

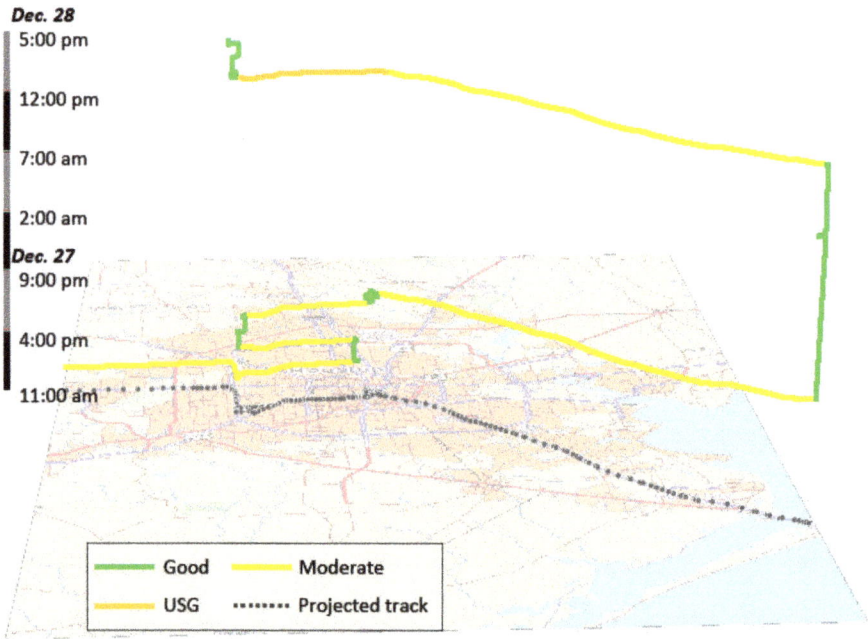

**Figure 4.** The volunteer's space-time path with personal AQIs in Houston, Texas on 27 and 28 December 2010.

The volunteer's real-time personal $O_3$ exposure (*i.e.*, $O_3$ concentration in immediate surroundings), real-time AQIs, and real-time $O_3$ intake rates are plotted together in Figure 5. Note that the three measures were plotted to be compatible along the y-axis for visualization purpose. As AQIs are directly related to $O_3$ concentration, the variation of the volunteer's real-time personal AQIs is similar to that of the concentration of $O_3$. However, there may be a big difference between the pollution concentration level (*i.e.*, concentration of $O_3$) and the pollutant intake level (*i.e.*, $O_3$ intake rate), as shown in Figure 5 for the mid-afternoon on 27 December. The nature of individual activities during this time is the major cause for this difference. Given a pollutant concentration level, the different types of activities required different levels of air intake, leading to different pollutant intake levels. This clearly shows the importance of understanding the dynamics of individual-environment interaction along a space-time path, which is the $E_i$ in the qua-tuples of $I$ ($x_i$, $y_i$, $t_i$, $E_i$).

Figure 5 shows that the volunteer's $O_3$ intake rates were higher and more unstable during the daytime. The volunteer visited different places and conducted different types of daytime activities, including walking, running, and staying. These activities changed his breathing rate and resulted in increased and changing levels of $O_3$ intake. During the nighttime, the volunteer stayed indoor with low-level physical activities, leading to low and stable $O_3$ intake rates.

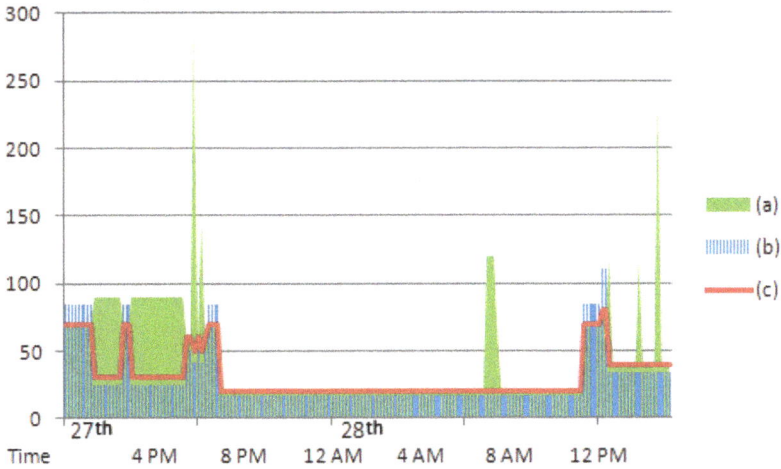

**Figure 5.** The volunteer's real-time air pollution exposure, intake, and AQI values during the 27–28 December 2010 data period in Houston, Texas. (**a**) $O_3$ intake rate ($10^{-8}$ L/min); (**b**) AQI; (**c**) measured $O_3$ concentration (ppb).

A total of six peaks of $O_3$ intake rate (*i.e.*, intake rates > $10^{-6}$ L/min) can be identified from Figure 5. Two peaks were occurred in the afternoon of 27 December, one peak in the morning of 28 December, one peak in the noon of December 28, and two peaks in the afternoon of December 28. By cross-referencing GPS trajectory data, the six peaks can be registered to specific positions in the space-time cube. Figure 6 highlights these corresponding six nodes as the high $O_3$ intake danger zones along the volunteer's space-time path. These zones signify the dangerous health risk segments in the individual's space-time behavior during the study period. Note that some of these danger zones fall onto the space-time segments where AQIs were not good, but others fall onto AQI safe segments along the volunteer's space-time path. The air quality at zone 1 and 2 was moderate (AQI color: yellow); the air quality at zone 3, 5, and 6 was good (AQI color: green); the air quality at zone 4 was unhealthy for sensitive groups (USG) (AQI color: orange). This manifests that even if the air quality is good, strenuous exercise may lead to inhalation of excessive air pollutants, and therefore exposure an individual to dangerous level of pollutant intake. On the other hand, when air pollutant concentration is relatively high and AQI is not good, maintaining inactive or less active physical activities so a minimum amount of pollutant entering into one's body may be an effective measure for alleviating the adverse health effects of air pollution.

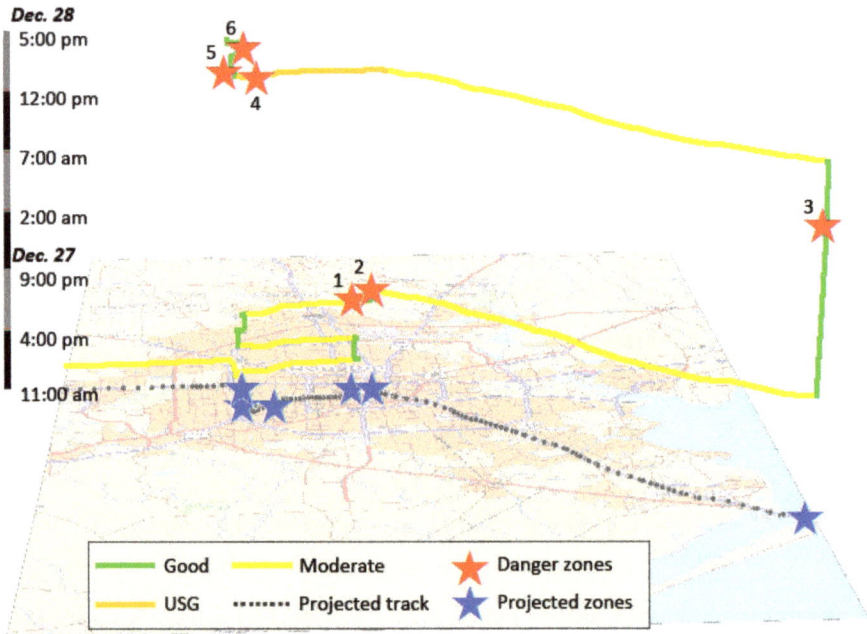

**Figure 6.** The personal air pollution danger zones for the volunteer in Houston, Texas on 27 and 28 December 2010.

## 4. Conclusions and Discussion

To assess personal health effects of air pollution, two steps are essential: to quantify personal ambient air pollution concentration, exposure duration and pollutant intake, and to evaluate the related personal health effects by considering personal space-time behavior. This paper showcases how an individual-level assessment of air pollution exposure and health effects can be evaluated through systematically integrating personal level data collection, space-time modeling, and geovisualization technologies. First, space-time trajectory information must be added to air pollution data in order to depict where, when, and how much an individual is exposed to air pollution. Real-time air pollution concentration data can be collected with ease using a portable air pollutant sensor. But they cannot be linked directly to personal space-time behavior. To solve this problem, the individual space-time trajectory data was collected by a portable GPS unit and was linked to air pollution data. Second, a 3D modeling and visualization approach was developed based on time geography framework to model a space-time cube, a space-time path within the cube, and the dynamics of individual-environment interaction along the path. The environment variation within a space-time cube and the dynamics of individual-environment interaction along a space-time path are not accounted for by the traditional time geography. This study models the space-time path as a collection of qua-tuples, $I\ (x_i,\ y_i,\ t_i,\ E_i)$, where $E_i$ represents environmental exposure at space-time point $i$. In the case of air pollution and individual health effect, the environmental exposure can be defined by the air pollutant concentration level and the pollutant intake rate (which is related to personal behavior and individual physical situation) along the space-time path. The approach of qua-tuples in space-time cube illustrates an effective measure to address the uncertain geography context problem (UGCoP) as identified by Kwan [31]. Third, based on the above two advancements, the proposed approach in this paper assesses personal level health effects of air pollution by considering air pollutant intake of an individual, which is determined by air pollution concentration, personal space-time trajectory, and personal activities throughout the trajectory. Air pollutant concentration levels in an individual's immediate surroundings were converted to space-time path AQI values (Figure 4); air pollutant intake was estimated using information from personal travel diary (Figure 5); personal air pollution health danger zones (Figure 6) were identified by considering air pollution and personal space-time behavior.

An individual's air pollution exposure changes through space and time, and it changes differently from his/her pollutant intake. In the case study, the $O_3$ concentration level for the volunteer's ambient air showed some regularity. When the volunteer drove along major roads, the $O_3$ concentration was medium to high level; when he stayed inside and walked/jogged in shopping/residential areas, the concentration remained low. This pattern may suggest the important role of traffic emission to elevated $O_3$ levels. Compared to $O_3$ concentration, the volunteer's $O_3$ intake rates varies largely (*i.e.*, ranging from $2 \times 10^{-7}$ L/min to $30 \times 10^{-7}$ L/min). Figures 5 and 6 revealed that the volunteer's $O_3$ intake had six peaks, indicating air pollution health danger zones. Five of the danger zones (zone 1, 2, 4, 5, and 6) are in the downtown area of Houston and one (zone 3) in Galveston, Texas. The downtown danger zones mostly highlight the volunteer's walking activities, and the Galveston danger zone represents the volunteer's running activity. The health danger zones represent the space-time-behavior segments along an individual's space-time trajectory where his/her personal health impact from air pollution was high. It is very important to understand that these danger zones may not fully agree with the personal AQI color zones. This is because AQI reflects concentration of pollutant in the ambient air, but the health danger zones are tailored to reflect the health effect from pollutant intake. Even though the pollutant concentration may not be as high, certain space-time activities may create personal danger zones due to the need for significantly elevated air intake.

The contributions of this study are mostly three-folds. First, time geography was extended to account for the dynamics of individual-environment interaction and was used as the framework for investigating the continuous change of personal exposure to and intake of air pollutant over space and time. Second, 3D maps were used to visualize personal exposure conditions. The 3D geovisualization modeled not only an individual's space-time path but also the changing dynamics of real-time personal AQI. Third, personal air pollution danger zones are defined and visualized along the personal space-time path to connect personal space-time behavior with personal health effects. Applying this approach to a future prediction of air pollution scenario, one can model personal air pollution danger zones for an individual's real or planned spatiotemporal activity trajectories. This can be used to assist an individual in selecting a spatiotemporal activity trajectory with the minimum adverse health effects. Therefore, the approach reported in this paper has a great potential in helping alleviate the adverse health effects of air pollution through managing individual-level space-time activities.

However, some limitations from the study should be noticed. First, the data accuracy for the empirical study in the paper is limited due to sensor precision. The increment of the portable $O_3$ sampler was limited to 10 ppb, and any variation below this level was not detectable by the sensor. Second, the volunteer's pollutant intake was estimated based on the past empirical findings about general population's air inhalation rate. Future research should consider using portable breath sensor that can automatically and accurately record an individual's breathe rate. Last, the limitation of the data collection means used in the reported pilot study must be recognized. It is impractical to expect the general public to carry both a GPS and a pollutant sensor while conducting daily activities. As pointed out by Fang and Lu [9], the research on real-time assessment of personal air pollution exposure and health effects are limited by the availability of sensors that can genuinely integrate location-detection and air pollutant sampling technologies. This technological limitation has been one major restriction for most of on-going or not-fully-implemented projects seeking to link air sensor to GPS (such as those reviewed in [19]). The focus of this paper is on the integration, analysis, and visualization of the related data for the purpose of analyzing personal health effects and health danger zones from air pollution. It is by no means the purpose of this study to showcase a GIS and air sensor integrated data collection approach.

This study pilots the investigation of personal air pollution exposure and intake, aiming at evaluating personal health effects of air pollution. Although only $O_3$ was investigated by the empirical study, the approach can be extended to other pollutants when proper sensors are available and health standard related AQIs are defined. Our next step of work will include continuous search for and design of an improved data collection means to support extending the individual-level health effects analysis and health danger zone modeling to the general public in order to serve people with varied backgrounds and different space-time behavior patterns.

**Author Contributions:** Yongmei Lu led the design of the research. Tianfang Bernie Fang conducted the fieldwork and data processing. Both authors contributed to the writing of the paper. Yongmei Lu finalized the manuscript for submission.

**Conflicts of Interest:** The authors declare no conflict of interest.

### References

1. EPA. Air Pollutants. Available online: http://www.epa.gov/ebtpages/airairpollutants.html (accessed on 21 August 2011).
2. Ostro, B.; Lipsett, M.; Reynolds, P.; Goldberg, D.; Hertz, A.; Garcia, C.; Henderson, K.D.; Bernstein, L. Long-term exposure to constituents of fine particulate air pollution and mortality: Results from the california teachers study. *Environ. Health Perspect.* **2010**, *118*, 363–369. [CrossRef] [PubMed]
3. EPA. *Air Quality Index—A Guide to Air Quality and Your Health*; U.S. Environmental Protection Agency, Office of Air Quality Planning and Stadards, Outreach and Information Division: Research Triangle Park, NC, USA, 2014.
4. EPA. What is Air Pollution? Available online http://www.epa.gov/airnow/airaware/day1.html (accessed on 21 August 2011).
5. EPA. *Technical Assistance Document for the Reporting of Daily Air Quality—The Air Quality Index (AQI)*; U.S. Environmental Protection Agency: Research Triangle Park, NC, USA, 2012.
6. AIRNow. Local Air Quality Conditions and Forecasts. Available online: http://www.airnow.gov/index.cfm?action=airnow.main (accessed on 21 August 2011).
7. EPA. Air Quality Index Report. Available online: http://iaspub.epa.gov/airsdata/adaqs.aqi?geotype=st&geocode=TX&geoinfo=st~TX~Texas&year=2008&sumtype=co&fld=gname&fld=gcode&fld=stabbr&fld=regn&rpp=25 (accessed on 21 August 2011).
8. Kwan, M.P. From place-based to people-based exposure measures. *Soc. Sci. Med.* **2009**, *69*, 1311–1313. [CrossRef] [PubMed]
9. Fang, T.B.; Lu, Y. Personal real-time air pollution exposure assessment methods promoted by information technological advances. *Ann. GIS* **2012**, *18*, 279–288. [CrossRef]
10. Duan, N. Models for human exposure to air pollution. *Environ. Int.* **1982**, *8*, 305–309. [CrossRef]
11. Lioy, P.J. Assessing total human exposure to contaminants. *Environ. Sci. Technol.* **1990**, *24*, 938–945. [CrossRef]
12. Monn, C. Exposure assessment of air pollutants: A review on spatial heterogeneity and indoor/outdoor/personal exposure to suspended particulate matter, nitrogen dioxide and ozone. *Atmos. Environ.* **2001**, *35*, 1–32. [CrossRef]
13. Weisel, C.P. Assessing exposure to air toxics relative to asthma. *Environ. Health Perspect.* **2002**, *110* (Suppl. 4), 527–537. [CrossRef] [PubMed]
14. Silverman, R.A.; Ito, K. Age-related association of fine particles and ozone with severe acute asthma in New York City. *J. Allergy Clin. Immunol.* **2010**, *125*, 367–373 e5. [CrossRef] [PubMed]
15. Holmes, J.R. How much air do we breathe? In *Measurement of Breathing Rate and Volume in Routinely Performed Activites*; Adams, W.C., Ed.; National Technical Information Service: Springfield, VA, USA, 1994.
16. Kwan, M.-P.; Ren, F. Analysis of human space-time behavior: Geovisualization and geocomputational approaches. In *Understanding Dynamics of Geographic Domains*; Hornsby, K., Yuan, M, Eds.; CRC Press: Boca Raton, FL, USA, 2008; pp. 93–113.
17. Shoval, N.; Isaacson, M. Sequence alignment as a method for human activity analysis in space and time. *Ann. Assoc. Am. Geogr.* **2007**, *97*, 282–297. [CrossRef]
18. Weber, J. Individual accessibility and distance from major employment centers: An examination using space-time measures. *J. Geogr. Syst.* **2003**, *5*, 51–70. [CrossRef]

19. Hägerstrand, T. What about people in regional science. *Pap. Reg. Sci. Assoc.* **1970**, *24*, 1–12. [CrossRef]

20. Miller, H. Modeling accessibility using space-time prism concepts within geographic information systems. *Int. J. Geogr. Inf. Syst.* **1991**, *5*, 287–301. [CrossRef]

21. Langran, G.; Chrisman, N. A framework for temporal geographic information. *Cartographica* **1988**, *25*, 1–14. [CrossRef]

22. Kwan, M.-P. Interactive geovisualization of activity-travel patterns using three-dimensional geographical information systems: A methodological exploration with a large data set. *Transp. Res. Part C: Emerg. Technol.* **2000**, *8*, 185–203. [CrossRef]

23. Shaw, S.-L.; Yu, H. A GIS-based time-geographic approach of studying individual activities and interactions in a hybrid physical-virtual space. *J. Transp. Geogr.* **2009**, *17*, 141–149. [CrossRef]

24. Chen, J.; Shaw, S.-L.; Yu, H.; Lu, F.; Chai, Y.; Jia, Q. Exploratory data analysis of activity diary data: A space-time GIS approach. *J. Transp. Geogr.* **2011**, *19*, 394–404. [CrossRef]

25. Lu, Y.; Liu, Y. Pervasive location acquisition technologies: Opportunities and challenges for geospatial studies. *Comput. Environ. Urban Syst.* **2012**, *36*, 105–108. [CrossRef]

26. Kraak, M.J. The space-time-cube revisited from a geovisualization perspective. In Proceedings of the Twenty-First International Cartographic Conference (ICC), Durban, South Africa, 10–16 August 2003; pp. 1988–1996.

27. Gatalsky, P.; Andrienko, N.; Andrienko, G. Interactive analysis of event data using spacetime cube. In Proceedings of the Eighth International Conference on Information Visualization, London, UK, 14–16 July 2004; pp. 145–152.

28. Demšar, U.; Virrantaus, K. Space-time density of trajectories: Exploring spatiotemporal patterns in movement data. *Int. J. Geogr. Inf. Sci.* **2010**, *24*, 1527–1542. [CrossRef]

29. Rossmo, K.; Lu, Y.; Fang, T.B. Spatial-temporal crime paths. In *Patterns, Prevention, and Geometry of Crime*; Andresen, M.A., Kinney, J.B., Eds.; Routledge: Cullompton, UK, 2012.

30. Fang, T.B.; Lu, Y. Constructing a near real-time space-time cube to depict urban ambient air pollution scenario. *Trans. GIS* **2011**, *15*, 635–649. [CrossRef]

31. Kwan, M.-P. The uncertain geographic context problem. *Ann. Assoc. Am. Geogr.* **2012**, *102*, 958–968. [CrossRef]

![isprs logo] International Journal of
**Geo-Information**

MDPI

Communication

# Use of the NASA Giovanni Data System for Geospatial Public Health Research: Example of Weather-Influenza Connection

**James Acker [1,*], Radina Soebiyanto [2,†], Richard Kiang [3,†] and Steve Kempler [4]**

[1] Goddard Earth Sciences Data and Information Services Center / Adnet Inc., Code 610.2, NASA Goddard Space Flight Center, Greenbelt, MD 20771, USA

[2] Goddard Earth Sciences Technology and Research (GESTAR), Universities Space Research Association, Columbia, MD 21046, USA; radina.p.soebiyanto@nasa.gov

[3] NASA Goddard Space Flight Center, Greenbelt, MD 20771, USA; richard.k.kiang@nasa.gov

[4] Goddard Earth Sciences Data and Information Services Center / NASA Goddard Space Flight Center, Greenbelt, MD 20771, USA; steven.j.kempler@nasa.gov

* Author to whom correspondence should be addressed; james.g.acker@nasa.gov; Tel.: +1-301-614-5435; Fax: +1-301-614-5268.

† These authors contributed equally to this work.

External Editors: Fazlay S. Faruque and Wolfgang Kainz

Received: 14 July 2014; in revised form: 20 October 2014; Accepted: 14 November 2014; Published: 10 December 2014

**Abstract:** The NASA Giovanni data analysis system has been recognized as a useful tool to access and analyze many different types of remote sensing data. The variety of environmental data types has allowed the use of Giovanni for different application areas, such as agriculture, hydrology, and air quality research. The use of Giovanni for researching connections between public health issues and Earth's environment and climate, potentially exacerbated by anthropogenic influence, has been increasingly demonstrated. In this communication, the pertinence of several different data parameters to public health will be described. This communication also provides a case study of the use of remote sensing data from Giovanni in assessing the associations between seasonal influenza and meteorological parameters. In this study, logistic regression was employed with precipitation, temperature and specific humidity as predictors. Specific humidity was found to be associated ($p < 0.05$) with influenza activity in both temperate and tropical climate. In the two temperate locations studied, specific humidity was negatively correlated with influenza; conversely, in the three tropical locations, specific humidity was positively correlated with influenza. Influenza prediction using the regression models showed good agreement with the observed data (correlation coefficient of 0.5–0.83).

**Keywords:** remote sensing; climate; weather; public health; disease; environment; atmosphere; ocean; biosphere; precipitation

---

## 1. Introduction

Investigation of connections between Earth's environment and public health issues can be considerably enhanced by the incorporation of remotely-sensed data. These investigations may comprise the examination of relationships between public health and primarily natural influences, such as meteorological and oceanic processes—an example would be the connection between water-borne diseases and heavy rainfall events, the latter potentially related to sea surface temperatures. Also included are processes affected by human activities, such as potentially harmful emissions into the atmosphere or water supply. An example of this relationship is the emission of sulfur dioxide

(SO$_2$) and nitrogen dioxide (NO$_2$) by fossil fuel combustion for energy production. Geographical setting, climatological baselines, and global teleconnections may also be included in research with remotely-sensed data that has public health implications.

The National Aeronautics and Space Administration (NASA) has acquired a rapidly growing archive of Earth remote sensing data, originating with the Landsat and Nimbus satellite missions in the 1970s and continuing with increasingly ambitious and technologically advanced missions to the present year, marked by the recent launches of the Global Precipitation Mission (GPM) and Orbiting Carbon Observatory-2 (OCO-2) satellites.

Since its inception in 2003, the NASA Geospatial Interactive Online Visualization ANd aNalysis Infrastructure (Giovanni) system provides access to a wide variety of NASA remote sensing data and other Earth science data sets, allowing researchers to apply selected data to a broad range of research topics. Currently hosted by the Goddard Earth Sciences Data and Information Services Center (GES DISC, Giovanni includes data from many different NASA missions and projects. An in-progress Advancing Collaborative Connections for Earth System Science (ACCESS) project titled "Federated Giovanni" will expand the data available in the system by including data from other NASA data centers.

This variety of data gives Giovanni marked potential for the investigation of different public health issues. One of Giovanni's primary attributes is ease-of-use; researchers who are generally unfamiliar with remote sensing data can use the system to find data that is applicable to their topic area and employ it. Only a relatively short investment of time and effort is required to become facile with the system. Correspondence with users in the public health research sector has indicated their high level of satisfaction with access to the data it provides, and the capability of determining whether or not remote sensing data can be used in their particular research area.

Giovanni provides remote sensing data alongside several different basic analytical capabilities, which include spatial maps of data variable values, difference maps, area-averaged time-series, animations, and vertical profiles of atmospheric variables. The mapping capability includes rapid averaging, so that mean values for months, seasons, or years can be visualized readily. All maps and plots generated by Giovanni can be immediately downloaded. Though it is not designed specifically as a data subsetting engine, for many data types Giovanni provides a relatively simple way to acquire spatially and temporally subsetted data, and it has been used for this purpose in numerous investigations. Giovanni is, ideally, a data exploration tool, allowing the performance of operations that used to require days and week for data acquisition and preparation to be performed in minutes, enabling more detailed analyses with considerably reduced time and effort. The Giovanni system is currently being transitioned from the current system, colloquially referred to as "Giovanni-3" [1], to a more flexible architecture, "Giovanni-4", that accelerates processing speed, adds new analysis capabilities, and which consolidates all of the data variables into a single search interface, rather than separate portals. The in-development Giovanni-4 system has not yet been described in a publication, but is available for use at the GES DISC Web site.

Missions, instruments, or projects providing data products available in Giovanni which are useful for public health research include: the Atmospheric Infrared Sounder (AIRS); the Tropical Rainfall Measuring Mission (TRMM); the Ozone Measuring Instrument (OMI); the Moderate Resolution Imaging Spectroradiometer (MODIS); the Modern Era Retrospective-analysis for Research and Applications (MERRA) project; the NASA Ocean Biogeochemical Model (NOBM); and both the North American Land Data Assimilation System (NLDAS) and the Global Land Data Assimilation System (GLDAS).

Although Giovanni is easy to use, many researchers and applied science professionals need guidance on how to find the appropriate datasets and how to interpret them. The NASA Applied Remote Sensing Training (ARSET) program provides online and in-person training for professional audiences, including health specialists, on how to use NASA resources and data, including data sets hosted at the GES DISC through Giovanni. Training modules with step by step instructions can be found online [2]. ARSET also has online resources that can supplement the use of Giovanni by non-specialists in remote sensing.

## 2. Data Parameters in Giovanni Relevant to Public Health

Data in Giovanni can be categorized with respect to its applicability to public health issues. In the following, three tiers of applicability will be presented: Tier 1, data that have a strong relationship to public health, and which are thus directly applicable in public health research; Tier 2, data that have indirect yet established relationships with an area of public health concern; and Tier 3, data that are related to weather or climate with an effect on public health and well-being.

### 2.1. Tier 1 Data Parameters

Tier 1 data types include:

- Precipitation
- Temperature
- Aerosol Optical Depth (AOD)
- Nitrogen Dioxide ($NO_2$)
- Carbon Monoxide (CO)
- Relative Humidity
- Cloud Cover

#### 2.1.1. Precipitation Data

Precipitation data finds wide application in public health research. Precipitation occurrence has frequently been associated with waterborne diseases, insect population outbreaks, and disease transmission modes (*i.e.*, shared water resources). Recent studies used Tropical Rainfall Measuring Mission (TRMM) daily data products to investigate the connection between rainfall and the location of cholera outbreaks in Haiti following a devastating earthquake [3]. Research on malaria transmission using remote sensing data frequently involves rainfall data. Malaria is a mosquito-borne disease, and since mosquitoes have an aquatic stage of their life cycle, mosquito populations are influenced by rainfall patterns. Kiang *et al.* [4] described research on malaria transmission patterns in Thailand, examining correlations with surface temperature, vegetation cover, and rainfall. Adimi *et al.* [5] described the potential for malaria risk prediction in Afghanistan. Both of these investigations accessed rainfall data products in Giovanni. Midekisa *et al.* [6] also used rainfall data to create early-warning models for malaria in Ethiopia.

Precipitation extremes also have public health effects—directly due to the danger posed by flood waters, subsequently due to damage to water utilities and freshwater sources affecting the water supply, and finally due to increased potential for disease outbreaks due to contaminated water. With regard to floods, Cools *et al.* [7] described the creation of a flash flood early warning system for Egypt that used precipitation data from Giovanni. Singh, Pandey, and Nathawat [8] used Giovanni to investigate the cause of the 2008 Kosi flood in India. GLDAS and NLDAS feature many different hydrological variables, including soil moisture and runoff in addition to precipitation. These variables can be used to study severe storms, snowmelt flooding, and drought intensity. In addition, cloud cover data can be correlated with changing precipitation patterns, as well as for tracking severe storms and weather fronts.

#### 2.1.2. Temperature Data

Temperature data, along with relative humidity, also can provide significant insight into public health concerns. Soebiyanto, Adimi, and Kiang [9] determined that temperature was a primary variable associated with seasonal influenza transmission. Surface temperature is a fundamental variable related to water resources, drought conditions, vegetation survival, insect overwintering survival, heat stress, and disease-vector species ranges. Giovanni has remotely-sensed land surface temperature data from the Moderate Resolution Imaging Spectroradiometer (MODIS), atmospheric temperature data from the Atmospheric Infrared Sounder (AIRS), model and assimilated model (GLDAS and NLDAS) temperature data, and high-resolution temperature data for specific regions. An example of such research was presented in Shen *et al.* [10], which described the pioneering data portal built for the Northern Eurasian Earth Science Partnership Initiative (NEESPI). Changes in this region, such as higher temperatures and increased fire outbreaks, were described.

### 2.1.3. Air Quality Data

Another area of public health concern is air quality. There are several regularly accessed variables related to air quality in Giovanni. Likely the most used are Aerosol Optical Depth (AOD) data products, which are acquired by MODIS and the Ozone Measuring Instrument (OMI). AOD indicates the optical clarity of the atmospheric air column, with higher values indicating more scattering and absorption by particles and chemicals in the atmosphere. Because of the direct relationship between AOD and some kinds of air pollution, particularly the frequently monitored PM2.5 and PM10 particle size fractions, AOD data variables have been primary resources in many different studies. Two examples are Li, Shao, and Buseck [11] on the effects of biomass burning aerosols on haze in Beijing, China, and Lu *et al.* [12] on sulfur dioxide emissions and trends in eastern Asia. AOD also has been used to track the regional impact of smoke from wildfires, which can be transported hundreds of miles from its source (Figure 1). Prados *et al.* [13] provides a comprehensive review of the use of air quality-related data sets in Giovanni.

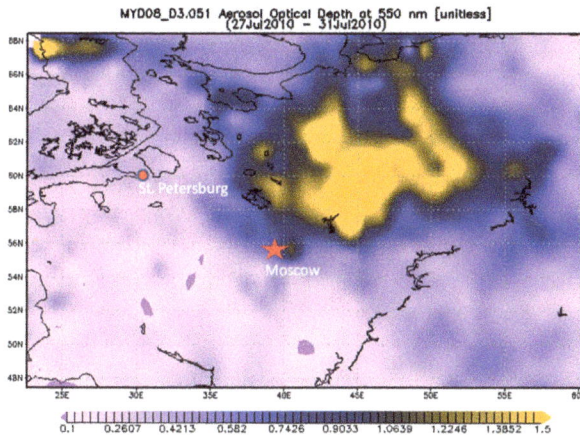

**Figure 1.** MODIS Aerosol Optical Depth (AOD) image showing the large area of elevated aerosol concentrations northeast of Moscow (yellow), stemming from massive wildfires that erupted in the hot summer of 2010. The daily AOD data was acquired for the period 27–31 July 2010, and averaged over this time period with Giovanni.

OMI is also an important source of other atmospheric chemistry data. The potential health significance of stratospheric ozone depletion is well-known, and OMI ozone data are integral to that research. OMI also provides a useful nitrogen dioxide ($NO_2$) data product, which can be used to track wildfire locations and movement, as well as air pollution sources resulting from the combustion of fossil fuels. Sergei Sitnov has been a prolific user of Giovanni, using the system to publish several papers on $NO_2$ and air quality in Russia. One such study looked at the weekly pattern of air quality and its relationship to meteorology in the environs of Moscow [14]. Another air quality indicator chemical species is carbon monoxide (CO), acquired by AIRS.

*2.2. Tier 2 Data Parameters*

Tier 2 health-related variables in Giovanni are:

- Chlorophyll concentration (phytoplankton)
- Euphotic Depth
- Sea Surface Temperature
- Ozone (O3) Erythemal Ultraviolet (UV) Daily Dose
- Normalized Difference and Enhanced Vegetation Indices (NDVI/EVI)
- Soil Moisture

2.2.1. Ocean Data

It may not be immediately apparent why oceanic phytoplankton chlorophyll concentrations are useful for health-related research. However, this data type actually has one of the longest associations with public health of any that has been provided by the GES DISC. This is due to the fact that *Vibrio cholerae*, the bacterial species responsible for cholera, has a stage in its life cycle when it infests copepods, a zooplankton species that feeds on phytoplankton. Thus, flood-related blooms of phytoplankton can provide a fertile ground for the proliferation of copepods and *V. cholerae*. Coastal Zone Color Scanner (CZCS) data were used in the 1980s to examine a cholera outbreak related to a phytoplankton bloom in the Bay of Bengal. These data in Giovanni can be used for cholera research, and to examine vectors of seafood contamination ("red tides" and other Harmful Algal Blooms, HABS), fish mortality, and severe storm effects. The use of ocean remote sensing data to study cholera outbreaks has been described previously [15–17].

Phytoplankton patterns also are related to fishery success or failure. Because fish constitute the major protein source for many coastal populations, these data too can have public health ramifications. Blooms also can indicate where someone should not fish; Van Holt showed that shellfish in areas with consistently elevated chlorophyll concentrations have more undesirable organisms clinging to their shells than in lower-chlorophyll zones [18]. Euphotic depth, a measure of water clarity, has been used for water quality studies and reports, and can indicate offshore flood effects. Sea surface temperature (SST) is directly related to water quality and phytoplankton growth, but indirectly it is related to coastal precipitation, storms, flooding, and the health of coral reefs. The Caribbean SERVIR (Sistema Regional de Visualización y Monitoreo) program used MODIS SST from Giovanni extensively in their research report "Sea Surface Temperature Trends in the Caribbean Sea and eastern Pacific Ocean" [19], published in 2011 to provide a baseline study for the impacts of these events on the population of countries in Central America, northern South America, and the Caribbean Sea.

2.2.2. Ozone Data

As noted earlier, OMI data is an obvious choice to examine stratospheric ozone depletion and the Antarctic "ozone hole" depth and extent. But the Erythemal Daily Dose data product, which describes the impact of ultraviolet radiation exposure on humans, has been used in some unique ways. Serrano, Cañada, and Moreno [20] used this data product to quantify the dangers to youth skiers of significant exposure to ultraviolet radiation.

2.2.3. Vegetation Indices

NDVI and EVI, both indices of vegetation greenness and ground cover, are also potentially useful data types for health research, as is soil moisture. These indices indicate the extent and intensity of drought, and thus are related to water resources and agricultural success. Kiang *et al.* [3] employed the vegetation indices in modeling malaria occurrence in Thailand, as they are relevant to land use and mosquito breeding environments. High resolution (5.6 kilometer) NDVI and EVI data are currently available in the Monsoon Asia Integrated Regional Study (MAIRS) high resolution monthly data portal.

## 2.3. Tier 3 Data Parameters

Tier 3 data types may be related to weather and climate, with effects on public health and well-being. Many of these data types measure quantities that are important to water resources:

- Snow Depth
- Snow Mass
- Snowfall Rate
- Snowmelt
- Fractional Snow Cover
- Snow/Ice Frequency
- Wind Speed
- Runoff

The current drought besetting western states of the United States, which some meteorologists describe as commencing in the year 2000, is having observable effects on snow in the mountain ranges, particularly those of California. As this will have ramifications for the management of water resources, and also impact wetland areas, the use of Giovanni to monitor such changes may be warranted. Furthermore, heavy snows can lead to floods, which may be predictable from snow depth data and observable with runoff data. Trends in snow parameters also may be indicators of climate change impacts and shifts in freeze and melt timing. A Giovanni time-series prepared for the NASA Data Investigations for Climate Change Education (DICCE) project [21] demonstrated how Giovanni could be used by teachers and students in New Mexico. Figure 2 shows a 1979–2010 monthly snow mass time-series for the mountainous area of northern New Mexico, a major source area for the Rio Grande River. Reduced snow mass from 1995–2005 is clearly visible.

The data tiers described above are necessarily broad classifications. Data types can have varying relevance to particular diseases, and research conducted on the connections between diseases and environmental factors must consider the spectrum of potential relationships. For example, snow depth or mass is rarely important for malaria incidence, but it is an important variable for malaria in Afghanistan. NDVI is not related to influenza, but is important for many vector-borne diseases, while AOD can be significant for respiratory diseases (such as influenza), but not for vector-borne diseases. Regional (particularly coastal) SST has been shown to be very relevant to cholera, while basin-scale SST is related to rainfall, and thus may have a relationship to diseases with a precipitation or hydrological connection.

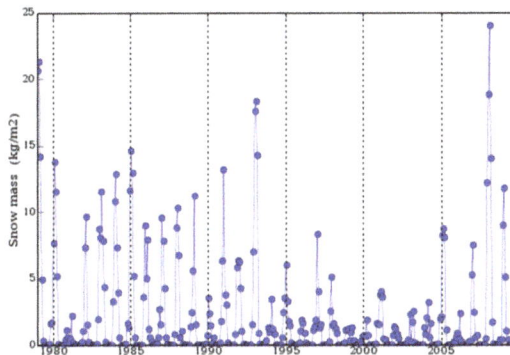

**Figure 2.** Monthly time-series of Modern Era Retrospective-analysis for Research and Applications (MERRA) snow mass data, plotted with Giovanni, for the central mountainous region of northern New Mexico, USA.

### 3. Influenza Example

The following example demonstrates how Giovanni was integral to the use of remote sensing data in studying the relationship between influenza and meteorological parameters, and further shows the capability of these parameters in predicting influenza activity. The burden of influenza, and how it is related to meteorological conditions, will be described first.

Influenza is an acute respiratory infection that can rapidly spread worldwide in seasonal epidemics. It approximately infects 5%–15% of the world population and causes up to 500,000 deaths each year [22]. In the United States, the economic cost of influenza epidemic is estimated to be around US$71–167 billion per year [22]. The epidemic timing varies across latitude, further suggesting the role of meteorological and environmental factors on influenza transmission. In the temperate region, influenza epidemic occurs during the winter time [23,24]. However, the seasonality and pattern of influenza epidemics in the tropics are less defined: from year-round high influenza activity, peaks that coincide with rainy seasons, to multiple peaks in a year [23,25–27]. Animal and laboratory studies have indicated that low temperature and humidity—consistent with wintertime conditions—provide suitable conditions for efficient transmission and longer virus survivorship [28,29]. In the tropics, rainfall is often associated with higher influenza activity although the direct causal relationship remains unclear. It is postulated that rainfall promotes indoor crowding that in turn, increases the probability for aerosol- and contact transmission [30].

In this study, influenza occurrence in five countries with either temperate or tropical climates was analyzed. The countries we studied were the Netherlands and New Zealand in temperate climate zones, and the Philippines, Vietnam and Sri Lanka in tropical climate zones. Influenza data was obtained from the World Health Organization Flu Net [31] for each country. Data was obtained for at least 3 years (Figure 3). Precipitation data was obtained from NASA's TRMM via Giovanni (TRMM 3B42 product). Briefly, the TRMM 3B42 product combines precipitation estimates from TRMM and other satellites as well as gauge analysis to produce daily precipitation at finer scale [32]. Near surface (2 m) specific humidity was obtained from Global Land Data Assimilation System (GLDAS), also archived in Giovanni. The Giovanni system was used to automatically download and subset the data based on the rectangular boundary of the study area. Output from Giovanni was an ASCII file of the aforementioned geophysical parameters, tagged with latitude, longitude and time. Spatial and temporal averaging was then performed on the retrieved data, as described below. The ASCII output format made it easier to do post-processing with statistical software (R), where we developed our model. Hence, the Giovanni system allowed us to effectively retrieve the dataset without the need to download or store large-sized HDF files. Ground stations were the source of minimum temperature data [33]. Due to the limitations of TRMM spatial coverage, precipitation data for the Netherlands was obtained from ground stations.

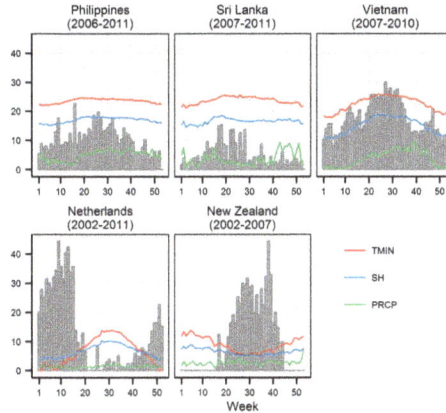

**Figure 3.** Weekly influenza positive (in %) and meteorological parameters averaged across study period. Bar plot shows the percentage of influenza positive. TMIN is minimum temperature (°C), SH is Specific Humidity (g/kg) and PRCP is precipitation (1 cm).

The weekly proportion of respiratory samples that tested positive for influenza acted as an influenza activity indicator. Logistic regressions were then developed for each study location, with minimum temperature, precipitation and specific humidity (averaged from current to previous 3 weeks) as covariates. The previous two weeks of influenza activity, and a third order polynomial function of the week number, were also included as covariates. Backward variable selection was then applied to obtain a parsimonious model (a model with as few covariates as possible). A more detailed description of the model can be found in Soebiyanto *et al.* [25].

Figure 3 plots the weekly influenza activity and meteorological parameters, averaged across the study period. Out of the three tropical study locations, Vietnam had larger variability in both minimum temperature and specific humidity. In the temperate locations, the Netherlands had larger variability in these two parameters (as compared to New Zealand). Precipitation in the tropical locations showed varying seasonality, while it was evenly distributed throughout the year in the temperate locations. The plots of the influenza positive proportion (Figure 3) showed that influenza peaks during wintertime in the temperate locations: around February–March in the Netherlands (Northern Hemisphere) and July–August in New Zealand (Southern Hemisphere). At this time, both temperature and specific humidity were also at their minimum values (Figure 3). In the tropics, the seasonality was not as well-defined. On average in the tropics, higher influenza activity appears to be associated with higher temperature, specific humidity, and precipitation values.

Results from the logistic regression models (Table 1) indicated that minimum temperature was inversely associated ($p < 0.05$) with influenza activity in Sri Lanka (Odds Ratio (OR) = 0.59, 95% Confidence Interval (CI) = 0.39–0.90). Specific humidity was significantly associated ($p < 0.05$) with influenza activity in all locations with varying relationships. Proportional associations were found in all three tropical locations (OR range of 1.13–1.47), while inverse associations were found in the two temperate locations (OR = 0.79 (0.67–0.95) in the Netherlands, and OR = 0.41 (0.29–0.58) in New Zealand). Here, proportional association indicates that an increase in the specified meteorological parameter was associated with an increase in influenza activity. Inverse association indicates an increase in a meteorological parameter was associated with a decrease in influenza activity. Precipitation, meanwhile, was not significantly associated ($p > 0.05$) with influenza activity in any of the locations.

**Table 1.** Multivariate regression between influenza positive proportion and meteorological parameters. Bold font indicates significance at α = 0.05 levels, RMSE indicate root mean squared error and Corr. Coeff. is the correlation coefficient between the observed and predicted influenza positive proportion. The models were adjusted for previous weeks' influenza activity, seasonality and other possible nonlinear relationships (modeled as a polynomial function, up to degree of 3, of the week number). When OR is not shown, the variable is not selected by the backward selection (not included in the final model).

| | Odds Ratio (95% Confidence Interval) | | | Model Performance | | |
| | Min. Temp. | Precipitation | Specific Humidity | Training | Prediction | |
| | (°C) | (mm) | (g/kg) | RMSE | RMSE | Corr. Coeff |
|---|---|---|---|---|---|---|
| Philippines | | | 1.13 (1.07, 1.19) | 0.064 | 0.048 | 0.831 |
| Sri Lanka | 0.59 (0.39, 0.90) | | 1.47 (1.11, 1.97) | 0.048 | 0.055 | 0.503 |
| Vietnam | | | 1.15 (1.09, 1.20) | 0.054 | 0.079 | 0.730 |
| Netherlands | | | 0.79 (0.67, 0.95) | 0.139 | 0.136 | 0.803 |
| New Zealand | | 1.00 (0.99,1.01) | 0.41 (0.29, 0.58) | 0.147 | 0.141 | 0.618 |

The inverse relationship between influenza activity and minimum temperature in Sri Lanka, and specific humidity in the temperate locations were consistent with experimental studies that indicated such conditions (low temperature and humidity) were suitable for longer influenza virus survivorship and more efficient transmission [28,29]. Findings on specific humidity in the tropical locations were in contrast to those in the temperate locations. These were consistent with other studies in the tropics [23,26,27]. The proportional association with specific humidity may indicate an indirect relationship with influenza activity, similar to precipitation. Indoor public places may provide opportunities for crowding when it rains or when humidity is high, and thus may enhance contact, aerosol, and droplet transmission.

We only find an association with minimum temperature in Sri Lanka, and not in the rest of the tropics and temperate study locations. In the temperate region, minimum temperature is often highly correlated with specific humidity. Hence, minimum temperature could be related to influenza in a similar fashion as specific humidity, but our model did not select this parameter, as it may not give the best model performance. This was consistent with another study showing that the relationship between influenza and temperature in the temperate United States was not as statistically strong as that of influenza and absolute humidity [24]. Meanwhile, temperature in the tropics typically remains relatively similar throughout the year without strong seasonality pattern. Therefore it was not associated with influenza, which was also observed in another study for tropical regions [23].

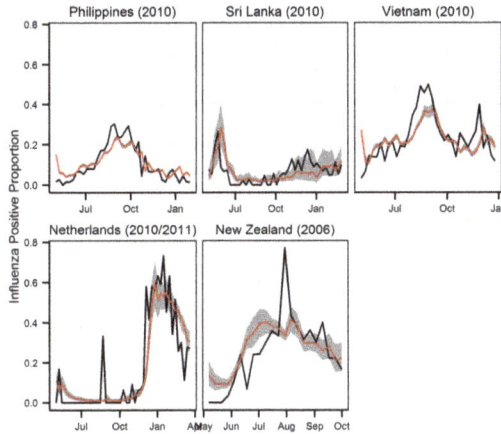

**Figure 4.** Regression models prediction of influenza positive proportion during the indicated period. The black line is the observed data (validation dataset, not used in training the models), and the red line is the model prediction with grey shades indicating the 95% Confidence Interval (CI).

The resulting models (one model in each study location) were then used to predict influenza activity during the final year of the data (Figure 4). These data sets were not used in training the models. The estimated influenza activity could be seen to reasonably follow the observed curves. Root mean squared errors (RMSE, Table 1) were less than 0.15 (influenza activity in this study was expressed in proportion). The correlation coefficients (Table 1) between the estimated and observed influenza activity showed a good agreement between the two (mean = 0.7, range = 0.5–0.83).

In conclusion, this analysis showed specific humidity as an important determinant for influenza activity across the climate zones. In temperate climate zones, influenza activity increased with decreasing specific humidity; the reverse was observed in the tropical locations. The former is consistent with low temperatures and low humidity occurring in winter, when influenza activity is elevated in temperate climate zones. It was also demonstrated, in Figure 4, that regression models which include meteorological, seasonal, and autoregressive inputs can be used to predict influenza activity relatively well. Hence, it is possible to use projected influenza activity as a guide for planning future prevention efforts. Short-term weather forecasts can be used to estimate influenza in the following week, whereas climate models can be used to assess how influenza activity and timing may change with climate over the next decades.

## 4. Conclusions

The multitude of remotely-sensed data parameters available in Giovanni, provide many potential research opportunities for examination of relationships between environmental factors and public health issues. Our example, an investigation of possible relationships between influenza activity and three variables—temperature, specific humidity, and precipitation—demonstrates how having such data types readily available for rapid subsetting and download in Giovanni, can enable the research process for researchers who are not necessarily familiar with the details of satellite remote sensing. This example does not fully demonstrate the capabilities of Giovanni's basic analytical functions, such as data variable maps, time-series, or Hovmöller diagrams, which can also be used in research. With due attention to the possibility of non-causal statistical correlations between environmental data and public health data, Giovanni provides an easy and rapid way to access and use NASA's Earth science resources in the health science sector. Furthermore, as climate change results in differing patterns of disease transmission and occurrence, Giovanni's analytical capabilities can be exploited to observe related changes in environmental factors.

**Acknowledgments:** Richard Kiang and Radina P. Soebiyanto are grateful for the support from NASA Public Health Applications in initiating the global influenza feasibility project.

**Author Contributions:** James Acker was responsible for the description of Giovanni and the generation of the figures. Radina Soebiyanto performed the influenza research in collaboration with Richard Kiang.

**Conflicts of Interest:** The authors declare no conflict of interest.

## References

1.  Berrick, S.; Leptoukh, G.; Farley, J.D.; Rui, H. Giovanni: A web service workflow-based data visualization and analysis system. *IEEE Trans. Geosci. Remote Sens.* **2009**, *47*, 106–113. [CrossRef]
2.  ARSET Web Site. Available online: http://arset.gsfc.nasa.gov (accessed on 5 September 2014).
3.  Rebaudet, S.; Gazin, P.; Barrais, R.; Moore, S.; Rossignol, E.; Barthelemy, N.; Gaudart, J.; Boncy, J.; Magloire, R.; Piarroux, R. The dry season in Haiti: A window of opportunity to eliminate cholera. *PLoS Curr.* **2013**. [CrossRef]
4.  Kiang, R.; Adimi, F.; Soika, V.; Nigro, J.; Singhasivanon, P.; Sirichaisinthop, J.; Leemingsawat, S.; Apiwathnasorn, C.; Looareesuwan, S. Meteorological, environmental remote sensing and neural network analysis of the epidemiology of malaria transmission in Thailand. *Geospatial Health* **2006**, *1*, 71–84. [PubMed]
5.  Adimi, F.; Soebiyanto, R.P.; Safi, N.; Kiang, R. Towards malaria risk prediction in Afghanistan using remote sensing. *Malar. J.* **2010**, *9*. [CrossRef] [PubMed]
6.  Midekisa, A.; Senay, G.; Henebry, G.M.; Semuniguse, P.; Wimberly, M.C. Remote sensing-based time series models for malaria early warning in the highlands of Ethiopia. *Malar. J.* **2012**, *11*, 165–181. [CrossRef] [PubMed]
7.  Cools, J.; Vanderkimpen, P.; El Afandi, G.; Abdelkhalek, A.; Fockedey, S.; El Sammany, M.; Abdallah, G.; El Bihery, M.; Bauwens, W.; Huygens, M. An early warning system for flash floods in hyper-arid Egypt. *Nat. Hazards Earth Syst. Sci.* **2012**, *12*, 443–457. [CrossRef]
8.  Singh, S.K.; Pandey, A.C.; Nathawat, M.S. Rainfall variability and spatio-temporal dynamics of flood inundation during the 2008 Kosi flood in Bihar State, India. *Asian J. Earth Sci.* **2011**, *4*, 9–19. [CrossRef]
9.  Soebiyanto, R.P.; Adimi, F.; Kiang, R.K. Modeling and predicting seasonal influenza transmission in warm regions using climatological parameters. *PLoS One* **2010**. [CrossRef]
10. Shen, S.; Leptoukh, G.; Loboda, T.; Csiszar, I.; Romanov, P.; Gerasimov, I. The NASA NEESPI data portal to support studies of climate and environmental changes in non-boreal Europe. In *Regional Aspects of Climate-Terrestrial-Hydrologic Interactions in Non-boreal Eastern Europe*; Pavel, G., Sergiy, I., Eds.; Springer: Berlin, Germany, 2009; pp. 9–16.
11. Li, W.J.; Shao, L.Y.; Buseck, P.R. Haze types in Beijing and the influence of agricultural biomass burning. *Atmos. Chem. Phys.* **2010**, *10*, 8119–8130. [CrossRef]
12. Lu, Z.; Streets, D.G.; Zhang, Q.; Wang, S.; Carmichael, G.R.; Cheng, Y.F.; Wei, C.; Chin, M.; Diehl, T.; Tan, Q. Sulfur dioxide emissions in China and sulfur trends in East Asia since 2000. *Atmos. Chem. Phys.* **2010**, *10*, 6311–6331. [CrossRef]
13. Prados, A.I.; Leptoukh, G.; Lynnes, C.; Johnson, J.; Rui, H.; Chen, A.; Husar, R.B. Access, visualization, and interoperability of air quality remote sensing data sets via the Giovanni online tool. *IEEE J. Sel. Top. Earth Obs. Remote Sens.* **2010**, *3*, 359–370. [CrossRef]
14. Sitnov, S.A. Weekly cycle of meteorological parameters over Moscow region. *Dokl. Earth Sci.* **2010**, *431*, 507–513. [CrossRef]
15. Lobitz, B.M.; Beck, L.R.; Huq, A.; Wood, B.; Fuchs, G.; Faruque, A.S.G.; Colwell, R.R. Climate and infectious disease: Use of remote sensing for detection of *Vibrio cholerae* by indirect measurement. *Proc. Natl. Acad. Sci. USA* **2000**, *97*, 1438–1443. [CrossRef] [PubMed]
16. Emch, M.; Feldacker, C.; Yunus, M.; Streatfield, P.K.; Thiem, V.; Canh, D.; Mohammad, A. Local environmental predictors of Cholera in Bangladesh and Vietnam. *Am. J. Trop. Med. Hyg.* **2008**, *78*, 823–832. [PubMed]
17. De Magny, G.; Murtugudde, R.; Sapianob, M.; Nizam, A.; Brown, C.; Busalacchi, A.; Yunus, M.; Nair, G.; Gil, A.; Calkins, J.; *et al.* Environmental signatures associated with Cholera Epidemics. *Proc. Natl. Acad. Sci. USA* **2008**, *105*, 17676–17681. [CrossRef] [PubMed]

18. Van Holt, T. Fisher success and adaptation to plantation systems in Chile. In *In Proceedings of the 2012 Gregory G. Leptoukh Online Giovanni Workshop*. Available online: http://disc.sci.gsfc.nasa.gov/giovanni/additional/newsletters/proceedings_2012_leptoukh_Giovanni_online_workshop (accessed on 10 November 2014).

19. Cherrington, E.A.; Hernandez, B.E.; Garcia, B.C.; Clemente, A.H.; Oynela, M.O. Sea Surface Trends in the Caribbean Sea and Eastern Pacific Ocean. Available online: http://issuu.com/cathalac/docs/servir_oceans_eng (accessed on 8 December 2014).

20. Serrano, M.A.; Cañada, J.; Moreno, J.C. Erythemal ultraviolet solar radiation doses received by young skiers. *Photochem. Photobio. Sci.* **2013**, *12*, 1976–1983.

21. SRI Inc. Data-Enhanced Investigations for Climate Change Education. Available online: http://dicce.sri.com/ (accessed on 5 September 2014).

22. WHO (2009) Influenza (Seasonal)—Fact Sheet No 211. Available online: http://www.who.int/mediacentre/factsheets/fs211/en/ (accessed on 14 June 2013).

23. Tamerius, J.D.; Shaman, J.; Alonso, W.J.; Bloom-Feshbach, K.; Uejio, C.K.; Andrew Comrie, A.; Viboud, C. Environmental predictors of seasonal influenza epidemics across temperate and tropical climates. *PLoS Pathog.* **2013**, *9*, e1003194. [PubMed]

24. Shaman, J.; Pitzer, V.E.; Viboud, C.; Grenfell, B.T.; Lipsitch, M. Absolute humidity and the seasonal onset of influenza in the continental United States. *PLoS Biol.* **2010**, *8*, e1000316. [CrossRef] [PubMed]

25. Soebiyanto, R.P.; Clara, W.; Jara, J.; Castillo, L.; Sorto, O.R.; Marinero, S.; De Antinori, M.E.B.; McCracken, J.P.; Widdowson, M. -A.; Kiang, R.K. The role of temperature and humidity on seasonal influenza in tropical areas: Guatemala, El Salvador and Panama, 2008–2013. *PLoS One* **2014**, *9*, e100659. [CrossRef] [PubMed]

26. Chadha, M.S.; Broor, S.; Gunasekaran, P.; Potdar, V.A.; Krishnan, A.; Chawla-Sarkar, M.; Biswas, D.; Abraham, A.M.; Jalgaonkar, S.V.; Kaur, H.; *et al.* Multisite virological influenza surveillance in India: 2004–2008. *Influenza Other Respi Viruses* **2012**, *6*, 196–203. [CrossRef]

27. Dosseh, A.; Ndiaye, K.; Spiegel, A.; Sagna, M.; Mathiot, C. Epidemiological and virological influenza survey in Dakar, Senegal: 1996–1998. *Am. J. Trop. Med. Hyg.* **2000**, *62*, 639–643. [PubMed]

28. Lowen, A.C.; Mubareka, S.; Steel, J.; Palese, P. Influenza virus transmission is dependent on relative humidity and temperature. *PLoS Pathog.* **2007**, *3*, 1470–1476. [CrossRef] [PubMed]

29. Polozov, I.V.; Bezrukov, L.; Gawrisch, K.; Zimmerberg, J. Progressive ordering with decreasing temperature of the phospholipids of influenza virus. *Nat. Chem. Biol.* **2008**, *4*, 248–255. [CrossRef] [PubMed]

30. Lofgren, E.; Fefferman, N.H.; Naumov, Y.N.; Gorski, J.; Naumova, E.N. Influenza seasonality: Underlying causes and modeling theories. *J. Virol.* **2007**, *81*, 5429–5436. [CrossRef] [PubMed]

31. World Health Organization FluNet. Available online: http://www.who.int/influenza/gisrs_laboratory/flunet/en/ (accessed on 14 June 2013).

32. Huffman, G.; Adler, R.F.; Bolvin, D.T.; Gu, G.; Nelkin, E.J.; Bowman, K.P.; Hong, Y.; Stocker, E.F.; Wolff, D.B. The TRMM multisatellite precipitation analysis (TMPA): Quasi-global, multiyear, combined-sensor precipitation estimates at fine scales. *J. Hydrometeorol.* **2007**, *8*, 38–55. [CrossRef]

33. National Climatic Data Center Global Surface Summary of Day. Available online: http://www.ncdc.noaa.gov/oa/ncdc.html (accessed on 14 June 2013).

International Journal of
*Geo-Information*

isprs

MDPI

*Article*

# Mapping Entomological Dengue Risk Levels in Martinique Using High-Resolution Remote-Sensing Environmental Data

Vanessa Machault [1,*], André Yébakima [2], Manuel Etienne [2], Cécile Vignolles [3], Philippe Palany [4], Yves M. Tourre [5], Marine Guérécheau [1] and Jean-Pierre Lacaux [1]

[1] Laboratoire d'Aérologie, Observatoire Midi-Pyrénées (OMP), Université Paul Sabatier, Toulouse 31400, France; marine.guerecheau@telespazio.com (M.G.); jean-pierre.lacaux@aero.obs-mip.fr (J.-P.L.)

[2] Service de Démoustication et de Lutte Anti-vectorielle, Conseil Général de la Martinique/Agence Régionale de Santé (SD-LAV), Fort-de-France, Martinique 97262, France; yebakima@cg972.fr (A.Y.); manuel.etienne@cg972.fr (M.E.)

[3] Direction de la Stratégie et des Programmes/Terre-Environnement-Climat, Centre National d'Etudes Spatiales (CNES), Toulouse 31400, France; cecile.vignolles@cnes.fr

[4] Météo-France Direction Inter-Régionale Antilles-Guyane, Fort-de-France, Martinique 97888, France; philippe.palany@meteo.fr

[5] Lamont-Doherty Earth Observatory (LDEO) of Columbia University, Palisades, New York, NY 10964-1000, USA; yvestourre@aol.com

* Author to whom correspondence should be addressed; vanessamachault@yahoo.com.br.

External Editors: Fazlay S. Faruque and Wolfgang Kainz

Received: 10 September 2014; in revised form: 6 November 2014; Accepted: 25 November 2014; Published: 10 December 2014

**Abstract:** Controlling dengue virus transmission mainly involves integrated vector management. Risk maps at appropriate scales can provide valuable information for assessing entomological risk levels. Here, results from a spatio-temporal model of dwellings potentially harboring *Aedes aegypti* larvae from 2009 to 2011 in Tartane (Martinique, French Antilles) using high spatial resolution remote-sensing environmental data and field entomological and meteorological information are presented. This tele-epidemiology methodology allows monitoring the dynamics of diseases closely related to weather/climate and environment variability. A Geoeye-1 image was processed to extract landscape elements that could surrogate societal or biological information related to the life cycle of *Aedes* vectors. These elements were subsequently included into statistical models with random effect. Various environmental and meteorological conditions have indeed been identified as risk/protective factors for the presence of *Aedes aegypti* immature stages in dwellings at a given date. These conditions were used to produce dynamic high spatio-temporal resolution maps from the presence of most containers harboring larvae. The produced risk maps are examples of modeled entomological maps at the housing level with daily temporal resolution. This finding is an important contribution to the development of targeted operational control systems for dengue and other vector-borne diseases, such as chikungunya, which is also present in Martinique.

**Keywords:** dengue; remote-sensing; risk mapping; *Aedes aegypti*; medical entomology

---

## 1. Introduction

Dengue is an infectious disease caused by one of the four serotypes (DEN-1 to DEN-4) of the dengue virus. It is transmitted by the bite of infected *Aedes* female mosquitoes and primarily occurs in urban areas. Even if the mortality rate is low among human populations, dengue is considered as one of the most important mosquito-borne viral disease due to its extensive geographic spread (125 endemic countries) with 50 to 200 million annual infections [1].

In Martinique (French Antilles), six dengue epidemic waves occurred during the last 20 years. More than 41,000 clinical cases were reported during the penultimate epidemic in 2010, accounting for approximately 10% of the island population. In this region, the *Aedes aegypti* mosquito is the single identified vector for the transmission of dengue virus to date. This mosquito breeds mostly in artificial domestic or peridomestic containers filled with clean water with little organic debris and low concentrations of inorganic nutrients [2,3]. In Martinique, potential breeding sites include (i) flower pots with saucers, detritus, and debris, abandoned cars and tires, badly maintained gutters, discarded old domestic appliances or pools that may be all filled-up naturally with rainfall; (ii) containers such as drum barrels that may be deliberately placed under gutters or in yards to collect rain water for watering/cleaning purposes; (iii) rarely, containers that can be artificially filled-up when watering. Antivectorial and mosquito nuisance control are managed by a public organization (*Service de Démoustication et de Lutte Antivectorielle*, SD-LAV). Since 1991, the SD-LAV has collected information on dengue vectors with an additional effort during outbreaks.

No specific treatment is available for dengue, and no operational vaccine is currently available [4]. Controlling virus transmission thus mainly consists of integrated vector management: (i) information is provided to the inhabitants to avoid the creation of potential larval habitats; (ii) source reduction occurs via the physical destruction of the potential/positive breeding sites; and (iii) insecticide spraying mainly occurs during epidemics. It is also important to note that it has been shown that for dengue vectors, "targeting only the most productive water container types (roughly half of all water holding container types) was as effective in lowering entomological indices as targeting all water holding containers at lower implementation costs" [5]. In this context, good knowledge of the entomological conditions in a given area and during a given period of time is a prerequisite for implementing efficient control. Unfortunately, entomological data are seldom collected longitudinally, and available data often provide only a snapshot of a rather continuous phenomenon. Risk maps at appropriate scales can provide surrogate data and valuable information regarding a spatio-temporal evaluation of entomological risk. Mapping at global scales may co-exist with fine scale or local mapping to establish local control strategies.

From global/regional to local scales, the heterogeneity of spatial and temporal distributions of dengue vectors/cases is partly led by weather/climate conditions (e.g., rainfall amount, relative humidity, and temperature), environment/landscape (e.g., vegetation and soil types) or human activities (e.g., transportation, urbanization, and waste management). Subsequent modeling of entomological/epidemiological dengue risk may benefit from the use of remote-sensing (RS) information that may provide appropriate ecological, meteorological, and geographic outputs. Choices may be made among available satellite products at various temporal, spectral, and spatial scales. In recent years, satellite products have been used to map numerous vector-borne diseases [6–10], and these products were recently proven to provide useful information for modeling *Aedes aegypti* or *Aedes albopictus* distribution [11–19], human dengue cases distribution [20,21] or the potential for future dengue vector or disease expansion [22–24].

The present study involves mapping the risk of the presence of *Aedes aegypti* immature stages around houses. This condition is necessary but not sufficient for dengue fever emergence given the numerous factors interplaying between the presence of *Aedes* larvae and dengue cases (rate of adult mosquito emergence, human-vector contact, human population movements, and acquired immunity to circulating serotype). Indeed, it has been argued that "mapping and spatial modeling based on mosquito presence or abundance data should be viewed as only representing potential dengue risk" [25]. Therefore, the larval maps of the present study will be referred to as "entomological dengue risk maps". Recipients of such maps should include the vector managing units that subsequently focus control interventions in places and times where/when the risk of vector presence is highest.

The practical and conceptual approach of tele-epidemiology could then be applied to the spatio-temporal mapping of entomological dengue risk in urban settings in Martinique. It has been developed and patented by the French Spatial Agency (CNES) and its partners [26,27]. The approach involves monitoring and studying the spatio-temporal dynamics of human and animal diseases that are closely related to weather/climate and environment variability. It relies on the identification of an experimental unit, which serves as the "object" that must be identified/characterized to properly assess the levels of risk. This unit is based on the sound knowledge of the biological and physical processes that underline the presence/densities of immature and adult vectors. It is thus widely dependent upon the disease being investigated. For example, this experimental unit is a pond (~1 ha) when studying Rift Valley Fever entomological risk [28] and a water body or aggregates of small water bodies (~0.1 ha) when studying urban malaria entomological risk [29]. Then, appropriate choices of satellite data and dynamic models must be assessed along with extensive use of *in situ* measurements.

Three observations underpinned the present study. Firstly, if the potential *Aedes aegypti* breeding habitats could not be directly detected using satellite images even at very high spatial resolution, their specific environment could be mapped. Indeed, the dwelling and yard conditions may reflect local habits regarding the maintenance of private yards/gardens and their close surroundings that may be associated with the presence of containers that retain water. Secondly, estate characteristics (*i.e.*, shading and tidiness of a house and its yard) have been identified as determinants of the presence/abundance of *Aedes aegypti* immature stages and eggs [18,30–32]. Thus, the characterization of fine-scale environments could provide information on the risk factors for the presence of dengue vector immature stages in areas where containers are present. Thirdly, meteorological conditions mainly drive the temporal dynamics for container filling as well as entomological dynamics (e.g., eggs hatching and larvae development). Consequently, the experimental unit here has been defined as the house with its nearby environment studied on one specific date. The state of such units was then described with details at the field level (*i.e.*, ground entomological investigations), meteorological level (high temporal resolution ground observation station), and environmental level (*i.e.*, high spatial resolution RS data). The main objective here was to model in space and time the houses considered to be so-called "positive" for *Aedes aegypti* immature stages from 2009 to 2011 in Tartane (Martinique, French Antilles) by using RS environmental data and field meteorological information. This modeling was performed to produce high spatio-temporal resolution dengue entomological risk maps.

## 2. Multi-Disciplinary Data and Methods

### 2.1. Studied Site and Period

The city of Tartane (14°45'29.24"N, 60°55'10.56"W) belongs to the Caravelle Peninsula, which is located northeast of Martinique. This area has historically served as a fishing cove with small and low-rise dwellings surrounded by small gardens or yards. The city center is near the seaside, whereas other sections are located uphill. The population is made up of ~3000 inhabitants. The area is a tourist attraction and includes many vacation accommodations. The studied site is approximately 8 km² (Figure 1). Several dengue outbreaks, including one that occurred in 2010, have started in Tartane, possibly due to favorable entomological conditions [33]. Although dengue epidemics "typically" last from July to December, viral circulation in 2010 was noted in early February, peaked in June and lasted until the end of the year. The studied period thus included this epidemic and ran from June 2009 to August 2011.

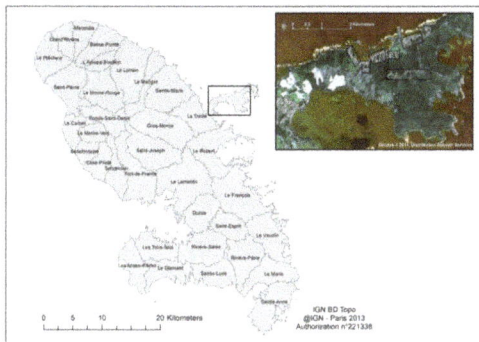

**Figure 1.** The Martinique Island, the studied area, and the six studied sections (black rectangles numbered 1 to 6) on the Tartane Peninsula.

### 2.2. Entomological Data

The SD-LAV conducts entomological surveillance in Martinique and thus regularly records mosquito information in the houses of the island municipalities [34]. All of the available records were selected from the SD-LAV databases for the studied period and area. Each record was associated with a given inhabited house visited on one of the following dates: 19/06/2009, 01/10/2009, 22/10/2009, 09/11/2009, 22/02/2010, 23/02/2010, 05/03/2010, 23/08/2011, 29/09/2010, and 02/12/2012. The records contained information regarding the number and type of domestic and peridomestic containers as well as the presence of *Aedes aegypti* immature stages (*i.e.*, all larvae stages and pupae). As the records did not contain geographic coordinates, only the records that could have been retrospectively plotted with a global positioning system (GPS) were maintained in the final database. Plotting was performed during July 2012 by the operators who conducted the 2009–2011 ground surveys and was based on the recorded name of the inhabitant and a deep knowledge of the area. Types of domestic or peridomestic containers include drum barrels, tanks, waste-bins, flower pots and saucers, gutters, tires, discarded appliances, and pools. Sampled houses are positioned in Figure 2. From their spatial distribution, six sections were identified as shown in Figure 1.

**Figure 2.** Map of the sampled houses.

The ten different dates covered different seasons of the year. A total of 117 houses were visited (*i.e.*, 88 houses visited once, 18 twice, 10 thrice and one visited four times), representing approximately 12% of the total number of houses in the studied area. Thus, the number of observations, *i.e.*, experimental units, was 158. A total of 88 out of those 158 experimental units were positive for the presence of peridomestic water-filled containers (from one to 11 containers per house at a given date). Thirty experimental units were *Aedes* larvae-positive (19% of all experimental units and 34% of all water-positive experimental units). Among all types of water-filled containers, large containers and drum barrels were most frequently identified as *Aedes* larvae positive with 57% and 51% positive observations, respectively. These containers were noted at 82% of the total number of positive breeding sites.

### 2.3. Meteorological Data

In Martinique, the summer rainy season (July to November) is characterized by frequent and heavy rainfall and maximum temperatures of approximately 32 °C, whereas the dry season (February to April) exhibits maximum temperatures of approximately 30 °C. These seasons are separated by two intermediate seasons. Rainfall amounts are heterogeneous on the island, ranging from approximately 1500 mm to greater than 4000 mm in the mountainous area. The year 2009 was exceptionally hot with minimal rainfall except on the Atlantic coastline. In 2010, temperatures were also high particularly during February and March. February was almost completely dry. Heavy precipitation started in early June followed by a very dry period. Again, the year 2011 experienced hot temperatures. However, 2011 was the wettest year of the 2009–2011 period with basically no dry season.

Daily temperature and humidity (minimum, maximum, and mean values) as well as precipitation amounts were provided for the present study by Météo-France. They were recorded at the observing station located in the Caravelle Peninsula. Yearly precipitation levels recorded during 2009, 2010, and 2011, were 948 mm, 1408 mm, and 1823 mm, respectively. Several variables were calculated from the raw data, added to the entomological database, and matched according to the date of the ground surveys as follows:

- total rainfall amount for the 2-, 3-, 4-, ..., to 30-day period before each entomological ground investigation date;
- average of temperature and relative humidity for the 2-, 3-, 4-, ..., to 15-day period before each entomological ground investigation date.

### 2.4. Satellite Images and Environmental Data

A Geoeye-1 optical image with clear sky was acquired on 13/03/2011. Data included four spectral bands at a 0.41-m spatial resolution (blue, green, red, and near infrared). The image was projected in WGS 84, UTM Zone 20 N and geometrically corrected using the 50-m spatial resolution elevation map (IGN BD ALTI®) from the French National Geographic Institute IGN (*Institut National de l'Information Géographique et Forestière*). Image processing was performed using ENVI 4.8 and ENVI EX (Exelis Visual Information Solutions). Other available geographic data included IGN topographic map (IGN BD TOPO®) and cadastral map (IGN BD ADRESSE®). Slope and objects height maps at 1-m spatial resolution were available through Litto3D® (IGN, *Service Hydrographique et Océanographique de la Marine, Direction de l'Environnement, de l'Aménagement et du Logement—Martinique, Agence des Aires Marines Protégées*), which was produced from airborne LIDAR measurements.

Three vegetation and soil indicators were derived from the Geoeye-1 image (Table 1).

**Table 1.** Environmental indicators calculated from the Geoeye-1 image at 0.41-m spatial resolution.

| Environmental Indicator | Spectral Bands Combination * | Description |
|---|---|---|
| **NDVI** (Normalized Difference Vegetation Index) [35,36] | $\frac{NIR-red}{NIR+red}$ | Values increase with the presence and density of vegetation. A value superior to 0.2 typically corresponds to a vegetated area. Negative values indicate non-vegetated features, such as barren surfaces (rocks and soils), water, built-up areas or asphalt. |
| **NDWI Mac Feeters** (Normalized Difference Water Index) [37] | $\frac{green-NIR}{green+NIR}$ | It delineates open water features while eliminating the presence of soil and terrestrial vegetation features. Values increase with the presence of water and decrease with the presence of vegetation. |
| **ANDWI** (Adapted NDWI Mac Feeters Index) | $\frac{blue-NIR}{blue+NIR}$ | Using the blue band, this adapted NDWI Mac Feeters maximizes the detection of water. |

* NIR: Near infrared; SWIR: Short wave infrared.

A three-step classification procedure was applied to produce the Land Use and Land Cover (LULC) map of the studied area. Firstly, a supervised maximum likelihood pixel-based classification was performed in ENVI 4.8 based on a set of training areas covering 5.4% of the total surface of the image. For each identified LULC class, a set of training polygons were digitized by an operator who photo-interpreted the Geoeye-1 image. The spectral signature of each class was then built by the software. Each pixel was assigned to the class having the highest probability to be the correct one based on those spectral signatures. No exclusion threshold was defined, so every pixel of the studied zone was classified. Some validation regions were also digitized, and they covered 4.7% of the image total area. They were used to calculate the kappa coefficient, which provides a measurement of the classification accuracy. This coefficient was 0.91, indicating a good agreement between the resulting LULC classes and the validation areas. Secondly, as improvements could be expected in the classification accuracy of some elements of the landscape, an object-oriented classification was performed. Segmentation, merging objects, and implementation of rules (area, convexity, average values of bands...) were undertaken in the ENVI EX Feature Extraction module. The quality of this classification was assessed by photo-interpretation, and it was thus concluded that the roofs and the swimming pools were accurately classified. Thirdly, a final classification was produced by merging both pixel and object classifications using a decision tree. It included fourteen land-cover classes: *i.e.*, five for vegetation, including "trees", "sugar cane", "stubbles", "lawn", "sparsely vegetated soil", five different types of roofs, "sand", "asphalt", "swimming pools", and "sea/ocean". A merged class for all roof types was also created.

*2.5. Geographic Information System (GIS)*

A GIS was built using ArcGIS 10.0 (Environmental Research Systems Institute, Redlands, CA, USA) to characterize the experimental units defined in the Introduction. All 117 surveyed houses were plotted, and environmental indices, LULC map, and elevation data were added as geo-referenced layers. Each single house of the Caravelle Peninsula was isolated as an object based on the LULC map. The plot around each house was identified using the IGN cadastral map.

Environmental variables, *i.e.*, the minimum/maximum/mean for the three indices, the slope and the object height and the areas of each LULC class, were computed for each surveyed house. This was accomplished for each plot and for the 50-m and 100-m radius buffer zones around the individual houses. The Euclidian distance from the house to the first patch of each LULC class as well as the area

81

of houses (assumed equal to the area of their roofs) and plots were calculated. These data were merged with the entomological database for each house.

### 2.6. Modeling Strategy

The overall database included the entomological, environmental, and meteorological variables described above. Each record was associated with one house visited on one date, *i.e.*, with one experimental unit. The chosen scenario (Figure 3) included the following two steps that involved the investigation of driving environmental and meteorological factors:

- Step 1: the presence of one or several water-filled container(s) in the vicinity of a house at a given date, independently of the presence or not of *Aedes aegypti* immature stages. This involved the detection of the water-positive experimental units;
- Step 2: the presence of *Aedes aegypti* immature stages exclusively in the experimental units that held one or several water-filled container(s). This involved the detection of the *Aedes* larvae-positive experimental units. No reference to the larval density was included.

Given that the number of domestic water containers was very low in the area, only the peridomestic containers were considered.

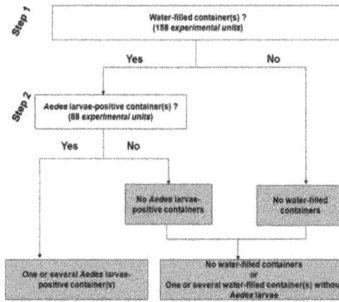

**Figure 3.** Scenario retained for dengue entomological risk mapping.

### 2.7. Statistical Analysis and Risk Mapping

Statistical analyses were performed using Stata 11 (Stata Corporation, College Station, TX, USA). Logistic regression analyses to explain the outcomes from both steps above were fitted at the experimental unit level using environmental and meteorological indicators as possible explanatory variables. For each model, uncorrelated variables with $p$-values < 0.25 from univariate analyses were candidates for multivariate analyses. Selection among the high number of co-linear variables, *i.e.*, meteorological and environmental variables that were created at several time and space scales, was performed by minimizing the AIC (the Akaike Information Criterion) in univariate analysis as well as by choosing variables with the best biological input. A manual backward stepwise selection procedure was applied in the final model to select variables with $p$-values < 0.05. The sampling scheme implied that some autocorrelations could exist between observations given that nearby observations could be more similar than distant ones due to the presence of more similar surroundings. In the case that the local environment was not fully considered by the explanatory variables, a random effect was added to the models at the level of the section. Model validity was assessed using the full sample and Receiver Operating Characteristic (ROC) curve (*i.e.*, representation of sensitivity against 1-specificity or true positive rate *versus* false positive rate, thereby providing the discriminative value of a test). It should be noted that the small amount of observations did not allow validity assessment of the models with a subset of observations. The cut-off value was chosen to maximize sensitivity and specificity.

Robustness was assessed using six sub-models of each final model fitted by separately omitting the experimental units of each section.

The linear equations derived from the final models allowed for the prediction of outcomes at the non-surveyed experimental units, *i.e.*, other Tartane houses and dates other than the dates of the surveys. We have chosen to undertake mapping for each day of the 2010 year to visualize the seasonal variability. The explanatory variables of the final models (Steps 1 and 2) were extracted from the GIS for every single building of the area. Meteorological independent variables that were significantly associated with the outcomes were calculated for each day of 2010. Then, the equation of Step 2 was applied to the buildings that were predicted as water positive in Step 1. The results were daily maps of the houses harboring *Aedes* larvae-positive container(s). These maps were merged into composite monthly maps that included the number of days for which each house was predicted to be *Aedes* larvae positive.

## 3. Results and Discussion

### 3.1. Results

#### 3.1.1. Step 1. Modeling the Water-Positive Experimental Units

In univariate analysis, several environmental and meteorological variables were significantly associated with the presence of water-filled container(s) in the experimental units. These environmental parameters included the area of the class "sparsely vegetated soil" in the plot, area of the class "tiled roof" in the plot, area of the class "swimming pool" within a 50-m buffer, slope of the plot, mean object height in the 100-m buffer (with positive sign), as well as area of the class "lawn" within the 50-m buffer, area of the class "sand" within the 50-m buffer and distance to the class "sparsely vegetated soil" (with negative sign). Some rainfall, temperature, and humidity variables were also positively or negatively associated with the presence of one or several water-filled container(s) in an experimental unit.

In multivariate analysis, the area of the class "sparsely vegetated soil" in the plot and the total rainfall during the 4-day period before the day of the field visit were positively associated with the outcome, whereas the area of the class "lawn" within the 50-m buffer around the house was negatively associated with the outcome (Table 2). The section random effect was not statistically significant in the final model. In addition, the six sub-models fitted by separately omitting the experimental units of each section provided similar coefficients compared with the final model. The area under the ROC curve was 0.72 (95% confidence interval: 0.64–0.80). A total of 103 experimental units out of 158 were correctly predicted (65%). The sensitivity was 63%, and the specificity was 69%. The positive predictive value was 71%, and the negative predictive value was 59%.

**Table 2.** Remote-sensing environmental and ground meteorological variables significantly associated with water-positive experimental units. Multivariate logistic regression analyses with section random effects are provided (Step 1).

| Explanatory Variables | Coefficient | 95% Confidence Interval | *p*-Value |
|---|---|---|---|
| **158 Experimental Units, 6 Sections** | | | |
| Surface of the class "sparsely vegetated soil" in the house plot (per 10 m²) | 0.10 | [0.02; 0.19] | 0.017 |
| Surface of the class "lawn" within the 50-m buffer around the house (per 100 m²) | −0.10 | [−0.17; −0.03] | 0.007 |
| Total rainfall during the 4-day period before field visit (per 10 mm) | 0.26 | [0.07; 0.46] | 0.007 |
| *section random effect* | | | *0.498* |

### 3.1.2. Step 2. Modeling the *Aedes* Larvae-Positive Experimental Units

In univariate analysis, several environmental and meteorological variables were associated with the presence of *Aedes* larvae-positive container(s) in the experimental units that held water-filled container(s). Those environmental variables included area of the class "tree" within the 50-m buffer, mean height of the houses within the 50-m buffer, mean NDVI within the 50-m buffer (with positive sign), as well as area of the class "asphalt" within the 50-m buffer, area of the class "swimming pool" within the 50-m buffer, and area of the class "tiled roof" within the 50-m buffer (with negative sign). Some rainfall, temperature, and humidity factors were also positively or negatively associated.

In multivariate analysis, the mean of the maximum humidity recorded during the 5-day period before the day of the ground investigation was positively associated with the outcome, whereas the area of the class "asphalt" within the 50-m buffer around the house was negatively associated (Table 3). The section random effect was not statistically significant in the final model. In addition, the six sub-models fitted by separately omitting the experimental units of each section provided very similar coefficients compared with the final model. The area under the ROC curve was 0.74 (95% confidence interval, 0.63–0.86). A total of 64 experimental units out of 88 were correctly predicted (73%). The sensitivity was 70%, and the specificity was 74%. The positive predictive value was 58%, and the negative predictive value was 83%.

**Table 3.** Remote-sensing environmental and ground meteorological variables significantly associated with *Aedes* larvae-positive experimental units among units that hold water-filled container(s). Multivariate logistic regression analyses with section random effects are provided (Step 2).

| Explanatory Variables | Coefficient | 95% Confidence Interval | *p*-Value |
|---|---|---|---|
| **88 Experimental Units, 6 Sections** | | | |
| Surface of the class "asphalt" within the 50-m buffer around the house (per 100 m$^2$) | −0.10 | [−0.17; −0.04] | 0.003 |
| Mean of maximum humidity during the 5-day period before field visit (per 1%) | 0.27 | [0.07; 0.48] | 0.008 |
| *section random effect* | | | *0.209* |

### 3.1.3. Application of the Scenario (Step 1 + Step 2)

Final predictions from the chosen scenario are displayed in Table 4. A total of 132 experimental units out of 158 were correctly predicted (84%). The sensitivity was 57%, and the specificity was 90%. The percentage of correctly classified predictions ranged from 67% to 92% depending on sections. The positive predictive value was 57%, and the negative predictive value was 90% (false positive rate = 43%; false negative rate = 10%).

**Table 4.** Final predictions of the scenario *versus* field data: number of *Aedes* larvae-positive experimental units *versus* number of *Aedes* larvae-negative experimental units.

| | | Prediction from Scenario | | |
|---|---|---|---|---|
| | | Number of *Aedes* Larvae-Negative Experimental Units | Number of *Aedes* Larvae-Positive Experimental Units | Total |
| **Field Data** | Number of *Aedes* larvae-negative experimental units | 115 | 13 | 128 |
| | Number of *Aedes* larvae-positive experimental units | 13 | 17 | 30 |
| | Total | 128 | 30 | 158 |

### 3.1.4. Predictive Entomological Risk Mapping in Tartane

Both steps of the scenario were successfully applied to every building (983) of the studied area at each day of the 2010 year to generate high spatio-temporal resolution entomological risk maps. The resulting composite monthly maps are displayed in Figure 4 and Supporting Information. None of the buildings was predicted as being *Aedes* larvae-positive for 100% of the days in 2010. A total of 126 buildings were predicted to have less than 10 *Aedes* larvae-positive days in the year, among which 28 were predicted to always be negative for *Aedes* larvae (0 positive days in the year). Maximum entomological risk was found in section 5 (see Figure 1, top right corner). The risk was decreased in sections 1, 2 and 4 (same risk) as well as section 6 and 3 (see Figure 1, top right corner) with yearly figures among the sections ranging from 77% to 15% *Aedes* larvae-positive experimental units. June and September 2010 exhibited the highest predicted entomological risk (45% and 44%, respectively, of the number of experimental units were predicted as *Aedes* larvae positive), whereas February and December were predicted with the lowest risk (17% and 18%, respectively, of the number of experimental units were predicted as *Aedes* larvae positive).

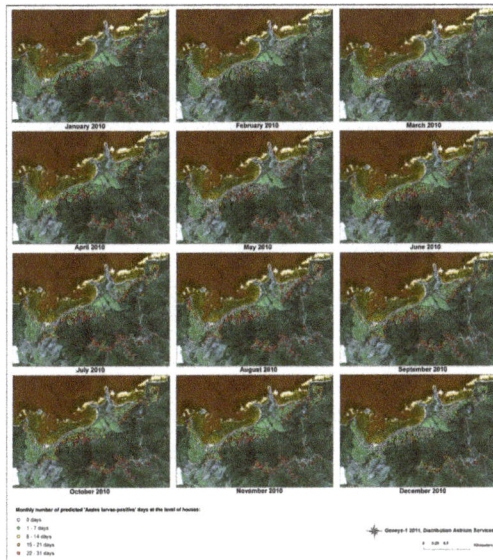

**Figure 4.** Monthly entomological risk maps from the modeling experiment based on data from January until December 2010. The number of predicted *Aedes* larvae-positive days for the 983 buildings within the studied area is provided (see color code at bottom left).

### 3.2. Discussion

In the present study, the practical and conceptual approach of tele-epidemiology, which was developed by the CNES and its partners, was used to generate spatio-temporal high-resolution entomological risk maps for the presence of *Aedes aegypti* immature stages in Tartane, Martinique. The experimental unit was the single house and its close surroundings studied at a specific date. This unit was proven to be an appropriate scale for mapping risk, which is consistent with the suggestion that entomological dengue risk would best be measured at the household scale [38]. Environmental and meteorological data, which are synergetic drivers of mosquito presence and density, were both included for modeling the presence of dengue vector larvae. Ecological data has also been used as a surrogate for sociological and behavioral information. The resulting maps are examples of modeled entomological risk maps at the dwelling level with daily temporal resolution. These maps highlight the spatio-temporal variability among the houses that contained *Aedes* larvae-positive container(s).

### 3.2.1. A Two-Step Approach

A two-step approach was performed to closely link entomological modeling to the biological, physical, and societal mechanisms that drive (i) the presence of water-filled containers and (ii) larval development. Each step involved distinct physical and biological mechanisms, so these steps were separately considered. The presence of containers around houses is related to population behavior and socio-economic/socio-cultural aspects, whereas the presence of larvae in those containers is associated with ecological and meteorological factors that impact their biological cycle.

Modeling of the houses harboring one or several water-filled container(s) at a given date was achieved using two predictors from the Geoeye-1 image and one field meteorological variable. Firstly, the surface of the class "sparsely vegetated soil" in the house plot was positively associated with the water-positive experimental units. Secondly, the surface of the class "lawn" within the 50-m buffer around the house was negatively associated with the outcome. Both variables represent garden maintenance given that houses surrounded by lawn are likely to be well maintained. On one hand, containers or waste-bins that could be filled with precipitation should occur less frequently in well-maintained environments. On the other hand, an individual's socio-economic level may reflect the need to collect rainfall water in drum barrels to save money. In contrast, collecting rainfall may also be related to environmental-friendly behavior. Thirdly, the total rainfall during the 4-day period before the field visit was logically a risk factor for the presence of water-filled container(s) given that most of the containers, either intentionally or not, were filled with rainfall. Artificial filling of containers, which rarely happens, was not considered in this study. Nevertheless, large containers and drum barrels, which serve as major *Aedes* larval sites (82% of the total number of positive sites), were not artificially filled and were correctly accounted for in the analysis.

Modeling of the houses harboring one or several *Aedes* larvae-positive container(s) among the sampled houses having water-filled container(s) was achieved using one predictor from the Geoeye-1 image and one ground meteorological variable. Firstly, the surface of the class "asphalt" within the 50-m buffer around the house was negatively associated with the outcome. This variable was strongly inversely correlated with the NDVI, which evaluates the density of vegetation. Many studies have highlighted the association between shade potentially provided by vegetation and the presence of *Aedes* immature stages [30,39,40]. Shade may lower the very high water temperatures that have been specifically recognized as a negative factor for the presence of *Aedes aegypti* larvae [3]. Vegetation may also provide nutrients for larvae via leaves falling in the water [41], whereas nectar could serve as food for adult mosquitoes [42]. In Martinique, *Aedes aegypti* is mainly endophilic, so vegetation probably does not serve as a resting site. Although the scale of the associations between shade and *Aedes aegypti* have been established at 2–3 m using ground data [40], the results of the present work corroborate previous studies that used remote-sensing images. Indeed, the presence of trees within a 30-m radius buffer zone was associated with adult *Aedes aegypti* abundance in Arizona [18], and niche modeling using Landsat 7 images (with 30-m spatial resolution) predicted the areas suitable for *Aedes aegypti* breeding sites in Colombia [11]. Nonetheless, the class "asphalt" likely provided more information than a NDVI variable given that this class was retained in the final statistical model instead of any NDVI variable. Indeed, asphalt surroundings, which are related to high temperatures and evaporation, may also be unfavorable for allowing the breeding sites to persist long enough to be suitable for full cycle larval development. Secondly, the mean of maximum humidity for the 5-day period before the entomological record was positively associated with the presence of *Aedes* larvae-positive container(s) in the experimental units. The association was robust given that the variables for the 2- to 14-day period before the entomological record were also significantly associated with this outcome. Similar results were noted in Brazil [43] and Australia [44]. Humidity is positively correlated with precipitation and negatively correlated with evaporation. Thus, humidity may be related to an increased presence of water in the containers, thereby increasing the probability of complete larval development. This finding is consistent with previous works that have highlighted the association between rainfall and *Aedes aegypti* [45–48] or included rainfall data as predictive variable for modeling dengue risk [24].

### 3.2.2. Scales and Resolutions

Very high spatial resolution remote-sensing environmental variables and high temporal resolution ground meteorological variables were included. The ecological factors were temporally static (*i.e.*, one value for the whole period), but they provided spatial dynamic as these variables were extracted from a very high spatial resolution satellite image (GeoEye-1 0.41 cm). The urban environment in Tartane remained quite stable during the studied period, justifying the use of one unique satellite image. On one hand, the majority of the LULC classes were appropriately defined by this unique image (buildings, asphalt, sea, swimming pools, sand, sugar cane, stubbles, and trees) given that they are not seasonally impacted. On the other hand, with regard to lawns, sparsely vegetated soil, and soil and vegetation indices that vary according the seasons, one unique image only provided a snapshot of the landscape. However, even static environmental data provide useful information for describing experimental units. Indeed, lawn and sparsely vegetated soils were identified as final explanatory variables at the first step of modeling, wherein environmental factors aimed to describe socio-economical and behavioral inputs. Using one unique image did not prevent from highlighting useful information; for example, the existence of a lawn during a dry season indicated that the garden was particularly well maintained, which reinforced the probability of the absence of water-filled containers. On the contrary, meteorological factors were spatially static (*i.e.*, one value for the whole studied area was obtained from a ground observation station). However, these factors were extracted at a fine temporal scale (daily), thereby providing temporal dynamics. Regarding the size of the studied area, it was difficult to obtain variability in meteorological data apart from installing ground devices in the different sections or accessing ground radar data (about 1 km spatial resolution). Indeed, even remotely sensed rainfall data (e.g., Tropical Rainfall Measuring Mission (TRMM) and Rainfall Estimation (RFE)) do not provide heterogeneity in such small areas due to their spatial resolution (0.25° for TRMM and RFE). Nevertheless, a recent study has revealed that it is crucial to consider the variability of rainfall amounts when modeling vector distribution, but the scales were beyond the size of the studied area [49].

### 3.2.3. Accuracy and Validity of Models

Given that the number of observations was limited, the validation was not based upon datasets different from those used to fit the models. The fact that 84% of the experimental units were correctly predicted is thus probably overestimated. The positive predictive value of the two-steps scenario was 57%, and the negative predictive value was 90% (false positive rate = 43% and false negative rate = 10%). From a larval control operational point of view, this scenario is very powerful to limit the amount of time spent on the ground by preventing the teams from visiting a large number of negative houses. However, the predictive ability to detect positive houses should be improved given that a large proportion of risky houses are missed.

Bias could have been introduced in the models given that ground data were not collected on the same dates for each section. In addition, some sections were preferentially investigated during rainy/dry seasons. Nevertheless, analysis of the resulting risk maps indicated that the sections that were followed mainly during rainy or dry seasons exhibited no particular pattern in terms of prediction of *Aedes aegypti* larvae presence. Indeed, although an entirely environmental model could not have been fitted due to this bias, the presence of meteorological variables in the models served as seasonal adjustment. In addition, the six sub-models which were fitted at both steps by separately omitting the experimental units of each section provided estimates similar to the final models. The latter indicates that none of the sections significantly modified the modeling results. Finally, the fact that the section effect was not significant at both steps of the analysis revealed that the explanatory variables managed to represent the environmental heterogeneity of the dataset.

### 3.2.4. From Entomological Risk Maps to Ground Control Actions

Models should be used to spatially and temporally prioritize prevention where the risk is the greatest [38]. On the one hand, entomological data are rarely collected in a routine fashion in a given area. When field studies are undertaken, they often provide only a snapshot of a continuous phenomenon. Risk maps are then expected to provide continuity for enhanced risk evaluation. On the other hand, maps that are available at the household level could facilitate the detection of "key premises" [50,51] for efficient control. Indeed, houses with an increased number of *Aedes* larvae-positive days in a month could be targeted for the destruction of breeding sites and dispersal of information to human populations to decrease the logistic burden. The resulting maps of the present study could subsequently be used as a tool for *Aedes aegypti* larvae control operational systems based on updated satellite images and meteorological information. Indeed, the equations resulting from the final models at both steps could be applied in the same area at other dates or even in other similar settings if the same LULC classes could be extracted from very high-resolution images. Although such image processing is time consuming, a unique image could be used to predict the entomological risk for several years in urban areas that are not experiencing rapid expansion. A prerequisite would involve testing the validity of any extrapolated predictions with new ground data. In any objective of using risk maps to assess dengue epidemiological risk, the association between the spatio-temporal distribution of immature stages and dengue disease should be evaluated in this area of Martinique. Indeed, a review paper highlighted various studies that have demonstrated this direct association as well as other studies that failed [25]. It is highly probable that additional layers of information should be added to those entomological maps, including pupal productivity or adult (infected) vector densities as well as human factors (acquired immunity, human-vector contact, population movement and distance to epidemic areas).

## 4. Conclusions

The present study revealed that environmental information at a fine spatial scale obtained using very high-resolution satellite images coupled with field meteorological data at a fine temporal scale were successfully highlighted as explanatory variables for the presence of *Aedes aegypti* larvae in Tartane, Martinique. Daily entomological predictive risk maps of the presence of *Aedes aegypti* larvae-positive container(s) were generated at the level of individual houses. As it has been often advocated, focusing interventions in the places and periods with maximum risk is paramount to enhanced allocation of limited resources and improved dengue control. In this context, such entomological risk maps may be considered as one of the tools available, and tele-epidemiology may be applied. Finally, the approach presented in this paper can be applied to assessments of the emerging chikungunya entomological risk levels in Martinique given that *Aedes aegypti* is also the vector for this disease [52].

**Acknowledgments:** The authors would like to thank David Flamanc from the DEAL Martinique (*Direction de l'Environnement, de l'Aménagement et du Logement*) for providing and permitting the full use of the Litto3D®data. The authors acknowledge the SERTIT (*SErvice Régional de Traitement d'Image et de Télédétection*; Hervé Yésou, Carlos Uribe, Claire Huber) for providing inputs for the definition of the very high-resolution indices. VM, CV and JPL thank Sanofi-Pasteur for supporting part of this study (images purchasing, data analysis). VM and JPL also acknowledge CNES (*Centre National d'Etudes Spatiales*) for support. This is an LDEO contribution #7843.

**Author Contributions:** VM, CV and JPL conceived and designed the experiments. VM, AY, ME, PP, MG, and JPL performed the experiments. VM, CV, MG, and JPL analyzed the data. VM, CV, YMT, and JPL contributed to the writing of the manuscript.

**Conflicts of Interest:** The founding sponsor had no role in the design of the study; the collection, analyses or interpretation of data; the writing of the manuscript, or the decision to publish the results.

# References

1. Murray, N.E.; Quam, M.B.; Wilder-Smith, A. Epidemiology of dengue: Past, present and future prospects. *Clin. Epidemiol.* **2013**, *5*, 299–309. [PubMed]

2. Christophers, S. *Aedes Aegypti (L.) The Yellow Fever Mosquito: Its Life History, Bionomics and Structure*; Cambridge University Press: Cambridge, UK, 1960.

3. Hemme, R.R.; Tank, J.L.; Chadee, D.D.; Severson, D.W. Environmental conditions in water storage drums and influences on *Aedes aegypti* in Trinidad, West Indies. *Acta Trop.* **2009**, *112*, 59–66. [CrossRef] [PubMed]

4. Wan, S.W.; Lin, C.F.; Wang, S.; Chen, Y.H.; Yeh, T.M.; Liu, H.S.; Anderson, R.; Lin, Y.S. Current progress in dengue vaccines. *J. Biomed. Sci.* **2013**, *20*. [CrossRef]

5. Tun-Lin, W.; Lenhart, A.; Nam, V.S.; Rebollar-Tellez, E.; Morrison, A.C.; Barbazan, P.; Cote, M.; Midega, J.; Sanchez, F.; Manrique-Saide, P.; *et al.* Reducing costs and operational constraints of dengue vector control by targeting productive breeding places: A multi-country non-inferiority cluster randomized trial. *Trop. Med. Int. Health* **2009**, *14*, 1143–1153. [CrossRef] [PubMed]

6. Stefani, A.; Dusfour, I.; Corrêa, A.P.S.A.; Cruz, M.C.B.; Dessay, N.; Galardo, A.K.R.; Galardo, C.D.; Girod, R.; Gomes, M.S.M.; Gurgel, H.; *et al.* Land cover, land use and malaria in the Amazon: A systematic literature review of studies using remotely sensed data. *Malar. J.* **2013**, *12*. [CrossRef] [PubMed]

7. Machault, V.; Vignolles, C.; Borchi, F.; Vounatsou, P.; Pages, F.; Briolant, S.; Lacaux, J.-P.; Rogier, C. The use of remotely sensed environmental data in the study of malaria. *Geospat. Health* **2011**, *5*, 151–168. [PubMed]

8. Yang, G.-J.; Vounatsou, P.; Zhou, X.-N.; Utzinger, J.; Tanner, M. A review of geographic information system and remote sensing with applications to the epidemiology and control of schistosomiasis in China. *Acta Trop.* **2005**, *96*, 117–129. [CrossRef] [PubMed]

9. Kalluri, S.; Gilruth, P.; Rogers, D.; Szczur, M. Surveillance of arthropod vector-borne infectious diseases using remote sensing techniques: A review. *PLoS Pathog.* **2007**, *3*, 1361–1371. [PubMed]

10. Bergquist, N.R. Vector-borne parasitic diseases: New trends in data collection and risk assessment. *Acta Trop.* **2001**, *79*, 13–20. [CrossRef] [PubMed]

11. Arboleda, S.; Jaramillo, O.N.; Peterson, A.T. Spatial and temporal dynamics of *Aedes aegypti* larval sites in Bello, Colombia. *J. Vector Ecol.* **2012**, *37*, 37–48. [CrossRef] [PubMed]

12. Neteler, M.; Roiz, D.; Rocchini, D.; Castellani, C.; Rizzoli, A. Terra and Aqua satellites track tiger mosquito invasion: Modelling the potential distribution of Aedes albopictus in north-eastern Italy. *Int. J. Health Geogr.* **2011**, *10*. [CrossRef]

13. Roiz, D.; Neteler, M.; Castellani, C.; Arnoldi, D.; Rizzoli, A. Climatic factors driving invasion of the tiger mosquito (*Aedes albopictus*) into new areas of Trentino, Northern Italy. *PLoS One* **2011**, *6*, e14800. [CrossRef] [PubMed]

14. Estallo, E.L.; Lamfri, M.A.; Scavuzzo, C.M.; Almeida, F.F.; Introini, M.V.; Zaidenberg, M.; Almiron, W.R. Models for predicting *Aedes aegypti* larval indices based on satellite images and climatic variables. *J. Am. Mosq. Control Assoc.* **2008**, *24*, 368–376. [CrossRef] [PubMed]

15. Rotela, C.H.; Espinosa, M.O.; Albornoz, C.; Lafaye, M.; Lacaux, J.-P.; Tourre, Y.M.; Vignolles, C.; Scavuzzo, C. Desarrollo de mapas predictivos de densidad focal de *Aedes aegypti* en la ciudad de Puerto Iguazú (Argentina), basados en información ambiental derivada de imágenes SPOT 5 HRG1. In Proceedings of XIII Simposio Latinoamericano de Percepcion Remota y Sistemas de Informacion Espacial (SELPER), Havana, Cuba, 22–28th September 2008.

16. Fuller, D.O.; Troyo, A.; Calderón-Arguedas, O.; Beier, J.C. Dengue vector (*Aedes aegypti*) larval habitats in urban environment of Costa Rica analysed with ASTER and QuickBird imagery. *Int. J. Remote Sens.* **2009**, *31*, 3–11. [CrossRef]

17. Vanwambeke, S.O.; Bennett, S.N.; Kapan, D.D. Spatially disaggregated disease transmission risk: Land cover, land use and risk of dengue transmission on the island of Oahu. *Trop. Med. Int. Health* **2011**, *16*, 174–185. [CrossRef] [PubMed]

18. Landau, K.I.; van Leeuwen, W.J. Fine scale spatial urban land cover factors associated with adult mosquito abundance and risk in Tucson, Arizona. *J. Vector Ecol.* **2012**, *37*, 407–418. [CrossRef] [PubMed]

19. Sarfraz, M.S.; Tripathi, N.K.; Tipdecho, T.; Thongbu, T.; Kerdthong, P.; Souris, M. Analyzing the spatio-temporal relationship between dengue vector larval density and land-use using factor analysis and spatial ring mapping. *BMC Public Health* **2012**, *12*. [CrossRef] [PubMed]

20. Van Benthem, B.H.; Vanwambeke, S.O.; Khantikul, N.; Burghoorn-Maas, C.; Panart, K.; Oskam, L.; Lambin, E.F.; Somboon, P. Spatial patterns of and risk factors for seropositivity for dengue infection. *Am. J. Trop. Med. Hyg.* **2005**, *72*, 201–208. [PubMed]

21. Rotela, C.; Fouque, F.; Lamfri, M.; Sabatier, P.; Introini, V.; Zaidenberg, M.; Scavuzzo, C. Space-time analysis of the dengue spreading dynamics in the 2004 Tartagal outbreak, Northern Argentina. *Acta Trop.* **2007**, *103*, 1–13. [CrossRef] [PubMed]

22. Neteler, M.; Metz, M.; Rocchini, D.; Rizzoli, A.; Flacio, E.; Engeler, L.; Guidi, V.; Luthy, P.; Tonolla, M. Is Switzerland suitable for the invasion of *Aedes albopictus*? *PLoS One* **2013**, *8*, e82090. [CrossRef] [PubMed]

23. ECDC. *Development of Aedes Albopictus Risk Maps ECDC Technical Report*; European Centre for Disease Prevention and Control: Stockholm, Sweden, 2009.

24. Rogers, D.J.; Suk, J.E.; Semenza, J.C. Using global maps to predict the risk of dengue in Europe. *Acta Trop.* **2014**, *129*, 1–14. [CrossRef] [PubMed]

25. Eisen, L.; Lozano-Fuentes, S. Use of mapping and spatial and space-time modeling approaches in operational control of *Aedes aegypti* and dengue. *PLoS Negl. Trop. Dis.* **2009**, *3*, e411. [CrossRef] [PubMed]

26. Vignolles, C.; Lacaux, J.P.; Tourre, Y.M.; Bigeard, G.; Ndione, J.A.; Lafaye, M. Rift Valley fever in a zone potentially occupied by Aedes vexans in Senegal: Dynamics and risk mapping. *Geospat. Health* **2009**, *3*, 211–220. [PubMed]

27. Vignolles, C.; Tourre, Y.M.; Mora, O.; Imanache, L.; Lafaye, M. TerraSAR-X high-resolution radar remote sensing: An operational warning system for Rift Valley fever risk. *Geospat. Health* **2010**, *5*, 23–31. [PubMed]

28. Lacaux, J.-P.; Tourre, Y.-M.; Vignolles, C.; Ndione, J.-A.; Lafaye, M. Classification of ponds from high-spatial resolution remote sensing: Application to Rift Valley fever epidemics in Senegal. *Remote Sens. Environ.* **2006**, *106*, 66–74. [CrossRef]

29. Machault, V.; Vignolles, C.; Pagès, F.; Gadiaga, L.; Tourre, Y.M.; Gaye, A.; Sokhna, C.; Trape, J.-F.; Lacaux, J.-P.; Rogier, C.; *et al.* Risk mapping of *Anopheles gambiae s.l.* densities using remotely-sensed environmental and meteorological data in an urban area: Dakar, Senegal. *PLoS One* **2012**, *7*, e50674. [CrossRef] [PubMed]

30. Tun-Lin, W.; Kay, B.H.; Barnes, A. The Premise Condition Index: A tool for streamlining surveys of Aedes aegypti. *Am. J. Trop. Med. Hyg.* **1995**, *53*, 591–594. [PubMed]

31. Nogueira, L.A.; Gushi, L.T.; Miranda, J.E.; Madeira, N.G.; Ribolla, P.E. Application of an alternative *Aedes* species (Diptera: Culicidae) surveillance method in Botucatu City, Sao Paulo, Brazil. *Am. J. Trop. Med. Hyg.* **2005**, *73*, 309–311. [PubMed]

32. Peres, R.C.; Rego, R.; Maciel-de-Freitas, R. The use of the Premise Condition Index (PCI) to provide guidelines for *Aedes aegypti* surveys. *J. Vector Ecol.* **2013**, *38*, 190–192. [CrossRef] [PubMed]

33. Etienne, M. Etude de la bioécologie d'Aedes Aegypti à la Martinique en relation avec l'épidémiologie de la dengue. Ph. D. Thesis, Université de Montpellier I, Montpellier, France, 22 June 2006.

34. Yebakima, A. Control of *Aedes aegypti* in Martinique. Contribution of entomology studies. *Bull. Soc. Pathol. Exot.* **1996**, *89*, 161–162. [PubMed]

35. Rouse, J.W.; Hass, R.H.; Schell, J.A.; Deering, D.W. Monitoring vegetation systems in the Great Plains with ERTS. In Proceedings of the Third ERTS Symposium, Washington, DC, USA, 10–14 December 1973; NASA SP 351. pp. 309–317.

36. Tucker, C.J. Red and photographic infrared linear combinations for monitoring vegetation. *Remote Sens. Environ.* **1979**, *8*, 127–150. [CrossRef]

37. McFeeters, S.K. The use of the normalised difference water index (NDWI) in the delineation of open water features. *Int. J. Remote Sens.* **1996**, *17*, 1425–1432. [CrossRef]

38. Scott, T.W.; Morrison, A.C. Vector dynamics and transmission of dengue virus: Implications for dengue surveillance and prevention strategies: Vector dynamics and dengue prevention. *Curr. Top. Microbiol. Immunol.* **2010**, *338*, 115–128. [PubMed]

39. Vezzani, D.; Albicocco, A.P. The effect of shade on the container index and pupal productivity of the mosquitoes *Aedes aegypti* and *Culex pipiens* breeding in artificial containers. *Med. Vet. Entomol.* **2009**, *23*, 78–84. [CrossRef] [PubMed]

40. Vezzani, D.; Rubio, A.; Velazquez, S.M.; Schweigmann, N.; Wiegand, T. Detailed assessment of microhabitat suitability for *Aedes aegypti* (Diptera: Culicidae) in Buenos Aires, Argentina. *Acta Trop.* **2005**, *95*, 123–131. [CrossRef] [PubMed]

41. Reiskind, M.H.; Greene, K.L.; Lounibos, L.P. Leaf species identity and combination affect performance and oviposition choice of two container mosquito species. *Ecol. Entomol.* **2009**, *34*, 447–456. [CrossRef] [PubMed]

42. Martinez-Ibarra, J.A.; Rodriguez, M.H.; Arredondo-Jimenez, J.I.; Yuval, B. Influence of plant abundance on nectar feeding by *Aedes aegypti* (Diptera: Culicidae) in southern Mexico. *J. Med. Entomol.* **1997**, *34*, 589–593. [PubMed]

43. Favier, C.; Degallier, N.; Vilarinhos Pde, T.; de Carvalho Mdo, S.; Yoshizawa, M.A.; Knox, M.B. Effects of climate and different management strategies on *Aedes aegypti* breeding sites: A longitudinal survey in Brasilia (DF, Brazil). *Trop. Med. Int. Health* **2006**, *11*, 1104–1118. [CrossRef] [PubMed]

44. Azil, A.H.; Long, S.A.; Ritchie, S.A.; Williams, C.R. The development of predictive tools for pre-emptive dengue vector control: A study of *Aedes aegypti* abundance and meteorological variables in North Queensland, Australia. *Trop. Med. Int. Health* **2010**, *15*, 1190–1197. [CrossRef]

45. Wee, L.K.; Weng, S.N.; Raduan, N.; Wah, S.K.; Ming, W.H.; Shi, C.H.; Rambli, F.; Ahok, C.J.; Marlina, S.; Ahmad, N.W.; *et al.* Relationship between rainfall and *Aedes* larval population at two insular sites in Pulau Ketam, Selangor, Malaysia. *Southeast Asian J. Trop. Med. Public Health* **2013**, *44*, 157–166. [PubMed]

46. Baruah, S.; Dutta, P. Seasonal prevalence of *Aedes aegypti* in urban and industrial areas of Dibrugarh district, Assam. *Trop. Biomed.* **2013**, *30*, 434–443. [PubMed]

47. Duncombe, J.; Clements, A.; Davis, J.; Hu, W.; Weinstein, P.; Ritchie, S. Spatiotemporal patterns of *Aedes aegypti* populations in Cairns, Australia: Assessing drivers of dengue transmission. *Trop. Med. Int. Health* **2013**, *18*, 839–849. [CrossRef] [PubMed]

48. Stewart Ibarra, A.M.; Ryan, S.J.; Beltran, E.; Mejia, R.; Silva, M.; Munoz, A. Dengue Vector Dynamics (*Aedes aegypti*) Influenced by Climate and Social Factors in Ecuador: Implications for targeted control. *PLoS One* **2013**, *8*, e78263. [CrossRef] [PubMed]

49. Guilloteau, C.; Gosset, M.; Vignolles, C.; Alcoba, M.; Tourre, Y.M.; Lacaux, J.-P. Impacts of satellite-based rainfall products on predicting spatial patterns of Rift Valley fever vectors. *J. Hydrometeorol.* **2014**, *15*, 1624–1635. [CrossRef]

50. Tun-Lin, W.; Kay, B.H.; Barnes, A. Understanding productivity, a key to *Aedes aegypti* surveillance. *Am. J. Trop. Med. Hyg.* **1995**, *53*, 595–601. [PubMed]

51. Chadee, D.D. Key premises, a guide to *Aedes aegypti* (Diptera: Culicidae) surveillance and control. *Bull. Entomol. Res.* **2004**, *94*, 201–207. [CrossRef] [PubMed]

52. Simon, F.; Savini, H.; Parola, P. Chikungunya: A paradigm of emergence and globalization of vector-borne diseases. *Med. Clin. North. Am.* **2008**, *92*, 1323–1343. [CrossRef] [PubMed]

isprs International Journal of
*Geo-Information*

MDPI

*Article*

# Improving Inland Water Quality Monitoring through Remote Sensing Techniques

Igor Ogashawara [1,2,*] and Max J. Moreno-Madriñán [3]

[1] Remote Sensing Division, National Institute for Space Research (INPE), Avenida dos Astronautas, 1758, São José dos Campos 12227-010, Brazil
[2] Department of Earth Sciences, Indiana University-Purdue University Indianapolis, 723 W. Michigan Street, SL118, Indianapolis, IN 46202, USA; igorogas@iupui.edu
[3] Department of Environmental Health, Fairbanks School of Public Health, Indiana University, 714 N. Senate Avenue, EF 206, Indianapolis, IN 46202, USA; mmorenom@iu.edu
* Author to whom correspondence should be addressed; igoroga@gmail.com; Tel.: +55-12-3208-6484.

External Editors: Fazlay S. Faruque and Wolfgang Kainz
Received: 9 June 2014; in revised form: 13 October 2014; Accepted: 30 October 2014;
Published: 14 November 2014

**Abstract:** Chlorophyll-*a* (chl-*a*) levels in lake water could indicate the presence of cyanobacteria, which can be a concern for public health due to their potential to produce toxins. Monitoring of chl-*a* has been an important practice in aquatic systems, especially in those used for human services, as they imply an increased risk of exposure. Remote sensing technology is being increasingly used to monitor water quality, although its application in cases of small urban lakes is limited by the spatial resolution of the sensors. Lake Thonotosassa, FL, USA, a 3.45-km$^2$ suburban lake with several uses for the local population, is being monitored monthly by traditional methods. We developed an empirical bio-optical algorithm for the Moderate Resolution Imaging Spectroradiometer (MODIS) daily surface reflectance product to monitor daily chl-*a*. We applied the same algorithm to four different periods of the year using 11 years of water quality data. Normalized root mean squared errors were lower during the first (0.27) and second (0.34) trimester and increased during the third (0.54) and fourth (1.85) trimesters of the year. Overall results showed that Earth-observing technologies and, particularly, MODIS products can also be applied to improve environmental health management through water quality monitoring of small lakes.

**Keywords:** chlorophyll-*a*; cyanobacteria biomass; empirical algorithms; remote sensing

## 1. Introduction

Besides traditional multiple uses of inland waters by mankind, urban lakes also provide services, such as storm-water buffer, waste removal and recreation [1]. Because of such multiple uses and services, the deterioration of water quality in urban lakes has been a serious ecological and social problem that can severely affect human health [2]. Indeed, the public health implications of surface water quality are among the main concerns related to aquatic systems. In inland waters in general, one of the greatest problems is the eutrophication process caused by the increase of nutrient inputs. This process has been affecting the ecological health of many shallow lakes worldwide [3], and urban lakes are not an exception in this category. This condition is enhanced by the increasing shortage of available water resources, growing urbanization and the lack of a water governance system, which affects mainly recently urbanized countries. These and other human-related factors, along with climate variations and rising temperatures, have been suggested as possible factors triggering an increasing trend in cyanobacteria abundance [4]. Among the problems caused by eutrophication, special interest is placed on the toxins and taste and odor production by some species of algae and cyanobacteria. Such toxins can enter the body via oral, dermal and inhalation routes through drinking water, freshwater food consumption and recreational water activities. Moreover, they can poison or even kill humans and animals that consume contaminated water and food [2,3]. Although low concentrations of toxins from cyanobacteria and algae may be adequately removed from the source water by conventional water treatment, the same may not be the case with high initial concentrations [5]. In Florida, an increased risk for liver cancer has been associated with residence within the area served by a surface water treatment plant as compared with residents living in contiguous areas [6].

The first scientific report of animal contamination from a harmful algal bloom (HAB) related to cyanobacteria occurred in 1878 by George Francis [7]. He reported a HAB in Lake Alexandrina (Australia) and observed that after animals drank its water, they were poisoned and rapidly died. The author even observed the time that cyanotoxins took to cause death in different animals: in sheep, from six to eight hours; in horses, from eight to twenty-four hours; in dogs, from four to five hours, and in pigs, from three to four hours. One of the first reported cases of human casualty associated with cyanobacteria and their toxins came about in 1996, in the city of Caruaru, PE, Brazil, where exposure through kidney dialysis led to the death of approximately fifty patients [8]. This disaster raised the awareness of water quality managers, environmental agencies, policy makers and the general public to the problem of HABs. Likewise, this incident warned about the need for a reliable and constant monitoring of HABs by environmental and public health programs. The World Health Organization's (WHO) Working Group on the Protection and Control of Drinking Water Quality identified cyanotoxins as an issue requiring urgent attention. In the United States, 22 hospitalizations out of two hundred and ninety-six cases were reported by the Center for Disease Control and Prevention (CDC) during the years 2009–2010 [9].

Phytoplankton primary production has been regarded as a reliable and accurate indicator to monitor eutrophication and HABs [3]. In turn, several research studies have confirmed that chlorophyll-*a* (chl-*a*) is a universally acknowledged indicator of phytoplankton abundance and trophic state due to its visible manifestation and for being part of the eutrophication process [10,11]. Thus, chl-*a* concentration has been long applied as an indicator for water quality monitoring of inland waters. Although chl-*a* is relatively easily measured in comparison to algal biomass, traditional methods consist of field water sampling and laboratory analysis, which are usually costly and time consuming [12]. Consequently, water quality monitoring is not conducted with enough regularity, which, in turn, may lead to limitations in the study of the environmental dynamics.

Because of these facts, there is rapidly growing interest in the application of satellite remote sensing technology in environmental management. The reasons for such interest are based on several advantages, such as: (1) the synoptic view of the satellite images, which allows the user to retrieve information from large geographic areas; (2) the acquisition of data from places that are otherwise difficult to access; (3) the temporal resolution, which can provide a historical dataset allowing the users to retrieve information from the past [13]. Consequently, remote sensing technology has been increasingly used to facilitate the decision-making process of environmental managers and policy makers.

Numerous studies have focused on deriving chl-*a* concentration information from remote sensing satellites in inland water bodies [11,12,14,15]. However, there are several limitations related mainly to the sensor resolutions required to monitor aquatic systems with such high spatial irregularities and often small areas. Ogashawara *et al.* [14] addressed the problem of spatial, temporal and spectral resolutions of satellite images in the monitoring of inland waters. Regarding spatial resolutions, the authors studied a small water body in which the use of a sensor with low spatial resolution caused high interference in the signal from adjacent (not targeted) features. The authors also discussed the importance of spectral resolution, since for turbid inland waters, the bio-optical models are usually based on the optical properties located at the red and near-infrared (NIR) regions of the spectrum [16–18]. These algorithms used the ratio of the chl-*a* reflectance peak around 700 nm (NIR) to the reflectance near 675 nm, which is the red chl-*a* absorption band [19,20]. The temporal resolution is

93

also a very important aspect of remote sensing estimation of chl-*a* in aquatic systems, because of the quick responses of chl-*a* to changes in the environment; thus, a good temporal resolution is crucial for consistent water quality monitoring.

The objective of this study was to assess the applicability of the Moderate Resolution Imaging Spectroradiometer (MODIS) daily product of surface reflectance (MOD09GA) to detect chl-*a* concentration in a small inland lake, using as a case Lake Thonotosassa, a suburban lake in Tampa, FL, USA, which has experienced episodic blooms of blue-green algae (cyanobacteria) [21]. We justify the use of the MOD09GA by its spatial resolution (500 m), which is higher than the spatial resolution of the spectral bands usually used in ocean color algorithms. Moreover, since this is a daily product, it could improve the temporal monitoring coverage for algal blooms in this aquatic system, which currently occurs on a monthly basis for three sampling stations by traditional methods. Besides ecological concerns for this lake's water quality, there are public health concerns related to its designated use for human recreation and the fact that the outflow of this lake discharges into the Hillsborough River and, ultimately, into the water reservoir that provides the drinking water supply to the City of Tampa.

The specific goals of this study were: (1) to evaluate the use of the most common ocean color algorithms applied on MODIS Level 0 products for the estimation of chl-*a* estimations; (2) to evaluate the use of 1-km spatial resolution MODIS spectral bands in the retrieval of chl-*a*; and (3) to propose an empirical algorithm for chl-*a* concentration estimation in Lake Thonotosassa with seasonal calibrations using the MOD09GA product. These goals were motivated by our intention to propose a more consistent monitoring of chl-*a* (both temporally and spatially) for public health management in this lake, further developing a methodology that could be applied to generate the appropriate algorithms to be used in similar lakes in other areas.

## 2. Materials and Methods

### 2.1. Study Site

The study site for this research is Lake Thonotosassa (28°03′N, 82°16′W), which is located in a suburban area in Hillsborough County, Florida, USA (Figure 1). The lake is supplied mainly by the runoff from surrounding citrus groves and from Baker Creek, an improved drainage canal originating in Dover, Florida [22]. The lake has a surface area of 3.45 km$^2$, with a mean depth of 3.5 m and a maximum depth of 5.1 m. The water temperature in Lake Thonotosassa varies from 14 to 34 °C during the year, with a mean of approximately 25 °C. Short-term thermal stratification (less than 5 °C from the surface to bottom) and hypolimnetic oxygen deficits occur in the deeper parts of the lake during the warmer months (May–October) [23]. Cowell *et al.* [22] showed that the lake was in advanced stages of eutrophy by analyzing its limnological characteristics. The authors observed that the eutrophication process occurred because of 15 years of artificial enrichment by organic wastes from domestic sewage and citrus processing plants. Their findings showed that inorganic nutrient levels were high, as well as oxygen deficits in the hypolimnion and at the mud-water interface. The phytoplankton community was large and dominated by cyanobacteria that generate a primary productivity rate comparable to those of grossly polluted lakes.

The high eutrophication levels in Lake Thonotosassa got the attention of policy makers, and consequently, the Environmental Protection Commission of Hillsborough County (EPCHC) has monitored its water quality since 1975. According to the Southwest Florida Water Management District (SFWMD) [21], the annual averages for total phosphorous concentrations for Lake Thonotosassa for the period of 1992–2000 varied from 0.3 to 0.77 mg/L. Average annual total nitrogen for the same period ranged from 1.8 to 4.1 mg/L. These high concentrations of nutrients led to high algal biomass (chl-*a* concentrations), which has ranged from 62 µg/L to 179 µg/L during the same period. Under current state standards, average chl-*a* levels greater than 20 µg/L are considered indicative of poor water quality for aquatic life in Florida lakes [24].

## 2.2. Dataset

### 2.2.1. Limnological Dataset

The water quality data from Lake Thonotosassa used in this study were collected and provided by the EPCHC. Data were collected monthly as part of routine water quality monitoring programs from three sampling stations, one located at the inlet, another at the middle and a third one at the outlet regions of the lake (see Figure 1 for the locations). Such a monitoring program also covers Tampa Bay and water bodies in its watershed. In this study, we used measurements of total phosphorus (TP), total nitrogen (TN) and chl-*a* concentration, which were collected during the time period between 2001 and 2011. Monthly water samples were analyzed by the EPCHC laboratory, which is a National Environmental Laboratory Accreditation Program (NELAP) certified laboratory. Water quality analyses were conducted using a combination of Environmental Protection Agency (EPA) and American Public Health Association (APHA) standard methods, following quality assurance and quality control (QA/QC) guidelines from both methodologies and all NELAP QA/QC rules. The TP concentration was determined following the EPA 365.4 methodology; the TN concentration were calculated by the sum of the total Kjeldahl nitrogen (TKN) and nitrate/nitrite nitrogen, where TKN was determined by EPA 351.2 and the nitrate/nitrite nitrogen concentration by Standard Methods 4500 $NO_3$ F (SM 4500 $NO_3$ F) [25]. The Chl-*a* concentration was determined by Standard Methods10200 H (SM 10200 H) [25], using acetone and a tissue grinder. More details on the methods of the laboratory analysis and sampling of *in situ* data have been described previously [18,26,27].

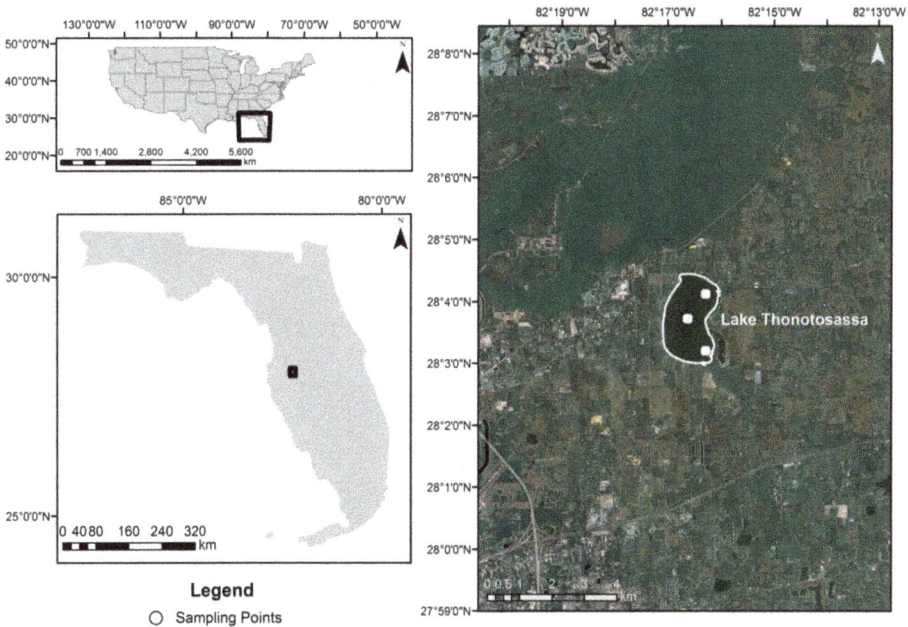

**Figure 1.** Location of Lake Thonotosassa in the State of Florida, USA, and the location of sampling points from the Environmental Protection Commission of Hillsborough County.

### 2.2.2. Remote Sensing Products

The remote sensing dataset consisted of two MODIS products. The following MODIS data were used: Level 0 MODIS data (L0_LAC) at a 1-km spatial resolution and Surface Reflectance Daily L2G at a 500-m spatial resolution (MOD09GA). MODIS L0 data were provided by NASA's Ocean Color

Products through their web portal [28]. MOD09GA data were provided by NASA's Earth Observing System Data and Information System (NASA/EOSDIS) through the Reverb web [29] portal. To process the L0 MODIS products to Level 3 products, we used the SeaWiFS Data Analysis System (SeaDAS) [30]. The MODIS Reprojection Tool (MRT) was used to re-project the MOD09GA dataset to the UTM coordinate system with the WGS-84 datum as the reference. The reflectance values were calculated using the scale factor of MOD09GA for the MOD09GA Collection 5 products acquired on the same day of the limnological collection from 2000 to 2011.

### 2.3. Ocean Color Algorithms Evaluation

Ocean color algorithms for MODIS usually use the spectral bands with 1-km spatial resolution. As is widely known, such a resolution is not useful for retrieving information from small inland aquatic systems, due to the mixing signal from other targets being different from the intended water body. Nevertheless, we used them for the purpose of comparison with our proposed algorithm. We evaluated three ocean color algorithms implemented in SeaDAS to show how MODIS 1-km bands could not retrieve geochemical data from Lake Thonotosassa.

### 2.3.1. Algorithms Used for Evaluation

We evaluated three existent chl-*a* algorithms, which are freely available from SeaDAS: the ocean color 3-band ratio (OC3M) [31], the Garver-Siegel-Maritorena model (GSM) [32] and the generalized inherent optical property (GIOP) [33]. OC3M is a fourth-order band ratio algorithm of remote sensing reflectance ($R_{rs}$), which can use two different band ratios: $R_{rs}443/R_{rs}547$ or $R_{rs}448/R_{rs}547$ [31]. The GSM is an optimized semi-analytical algorithm that simultaneously retrieves inherent optical properties (IOPs) from spectral measurements of normalized water leaving spectral radiance ($nL_w(\lambda)$) [32]. The GIOP model uses the spectral behaviors of several optically-active constituents (OACs) in the water column to apply in an inversion process. This process is based on finding the optimum set of eigenvalues between the modelled $R_{rs}$ and MODIS $R_{rs}$ using the Levenberg-Marquardt optimization scheme [33]. These three algorithms are presented and described in Table 1.

**Table 1.** Functional form of MODIS chl-*a* algorithms. OC3M, ocean color 3-band ratio; GSM, Garver-Siegel-Maritorena; GIOP, generalized inherent optical property.

| Algorithm | Reference | Functional Form |
|:---:|:---:|:---:|
| OC3M | [31] | $chl - a = 10^{(a_1 + a_2 R + a_3 R^2 + a_4 R^3 + a_5 R^4)}$ |
| GSM | [32] | $a_{phy, i} = chl - a \cdot a^*_{phy}$ |
| GIOP | [33] | $chl - a = \frac{a_{phy, ii}}{a^*_{phy}}$ |

From the table, $R$ is the chosen band ratio; $a_1$, $a_2$, $a_3$, $a_4$ and $a_5$ are coefficients from the polynomial equation with the following values: 0.2424, −2.7423, 1.8017, 0.0015 and −1.2280, respectively; $a_{phy,i}$ is the absorption coefficient of phytoplankton, which is substituted in the GSM model for the functional form; $a_{phy,ii}$ is the absorption coefficient of phytoplankton derived from the GIOP model (Equation (2)); and $a^*_{phy}$ is the average specific absorption coefficient of phytoplankton calculated from Morel [34] and implemented in SeaDAS.

The GSM uses Equation (1) to retrieve the chl-*a* concentration, absorption coefficient for dissolved and detrital materials ($a_{CDM}$) and the particulate backscatter coefficient ($b_{bp}$) at 443 nm. The parameters for the algorithm (Equation (1)) were obtained through simulated annealing, which is a global optimization technique [32].

$$L_{\omega N}(\lambda) = \frac{tF_0(\lambda)}{n_w^2} \sum_{i=1}^{2} g_i \left\{ \frac{b_{bw}(\lambda) + b_{bp}(\lambda_0)(\lambda/\lambda_0)^{-\eta}}{b_{bw}(\lambda) + b_{bp}(\lambda_0)(\lambda/\lambda_0)^{-\eta} + a_w(\lambda) + chla_{ph}*(\lambda) + a_{CDM}(\lambda_0)exp(-S(\lambda-\lambda_0))} \right\}^i \tag{1}$$

where $t$ is the sea-air transmission factor; $F_0(\lambda)$ is the extraterrestrial solar irradiance; $n_w$ is the index of the refraction of the water; $g_i$ is a fitting coefficient from Monte Carlo simulations of an idealized ocean by Gordon [35]. The GIOP algorithm uses the GSM algorithm [32] estimations of several inherent optical properties, such as: the $a_{phy}^*$, the specific absorption coefficient of non-algal particles ($a_{NAP}^*$), the specific absorption coefficient of colored detrital matter ($a_{CDM}^*$), the colored detrital matter absorption coefficient slope ($S_{CDM}$), the particle-specific backscattering coefficient ($b_{bp}$) and the backscattering coefficient slope ($S_{bp}$).

### 2.3.2. Level 0 MODIS data (L0_LAC)

The three previously described SeaDAS chl-*a* algorithms were evaluated using MODIS-Aqua L0 products. The products were atmospherically corrected by the Management Unit of the North Sea Mathematical Models (MUMM) algorithm using its default settings. This model for atmospheric correction was chosen because of its application for turbid waters, which is enhanced by the replacement of the usual assumption of zero water-leaving radiance in the NIR bands. Thus, it is substituted by the assumption of the spatial homogeneity of the reflectance ratio (748/869), which is used for aerosol and water reflectance within an image [36].

### 2.4. Algorithm Development

As the goal of this research, we developed an empirical algorithm for Lake Thonotosassa, FL, USA. The development process was divided into two parts: the band selection and the calibration and validation of the algorithm.

### 2.4.1. Band Selection

Once the evaluation of the three algorithms implemented with L0_LAC data was completed, we evaluated the use of MOD09GA data, which has a spatial resolution of 500 m, to estimate the chl-*a* concentration. In order to select the spectral bands from MOD09GA (Band 1 to Band 7) to be used in the algorithm, we firstly divided the images into 4 periods: January to March; April to June; July to September; and October to December. We used this seasonal distribution of the data, because aquatic systems respond differently to weather and seasonal conditions. Therefore, for each period, we analyzed the correlation among the spectral bands and chl-*a* concentration. We also evaluated the use of band ratios through the web tool, Interactive Correlation Environment (ICE), available at www.dsr.inpe.br/hidrosfera/ice [37]. This web tool builds a two-dimensional correlation plot of the radiometric measurement (*i.e.*, surface reflectance from the MOD09GA) and its relation to the interesting biogeochemical component (*i.e.*, chl-*a*). This two-dimensional correlation plot is important for band selection, because of its capability to cover all possible band ratios, making it a useful tool for the analysis of hyperspectral measurements with a large number of spectral bands.

### 2.4.2. Model Evaluation

We split the dataset in two groups: the first one using MOD09GA products from 2000 to 2007, which was used to calibrate the models; and the second using MOD09GA products from 2008 to 2011. Thus, we validated the calibration coefficients derived from a linear regression of the calibration dataset in the validation dataset. We also used error estimators, such as the bias, normalized bias (NBias), root mean squared error (RMSE) and normalized root mean squared error (NRMSE), which were calculated according to the equations in Table 2. The described methodology (Section 2) of this work is summarized in Figure 2, which shows a schematic flowchart of the entire work.

**Table 2.** Error estimators used in this study. NBias, normalized bias.

| Estimator | Formulas |
|---|---|
| Bias | $(y_i - x_i)$ |
| NBias | $\frac{(y_i - x_i)}{y_{i,\,max} - y_{i,\,min}}$ |
| RMSE | $RMSE = \sqrt{\frac{1}{n} \sum\limits_{i=1}^{n} (y_i - x_i)^2}$ |
| NRMSE | $NRMSE = \frac{RMSE}{y_{i,\,max} - y_{i,\,min}}$ |

Note: $y_i$ and $x_i$ are the measured and predicted chl-*a* concentration, respectively. In the *i*-th sample, $y_{i,max}$ and $y_{i,min}$ are the maximum and minimum chl-*a* concentrations, respectively.

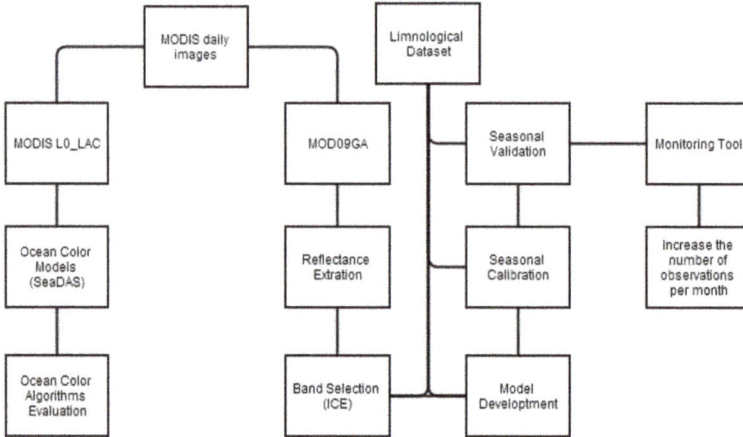

**Figure 2.** Flow chart of the methodology used to develop an empirical model for chl-*a* estimation in Lake Thonotosassa. SeaDAS, normalized bias; ICE, Interactive Correlation Environment.

## 3. Results and Discussion

### 3.1. Environmental Characteristics

The limnological variables in the water column for the 11 years of monthly analysis showed different responses in the three stations located in Lake Thonotosassa. The average chl-*a* concentration was lower (15.37 μg/L) at the inlet and higher (154.68 μg/L) at the outlet of the lake. The same pattern was observed for the average TN concentrations for the three sampling stations—a low (1051.07 μg/L) concentration at the entrance and a high (2785.69 μg/L) concentration at the outlet. For the average TP concentration the opposite pattern was found with a high average TP concentration (535.04 μg/L) at the entrance and a low concentration (329.20 μg/L) at the outlet. Table 3 summarizes the statistics for these variables. The mean ratio TN:TP was very low at the inlet (1.96) and almost four-times higher at the middle (8.82) and outlet (8.46) sampling points. Such increasing ratios of TN:TP along the transect between the inlet and the outlet indicate an increasing trend in the TN concentration, while a decreasing trend in the TP concentration along the same transect. Table 3 summarizes the statistics for the limnological variables used in this manuscript.

**Table 3.** Summary statistics for chl-*a* (µg/L), total nitrogen (TN) (µg/L), total phosphorus (TP) (µg/L) and the ratio TN:TP for the monthly water samples analysis of Lake Thonotosassa from 2000 to 2011.

|  | Inlet | Middle | Outlet |
|---|---|---|---|
|  | Mean ± SD (Min–Max) | Mean ± SD (Min–Max) | Mean ± SD (Min–Max) |
| chl-*a* | 15.37 ± 25.91 (0.3–163.54) | 146.48 ± 57.45 (4.4–373.8) | 154.68 ± 61.30 (2.2–339.4) |
| TN | 1051.07 ± 429.80 (31–3010) | 2675.10 ± 1,215.92 (843–9300) | 2785.69 ± 1,342.21 (978–9159) |
| TP | 535.04 ± 263.48 (58–1994) | 302.96 ± 158.46 (47–1030) | 329.20 ± 171.93 (40–1204) |
| TN:TP | 1.96 ± 1.15 (0.61–7.21) | 8.82 ± 4.25 (0.10–23.51) | 8.46 ± 4.32 (2.45–22.26) |

To relate these limnological analyses to cyanobacteria biomass (CBB), we applied an empirical algorithm developed by Beaulieu *et al.* [38] to predict CBB. Such an algorithm was developed based on data from approximately 1100 lakes from the entire continental United States. The authors divided the dataset according to basin type (shallow or deep lakes, as well as natural ones or reservoirs). For each of these basin types, it was possible to generate several relations, that of a shallow natural lake being the one corresponding to Lake Thonotosassa. The predictive models of CBB based on the shallow natural lake type and TN and TP concentrations (in µg/L) are shown in Equations (2) and (3), respectively.

$$CBB = -1.08 + 1.17 \, \log_{10} TN \tag{2}$$

$$CBB = 1.19 + 0.76 \, \log_{10} TP \tag{3}$$

Figure 3 shows the results of the application of both equations for the three sampling points for the estimation of CBB. Based on TN, Figure 3A shows a lower CBB concentration at the inlet (blue line) as compared with that at the middle (green line) and the outlet (red line). Figure 3B represents the CBB estimation based on TP, which shows the TP trend line at the inlet (blue line) being the highest during most of the study period. For both cases, an estimation of CBB concentration for any time point with TN and/or TP data was observed. Since we do not have *in situ* data for cyanobacteria, these results, together with the data on chl-*a* concentrations (Figure 3C) and the history of cyanobacteria bloom in the lake, strongly suggest the presence of cyanobacteria.

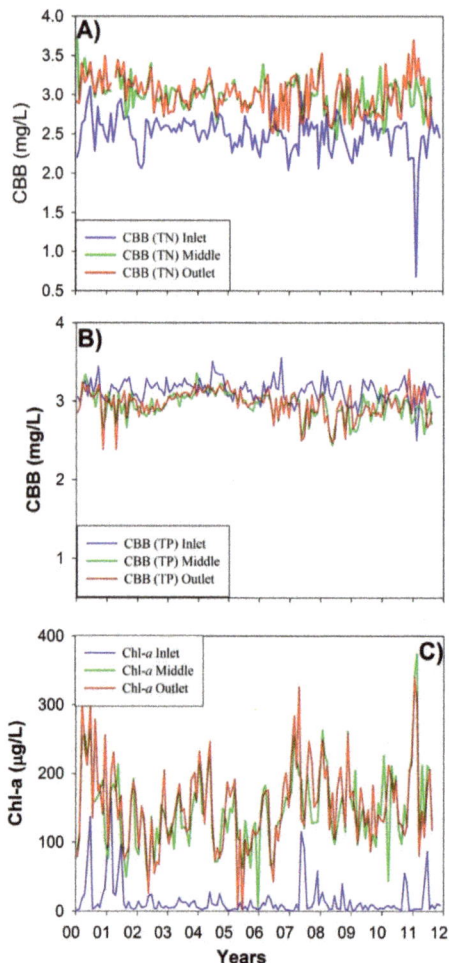

**Figure 3.** Time series of cyanobacteria biomass (CBB) estimations from (**A**) TN and (**B**) TP at the three sampling points, (**C**) data on measured chl-a concentrations.

However, Figure 3A,B shows different patterns for the cyanobacteria presence among the three regions. Figure 3A, based on Equation (2), showed a lower CBB at the inlet as compared with the other two monitoring sites, while Figure 3B shows the opposite. To better evaluate the presence of cyanobacteria in Lake Thonotosassa, we analyzed the TN: TP ratio (shown in Table 3), since TN:TP ratios have been associated with the concentration of cyanobacteria in lake water [5]. Although it has been proposed that low ratios of TN:TP lead to a higher cyanobacteria concentration [39,40], the opposite case has also been proposed, where low ratios of TN:TP may rather be the result of a high cyanobacteria presence, due to the possible ability of cyanobacteria to pump out phosphorus from enriched sediments [41]. Furthermore, other studies have supported alternative explanations, according to which other factors, such as nutrient variations, rather than low ratios of TN:TP, could be the main triggers of higher cyanobacteria concentration or toxic blooms [42].

In our study case, Table 3 shows the inverse relationship between the trends in TN and TP along the inlet-outlet transect and the direct relationship between the trends of chl-*a* and TN along it. This could in fact suggest that cyanobacteria are an important contributor in the overall chl-*a* concentration in the water column of the lake. Such a hypothesis could be explained as a consequence of the well-known ability of cyanobacteria to fix nitrogen [39]. Cyanobacteria would be favored over other species of phytoplankton by conditions of relatively limited nitrogen and abundant phosphorus in the water supply to the lake [43,44]. Depending on nutrient fluxes and the sink capacity of the sediments, it is possible that as water flows from the inlet to the outlet of the lake, the ratio between TN and TP in the water column gradually changes as cyanobacteria uptake TP from the solution and fix nitrogen in solution. Indeed, it can be noticed from Table 3 that both the TN:TP ratio and the chl-*a* concentration are lowest at the inlet, presumably because at that site, the inflow water with the lowest TN:TP ratio has just entered the lake and has not had enough retention time yet to affect the cyanobacteria; thus, the chl-*a* concentration is low. The middle and outlet sites monitor lake water with a longer retention time; thus, there is a longer opportunity for cyanobacteria to increase in abundance by taking advantage of the still low TN:TP ratio. Although such a ratio has increased by the time the water reaches the middle and outlet, it is still low and, therefore, still nitrogen limiting and, consequently, favorable for cyanobacterial growth. This could explain the higher chl-*a* concentration at the middle and outlet sites.

According to the nutrient limitation criteria suggested by Brezonik [45] and the results of Table 3, Lake Thonotosassa is a nitrogen-limited environment. Such criteria propose that lakes with TN:TP ratios less than 10 are nitrogen-limited, while those lakes with ratios greater than 30 are phosphorus-limited, and those ranging between 10 and 30 are balanced (both nutrients are limiting). This, along with the episodic blooms of cyanobacteria reported at the lake [21], agree with the associations reported in the literature between cyanobacteria and low TN:TP ratios [5,40]. Furthermore, the fact that both the TN: TP ratio and chl-*a* concentration increase together from the inlet to the middle does not coincide with the proposition of Xie *et al.* [41] that a low TN:TP ratio is a result of a high cyanobacteria concentration, in which case, a lower TN:TP ratio would be expected with more chl-*a*.

It seems plausible that the TN:TP ratio has indeed an effect on the cyanobacteria in the lake water, which would be mostly the case from the TN:TP ratio of the inflow water [43,44]. This highlights the importance for management plans to focus on the reduction of phosphorus inputs as a way to prevent cyanobacteria blooms and consequent public health issues. As we have indicated, since it is highly plausible that in our study case that the chl-*a* is importantly composed of cyanobacteria, the importance of monitoring chl-*a* in a more consistent manner as a way to monitor the effectiveness of such management plans is evident.

*3.2. Ocean Color Algorithms Performances*

Figure 4 shows the application results from the three algorithms, used for comparison, in a MODIS-Aqua L0 image acquired on 14 April 2006, over the State of Florida. As expected, none of the three algorithms were able to retrieve chl-*a* concentrations from Lake Thonotosassa. The OC3M was the only algorithm that could perform chl-*a* estimation in the MODIS L0 product (Figure 4). However, such an estimation was not useful for inland aquatic systems, mainly due to the size of the water body and consequent issues with spatial resolution. Accordingly, this algorithm has been reported previously to be able to retrieve some variation of chl-*a* in Tampa Bay, Florida, a body of water covering about 1000 km$^2$ at a high tide, but even in that case, its performance was considerably lower as compared with a 1-km spatial resolution MODIS chlorophyll fluorescence line height (FLH) product [18]. The use of GSM and GIOP could not retrieve any spatial variation of the chl-*a* concentration. These results agree with those found by Tilstone *et al.* [46], who evaluated the same algorithms for the eastern Arabian Sea coast and similarly found that OC3M had the best performance among the same three algorithms. These observations highlight the need to develop water quality products with a higher spatial resolution for the study and monitoring of small aquatic systems.

chlor_a [mg m^-3]

0.0  0.18  0.35  0.53  0.7  0.88  1.05  1.23  1.4  1.58  1.75  1.93  2.1  2.28  2.45  2.63  2.8  2.98  3.15  3.33  3.5  3.68  3.85  4.03  4.2  4.38

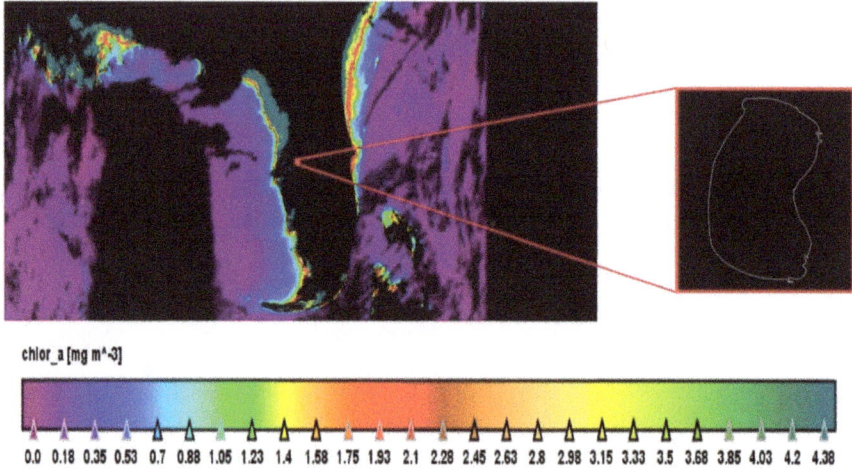

**Figure 4.** OC3M application using SeaDAS 7.02 on a MODIS-Aqua L0 product.

### 3.3. Locally-Tuned Algorithm

### 3.3.1. Band Selection

Upon confirming the unsatisfactory results from the MODIS-Aqua L0 1-km product, we used the MOD09GA product with a spatial resolution of 500 m to develop empirical models. However, the single spectral bands from the MOD09GA product were not suitable for water color studies either, since they cannot detect the small spectral variations required [14]. We compared chl-*a* concentration against MOD09GA reflectance to analyze their relationship using the reflectance from single bands. Table 4 shows the correlation for each spectral band of MOD09GA (days without cloud cover over Lake Thonotosassa) to the chl-*a* concentration per period of the year using the calibration dataset. As shown in Table 4, algorithms based on a single band were not useful for estimating the chl-*a* concentration in Lake Thonotosassa during the last six months of the year.

**Table 4.** Coefficient of determination ($R^2$) for chl-*a* ($\mu$g/L) and reflectance values from each band for four periods of the year: January to March (JFM), April to June (AMJ), July to September (JAS) and October to December (OND).

|  | JFM | AMJ | JAS | OND |
|---|---|---|---|---|
| **Band 1 (620–670 nm)** | 0.07 | 0.67 | 0.00 | 0.00 |
| **Band 2 (841–876 nm)** | 0.00 | 0.71 | 0.01 | 0.01 |
| **Band 3 (459–479 nm)** | 0.15 | 0.62 | 0.01 | 0.00 |
| **Band 4 (545–565 nm)** | 0.25 | 0.58 | 0.00 | 0.00 |
| **Band 5 (1230–1250 nm)** | 0.03 | 0.67 | 0.00 | 0.01 |
| **Band 6 (1628–1652 nm)** | 0.02 | 0.70 | 0.00 | 0.00 |
| **Band 7 (2105–2155 nm)** | 0.01 | 0.71 | 0.00 | 0.01 |

Different spectral bands were found to be the most correlated for each period. The highest correlations were observed in the April to June (AMJ) period, which is also the period of the year with higher chl-*a* concentrations (179.36 $\mu$g/L). For the periods with a low chl-*a* concentration, July to September (JAS) (126.83 $\mu$g/L) and October to December (OND) (132.12 $\mu$g/L), the $R^2$ was low. Finally, the January to March (JFM) period, which presented a higher $R^2$ as compared to JAS and OND, but lower $R^2$ as compared to AMJ, presented a chl-*a* concentration greater than that of JAS and

OND, but lower than that of AMJ (148.49 µg/L). These observations demonstrate the need for seasonal calibrations of the empirical algorithms.

To understand the physical principles behind the algorithm development, we plotted the average reflectance spectrum for each period (Figure 5). No relationship between chl-*a* concentration and the reflectance peak at Band 2 (in the NIR) was observed. The highest reflectance value was detected in the OND period, with an average chl-*a* concentration of 132.12 µg/L, followed in order by the JAS, AMJ and JFM periods with average chl-*a* concentrations of 126.83 µg/L, 179.36 µg/L and 148.49 µg/L, respectively. It was concluded from this lack of correlation that single bands were not useful for the estimation of chl-*a* concentration.

As a next step, band ratios among the seven spectral bands were tested. The importance of using band ratios lies in the fact that the specular reflection from water under wavy conditions gets suppressed by the ratio architecture of this type of algorithm, which cancels out the specular reflection from the two bands used in the ratio [47]. To analyze all possible band ratios, we used the ICE [37] to generate two-dimensional correlation plots of the $R^2$ between chl-*a* concentration and band ratio values. ICE was previous described in Section 2.4.1, and more information can be found in [37]. Figure 6 shows the plots for the four seasons analyzed using the calibration dataset.

As shown by the four 2D correlation plots in Figure 6, the best performances are obtained from the ratio between Band 1 and Band 4 or Band 4 and Band 1 (the highest $R^2$ on the 2D color correlation plot for all of the periods; Figure 6A–D). Band 1 is located in the red channel around 675 nm, where there is an important chl-*a* absorption feature. Another reflectance peak of chl-*a* is located at Band 4 in the green channel around 550 nm. With attention to this, the algorithm we proposed here for all four seasons is shown in Equation (4):

$$Chl - a \approx \frac{B_1 - B_4}{B_1 + B_4} \tag{4}$$

where $B_1$ is related to MOD09GA reflectance from spectral Band 1 and $B_4$ is related to MOD09GA reflectance from spectral Band 4.

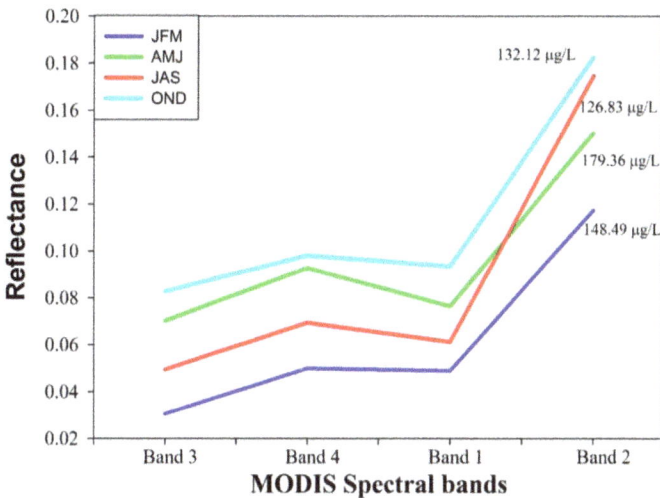

**Figure 5.** Average reflectance spectra of MOD09GA for each seasonal period.

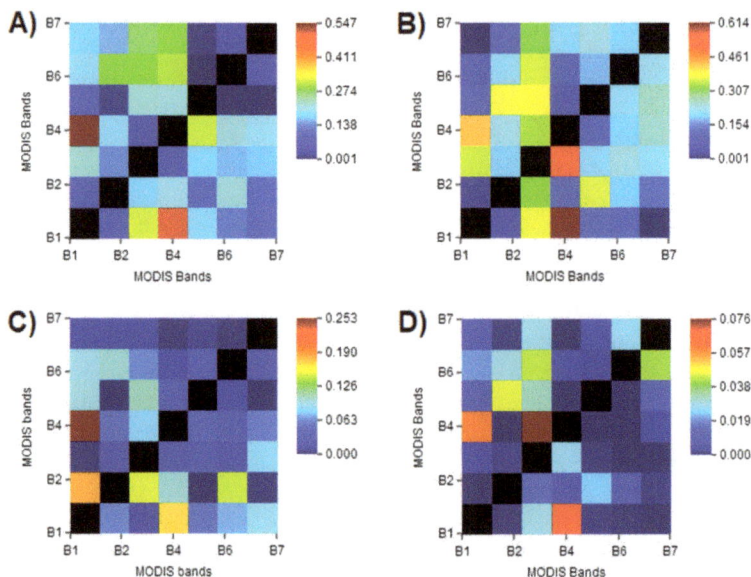

**Figure 6.** 2D correlation plots of MOD09GA spectral bands using ICE [37]. (**A**) JFM; (**B**) AMJ; (**C**) JAS; (**D**) OND.

### 3.3.2. Calibration and Validation

The algorithm was seasonally calibrated using MOD09GA data from 2000 to 2007. A linear calibration was performed for each period to retrieve its slope and intercept. Table 5 shows the calibrations coefficients, indicating an average performance for most of the periods, with $R^2$ of 0.53, 0.56 and 0.67 for the JFM, AMJ and JAS periods, respectively. Nevertheless, a clearly lower $R^2$ (0.06) was detected for the linear calibration in the OND period.

**Table 5.** Calibration coefficients for the algorithm for each period.

|  | $R^2$ | Slope | Intercept | $p$-Value |
|---|---|---|---|---|
| JFM | 0.53 | −426.81 | 144.78 | 0.003 |
| AMJ | 0.56 | −289.51 | 137.72 | 0.008 |
| JAS | 0.67 | 357.46 | 137.18 | 0.012 |
| OND | 0.06 | −148.15 | 125.87 | 0.440 |

It can be observed from the 2D correlation plots in Figure 6 that the relationship between reflectance and chl-*a* concentration in the OND period did not present any $R^2$ higher than 0.08. The same was observed by applying the proposed algorithm with a very low $R^2$ with a very high $p$-value (0.440). As can be appreciated from Table 5 and Figure 7, the models were better calibrated for the first three periods: JFM, AMJ and JAS. Figure 7C shows a positive slope during the JAS period, while all of the others figures (Figure 7A–C for periods JFM, AMJ and OND, respectively) present a negative slope. Such a negative slope can be explained by the algorithm structure, which is sensitive to the absorption of phytoplankton in the red channel, as well as the phytoplankton reflectance peak in the green channel. Conversely, a positive slope indicates an increase in the signal in the red channel, which is related to a cyanobacteria pigment known as phycocyanin (PC) with a fluorescence peak at 650 nm. Such a fluorescence peak is close to the center of MODIS Band 1. This circumstance causes an increase in the signal of this band at high cyanobacteria concentrations. Moreover, the JAS period is usually related to the summer, where several cyanobacteria cases are reported mainly due to the high temperature observed in this season.

The linear calibrations depicted in Figure 7 were used in the validation process with MOD09GA data, which covered the period from 2008 to 2011. The coefficients of the calibration (slope and intercept from Table 5) were applied on the surface reflectance data derived from the MOD09GA product. The error estimators in Table 2 were used to evaluate the performance of the model. Table 6 shows the error estimators for the four periods of the year.

**Table 6.** Errors estimators used in the validation dataset for each period.

|        | JFM   | AMJ    | JAS   | OND   |
|--------|-------|--------|-------|-------|
| Bias   | 38.58 | 91.46  | 52.94 | 25.53 |
| NBias  | 0.23  | 0.27   | 0.46  | 1.74  |
| RMSE   | 45.2  | 112.08 | 62.02 | 27.16 |
| NRMSE  | 0.27  | 0.34   | 0.54  | 1.85  |

The variation in the error estimator from period to period of each year can be appreciated in Table 6. Both NBias and NRMSE indicated lower error estimators for the periods with higher chl-*a* concentrations (JFM and AMJ) as compared with those periods with lower chl-*a* concentrations (JAS and OND). These errors could be related to the atmospheric correction of the MOD09GA product, which is based on the utilization of look-up tables (LUTs) of top of atmosphere (TOA) reflectance values [48]. This is needed to retrieve surface reflectance properties on the basis of given TOA reflectance values and atmospheric parameters [48]. Hence, the accuracy of using MOD09GA depends on several aspects, such as the accuracy of sensor calibration, input atmospheric parameters, LUTs and operational implementation of correction for bidirectional reflectance distribution function effects [49]. More details about the atmospheric correction used in the MOD09GA product can be found in Vermote *et al.* [48]. The NRMSE results, though, agree with Cowell *et al.* [22], which identified more abundant green algae (*Chlorophyceae*) in Lake Thonotosassa during the spring and early summer months, while blooms of blue-green algae (mainly *Anabaena spiroides*) were during the summer. This phytoplankton dynamic enhanced the accuracy for the first two trimesters, which received lower errors. During the third and fourth trimester, Cowell *et al.* [22] identified the occurrence of diatom blooms (mainly *Stephanodiscus hantzschii*). This could explain the high errors, since diatoms have a different spectral shape with higher reflectance values, and their steep spectrum decreases from 412 to 510 nm [50]. As shown by Ogashawara *et al.* [14], this range of estimator errors for chl-*a* concentration estimation from a MOD09GA product is reasonable. This assumption is mainly based on the spectral resolution of the seven first MODIS bands, which were not the ideal for water color studies, since they are not able to collect the signal from important sections of the water leaving radiance spectra.

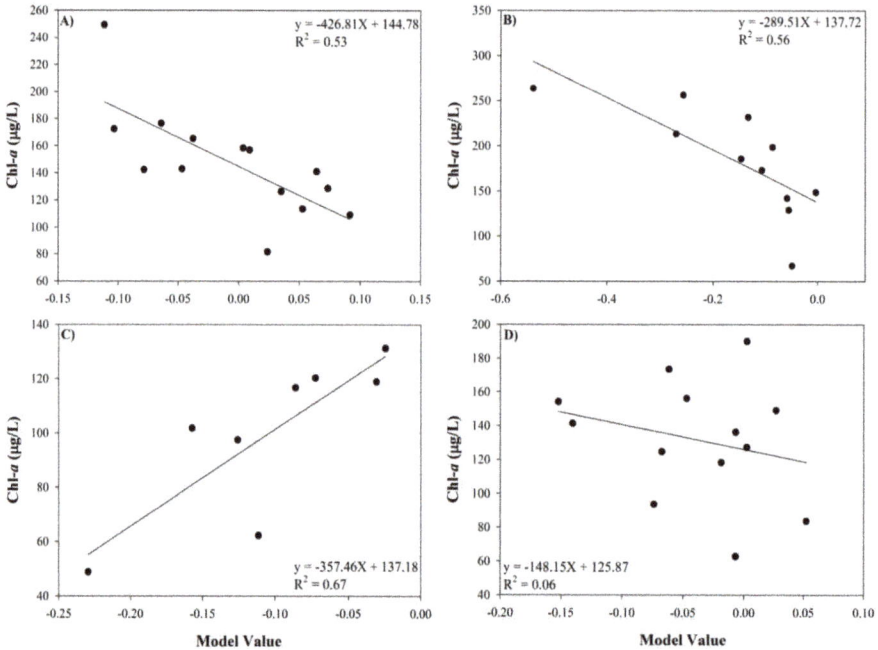

**Figure 7.** Linear regression plots of the calibration between model values and chl-*a* concentration for each period of the year. (**A**) JFM; (**B**) AMJ; (**C**) JAS; (**D**) OND.

Moreover, several studies using more appropriate MODIS spectral bands for water color analysis still have presented similar or poorer error estimator results. Le *et al.* [51] used MODIS Bands 11 (526–536), 12 (546–556) and 14 (673–683), which have a narrow range that meets the chl-*a*-specific spectral range of reflectance and absorption. Using these spectral bands, the authors produced an RMSE (%) of 36.5%, and they also attributed a value of 39.6% of the error to the red-green ratio algorithms as acceptable for estuaries.

### 3.4. Possible Applications

The development of empirical algorithms for the monitoring of chl-*a* concentration for Lake Thonotosassa can lead to several environmental and public health applications. Ogashawara *et al.* [14] demonstrated the usefulness of applying empirical algorithms to retrieve a time-series of chl-*a* concentration to improve the monitoring in regions with a considerable lack of data acquisition. Although Lake Thonotosassa is monitored monthly, the possibility of having more frequent data is important, because water quality parameters can change rapidly. Furthermore, a monitoring method based on fixed locations may not be representative of the mean chl-*a* concentration in the lake, as other locations of potential interest within the lake are not considered. Thus, the use of MOD09GA, which is a daily product, can improve the frequency of monitoring water quality, mainly algal blooms, further providing the feasibility to cover areas within the lake not included in the routine monitoring plan. As indicated by calculations according to Beaulieu *et al.* [38] and by the analysis of *in situ* data for TN:TP ratios, there is an important indication that cyanobacteria are a core component of the chl-*a* present in Lake Thonotosassa. As growing populations in urban and suburban areas increasingly rely on lakes for ecosystem services, such as recreation, aesthetics, culture and storm-water drainage and treatment [1,52], the use of practical time- and cost-effective methodologies to monitor water quality is accordingly becoming more imperative. The risk of cyanobacteria presence in urban lakes and consequent potential danger to public health amply justify the implementation of water quality monitoring programs. These programs are important, since human intoxication from cyanotoxins, such as microcystin, does not occur only through the ingestion of drinking water or food. It can also occur through recreational dermal contact during aquatic activities, such as practicing aquatic sports, bathing, swimming and diving [53]. Moreover, these activities also promote accidental ingestion of water, which is a concern, since ingestion of even a small quantity of cyanotoxins can have serious health consequences [54]. Therefore, in a broader scope, chl-*a* monitoring in general is an increasingly important need. The web tool procedures and remote sensing techniques used in this study were shown to be useful for applications to water chl-*a* monitoring or algal bloom identification in Lake Thonotosassa. Furthermore, the proposed approach and procedures can also be applied to develop customized remote sensing methodologies in other lakes with similar environmental conditions. This initiative can improve local water governance systems, as well as can be an important tool for environmental and public health managers. Additionally, the community as the major stakeholders could be trained in the use of remote sensing technology to be involved in the monitoring process in a sustainable low-cost approach.

## 4. Conclusions

Our analysis of *in situ* data of the changes in TN: TP ratios and chl-*a* concentration along the flow transect covering the three monitoring sites suggests the presence of cyanobacteria in Lake Thonotosassa. To validate these findings, an analysis through the TN and TP relations established by Beaulieu *et al.* [38] along with chl-*a* concentration highlighted the presence of cyanobacteria. Considering that some cyanobacteria species have the ability to produce toxins (which can be dangerous to human health through different routes of entry) along with the potential risk of exposure (as this lake is located in a suburban area and discharges into the Hillsborough River, which is the main source of municipal water supply for the City of Tampa), the importance of implementing monitoring programs in this water body is clear. Such a practice could anticipate and prevent related health issues in the community interacting with the lake.

Remote sensing techniques, such as bio-optical modeling, are an alternative to qualitatively monitoring of biological activity through chl-*a* concentration. Using MODIS-Aqua L0 products and SeaDAS, we implemented three ocean color algorithms that resulted in non-suitable products, due to the use of spectral bands with low spatial resolution, which were unable to identify small water bodies. Since they are global algorithms and originally modeled for oceanic waters, our empirical model especially designed for a small inland lake was shown to improve the results. We used MOD09GA (500 m) data to develop an empirical chl-*a* bio-optical model based on the band selection provided by 2D color correlation plots of $R^2$ between band ratios and measured chl-*a* concentration. We grouped 11 years of MOD09GA data by trimester and calibrated the algorithm for each of these time periods. The NRMSE for the first trimester of the year was 0.27, while for the second, third and fourth trimester, it was 0.34, 0.54 and 1.85, respectively. These results showed a greater error during the fourth trimester of the year, as compared with the other trimesters. This is probably explained by the presence of diatoms during this period of time, which shows the need for the development of filters to remove the effect of others water constituents. We also highlight here the need for seasonal calibration of bio-optical models, as seasonal variability in the aquatic system can change the environment dynamics. Thus, we have shown the applicability of remote sensing techniques to monitor water quality, even in small water bodies, such as Lake Thonotosassa. Nevertheless, it is clear that the understanding of environmental factors and their effects on seasonal variations are crucial for better algorithm calibration.

Although the spatial resolution from MODIS products is often not suitable for small water bodies, it is still possible to acquire daily data to retrieve a long-term time-series from one pixel that fits the required conditions. This can be useful to monitor the effects on water quality from human activities in the watershed and the progress of environmental management plans aimed at improving water quality. This improvement is needed, since an adequate monitoring system for aquatic environments may imply high costs that make difficult its implementation difficult. Moreover, immediate management options should be available to provide information for the population, especially for those who depend on the aquatic environment. Remote sensing techniques showed here a high potential for the environmental and public health management of inland aquatic systems. MODIS products constitute a great help in understanding environmental dynamics in order to propose better environmental and health policies concerning water resources. Thus, remote sensing can be an important tool for environmental decision makers, which will influence the environment and human health.

**Acknowledgments:** The authors express their appreciation to the Environmental Protection Commission of Hillsborough County (EPCHC) for providing the *in situ* water quality data and, particularly, to Rick Garrity and Richard Boler for facilitating the process of data sharing. MODIS data collection and processing were made possible through the efforts of MODIS Adaptive Processing System (MODAPS) services at the NASA Goddard Space Flight Center (NASA/GSFC) and NASA's Earth Observing System Data and Information System (NASA/EOSDIS). Also acknowledged is the SeaDAS Development group at NASA GSFC for the use of the SeaDAS software to process the MODIS imagery. Special acknowledgment is given to Steve Padgett Vasquez for introducing the authors.

**Author Contributions:** Igor Ogashawara and Max J. Moreno-Madrinan conceived of and designed the paper together. Igor Ogashawara is the main author of the paper, who performed the RS analyses. Igor Ogashawara and Max J. Moreno-Madriñán equally participated in the writing of the manuscript and limnological analysis. Max J. Moreno-Madriñán classified the *in-situ* dataset and revised the entire manuscript.

**Conflicts of Interest:** The authors declare no conflict of interest.

## References

1. Rodríguez, J.P.; Beard, T.D., Jr.; Bennett, E.M.; Cumming, G.S.; Cork, S.J.; Agard, J.; Dobson, A.P.; Peterson, G.D. Trade-offs across space, time, and ecosystem services. *Ecol. Soc.* **2006**, *11*, 1–14.
2. Pitois, S.; Jackson, M.H.; Wood, B.J. Sources of the eutrophication problems associated with toxic algae: An overview. *J. Environ. Health* **2001**, *64*, 25–32. [PubMed]
3. Mudroch, A. (Ed.) *Planning and Management of Lakes and Reservoirs, An Integrated Approach to Eutrophication*; UNEP International Environmental Technology Centre: Osaka, Japan, 1999.
4. O'Neil, J.M.; Davis, T.W.; Burford, M.A.; Gobler, C.J. The rise of harmful cyanobacteria blooms: The potential roles of eutrophication and climate change. *Harmful Algae* **2012**, *14*, 313–334. [CrossRef]
5. Kotak, B.G.; Zurawell, R.W. Cyanobacterial toxins in Canadian freshwaters: A review. *Lake Reserv. Manag.* **2007**, *23*, 109–122. [CrossRef]
6. Fleming, L.E.; Rivero, C.; Burns, J.; Williams, C.; Bean, J.A.; Shea, K.A.; Stinn, J. Blue green algal (cyanobacterial) toxins, surface drinking water, and liver cancer in Florida. *Harmful Algae* **2002**, *1*, 157–168. [CrossRef]
7. Francis, G. Poisonous Australian Lake. *Nature* **1878**. [CrossRef]
8. Azevedo, S.M.F.O.; Carmichael, W.W.; Jochimsen, E.M.; Rinehart, K.L.; Lau, S.; Shaw, G.R.; Eaglesham, G.K. Human intoxication by microcystins during renal dialysis treatment in Caruaru—Brazil. *Toxicology* **2002**, *181–182*, 441–446. [CrossRef] [PubMed]
9. Centers for Disease Control and Prevention (CDC). Recreational Water—Associated Disease Outbreaks—United States, 2009–2010. Available online: http://www.cdc.gov/mmwr/preview/mmwrhtml/mm6301a2.htm?s_cid=mm6301a2_w (accessed on 8 May 2014).
10. Jogensen, S.E.; Bendoricchio, G. *Fundamentals of Ecological Modeling*, 3rd ed.; Elsevier: New York, NY, USA, 2001.
11. Dall'Olmo, G.; Gitelson, A.A.; Rundquist, D.C.; Leavitt, B.; Barrow, T.; Holz, J.C. Assessing the potential of SeaWiFS and MODIS for estimating chlorophyll concentration in turbid productive waters using red and near-infrared bands. *Remote Sens. Environ.* **2005**, *96*, 176–187. [CrossRef]
12. Le, C.; Li, Y.; Zha, Y.; Sun, D.; Huang, C.; Lu, H. A four-band semi-analytical model for estimating chlorophyll a in highly turbid lakes: The case of Taihu Lake, China. *Remote Sens. Environ.* **2009**, *113*, 1175–1182. [CrossRef]
13. Hadjimitsis, D.G.; Clayton, C. Assessment of temporal variations of water quality in inland water bodies using atmospheric corrected satellite remotely sensed image data. *Environ. Monit. Assess.* **2009**, *159*, 281–292. [CrossRef] [PubMed]

14. Ogashawara, I.; Alcântara, E.H.; Curtarelli, M.P.; Adami, M.; Nascimento, R.F.F.; Souza, A.F.; Stech, J.L.; Kampel, M. Performance analysis of MODIS 500-m spatial resolution products for estimating Chlorophyll-*a* concentrations in Oligo- to Meso-Trophic waters case study: Itumbiara Reservoir, Brazil. *Remote Sens.* **2014**, *6*, 1634–1653. [CrossRef]

15. Gitelson, A.; Garbuzov, G.; Szilagyi, F.; Mittenzwey, K.H.; Karnielli, A.; Kaiser, A. Quantitative remote sensing methods for real-time monitoring of inland waters quality. *Int. J. Remote Sens.* **1993**, *14*, 1269–1295. [CrossRef]

16. Gons, H.J. Optical teledetection of chlorophyll a in turbid inland waters. *Environ. Sci. Technol.* **1999**, *33*, 1127–1132. [CrossRef]

17. Moses, W.J.; Gitelson, A.A.; Berdnikov, S.; Bowles, J.H.; Povazhnyi, V.; Saprygin, V.; Wagner, E.J.; Patterson, K.W. HICO-based NIR-Red models for estimating Chlorophyll-a concentration in productive coastal waters. *IEEE Geosci. Remote Sens.* **2014**, *11*, 1111–1115. [CrossRef]

18. Moreno-Madriñán, M.J.; Fischer, M.A. Performance of the MODIS FLH algorithm in estuarine waters: A multi-year (2003–2010) analysis from Tampa Bay, Florida (USA). *Int. J. Remote Sens.* **2013**, *34*, 6467–6483. [CrossRef]

19. Dall'Olmo, G.; Gitelson, A.A. Effect of bio-optical parameter variability on the remote estimation of chlorophyll-$\alpha$ concentration in turbid productive waters: Experimental results. *Appl. Opt.* **2005**, *44*, 412–422. [CrossRef] [PubMed]

20. Dall'Olmo, G.; Gitelson, A.A. Effect of bio-optical parameter variability and uncertainties in reflectance measurements on the remote estimation of chlorophyll-$\alpha$ concentration in turbid productive waters: Modeling results. *Appl. Opt.* **2006**, *45*, 3577–3592. [CrossRef] [PubMed]

21. Southwest Florida Water Management District. *Lake Thonotosassa Surface Water Improvement and Management (SWIM) Plan*; Southwest Florida Water Management District: Plant City, FL, USA, 2003.

22. Cowell, B.C.; Dye, C.W.; Ada, R.C. A synoptic study of the limnology of Lake Thonotosassa, Florida. Part I. effects of primary treated sewage and citrus wastes. *Hydrobiologia* **1975**, *46*, 301–345. [CrossRef]

23. Cowell, B.C.; Vodopich, D.S. Distribution and seasonal abundance of benthic macroinvertebrate in a subtropical Florida lake. *Hydrobiologia* **1981**, *78*, 97–105. [CrossRef]

24. Florida Department of Environmental Regulation. *Integrated Water Quality Assessment for Florida: 2012 305(b) Report and 303(d) List Update*; Florida Department of Environmental Regulation—Division of Environmental Assessment and Restoration: Tallahassee, FL, USA, 2012.

25. Clesceri, L.S.; Eaton, A.D.; Greenberg, A.E.; Franson, M.A.H. (Eds.) *Standard Methods for the Examination of Water and Wastewater*; American Public Health Association, American Water Works Association & Water Environment Federation: Washington, DC, USA, 1998.

26. Moreno-Madriñán, M.J. Analysis of limnological variables associated to water quality in lakes of Northwestern Hillsborough County, Florida. *Fla. Sci.* **2010**, *73*, 218–224.

27. Moreno-Madriñán, M.J. Analysis of the relationship between Submerged Aquatic Vegetation (SAV) and water trophic status of lakes clustered in Northwestern Hillsborough County, Florida. *Water Air Soil Poll.* **2011**, *214*, 539–546. [CrossRef]

28. Ocean Color Web. Available online: http://oceancolor.gsfc.nasa.gov/ (accessed on 3 February 2014).

29. EOS ClearingHouse (ECHO). Available online: http://reverb.echo.nasa.gov/ (accessed on 3 February 2014).

30. Fu, G.; Settle, K.; McClain, C.R. SeaDAS: The SeaWiFSData analysis system. In Proceedings of the 1998 Pacic Ocean Remote Sensing Conference, Qingdao, China, 28–31 July 1998; Secretariat: Beijing, China, 1998; pp. 73–77.

31. O'Reilly, J.E.; Maritorena, S.; O'Brien, M.C.; Siegel, D.A.; Toole, D.; Menzies, D.; Smith, R.C.; Mueller, J.L.; Mitchell, B.G.; Kahru, M.; *et al. SeaWiFS Postlaunch Calibration and Validation Analyses*; Part 3, Volume 11; National Aeronautics and Space Administration: Washington, DC, USA, 2000.

32. Maritorena, S.; Siegel, D.A.; Peterson, A.R. Optimization of a semianalytical ocean color model for global-scale applications. *Appl. Opt.* **2002**, *41*, 2705–2714. [CrossRef] [PubMed]

33. Franz, B.A.; Werdell, P.J. A generalized framework for modeling of inherent optical properties in ocean remote sensing applications. In Proceedings of the 2010 Ocean Optics, Anchorage, AK, USA, 27 September–1 October 2010.

34. Morel, A. Optical modeling of the upper ocean in relation to its biogenous matter content (case I waters). *J. Geophys. Res.* **1988**, *931*, 10749–10768. [CrossRef]

35. Gordon, H.R. Ocean color remote sensing: Influence of the particle phase function and the solar zenith angle. *EOS Trans. Am. Geophys. Union* **1986**, *14*, 1055.

36. Ruddick, K.; Ovidio, F.; Rijkeboer, M. Atmospheric correction of SeaWiFS imagery for turbid coastal and inland waters. *Appl. Opt.* **2000**, *39*, 897–912. [CrossRef] [PubMed]

37. Ogashawara, I.; Curtarelli, M.P.; Souza, A.F.; Augusto-Silva, P.B.; Alcântara, E.H.; Stech, J.L. Interactive Correlation Environment (ICE)—A statistical web tool for data collinearity analysis. *Remote Sens.* **2014**, *6*, 3059–3074. [CrossRef]

38. Beaulieu, M.; Pick, F.; Gregory-Eaves, I. Nutrients and water temperature are significant predictors of cyanobacterial biomass in a 1147 lakes data set. *Limnol. Oceanogr.* **2013**, *58*, 1736–1746. [CrossRef]

39. Smith, V.H. Low nitrogen to phosphorus ratios favor dominance by blue-green algae in lake phytoplankton. *Science* **1983**, *221*, 669–671. [CrossRef] [PubMed]

40. Orihel, D.M.; Bird, D.F.; Brylinsky, M.; Chen, H.; Donald, D.B.; Huang, D.Y.; Giani, A.; Kinniburgh, D.; Kling, H.; Kotak, B.G.; *et al.* High microcystin concentrations occur only at low nitrogen-to-phosphorus ratios in nutrient-rich Canadian lakes. *Can. J. Fish. Aquat. Sci.* **2012**, *69*, 1457–1462. [CrossRef]

41. Xie, L.; Xie, P.; Li, S.; Tang, H.; Liu, H. The low TN:TP ratio, a cause or a result of Microcystis blooms? *Water Res.* **2003**, *37*, 2073–2080. [CrossRef] [PubMed]

42. Downing, J.A.; Watson, S.B.; McCauley, E. Predicting Cyanobacteria dominance in lakes. *Can. J. Fish. Aquat. Sci.* **2001**, *58*, 1905–1908. [CrossRef]

43. Schindler, D.W. Evolution of phosphorus limitation in Lakes. *Science* **1977**, *195*, 260–262. [CrossRef] [PubMed]

44. Schindler, D.W.; Hecky, R.E.; Findlay, D.L.; Stainton, M.P.; Parker, B.R.; Paterson, M.J.; Beaty, K.G.; Lyng, M.; Kasian, S.E.M. Eutrophication of lakes cannot be controlled by reducing nitrogen input: Results of a 37-year whole-ecosystem experiment. *Proc. Natl. Acad. Sci. USA* **2008**, *105*, 11254–11258. [CrossRef] [PubMed]

45. Brezonik, P.L. Trophic state indices: Rationale for multivariate approaches. *Lake Reserv. Manag.* **1984**, *1*, 441–445. [CrossRef]

46. Tilstone, G.H.; Lotliker, A.A.; Miller, P.I.; Ashraf, P.M.; Kumar, T.S.; Suresh, T.; Ragavan, B.T.; Menon, H.B. Assessment of MODIS-Aqua chlorophyll-α algorithms in coastal and shelf waters of the eastern Arabian Sea. *Cont. Shelf Res.* **2013**, *65*, 14–26. [CrossRef]

47. Vincent, R.K.; Qin, X.; McKay, R.M.L.; Miner, J.; Czajkowski, K.; Savino, J.; Bridgeman, T. Phycocyanin detection from LANDSAT TM data for mapping cyanobacterial blooms in Lake Erie. *Remote Sens. Environ.* **2004**, *89*, 381–392. [CrossRef]

48. Vermote, E.F.; El Saleous, N.Z.; Justice, C.O. Atmospheric correction of visible to middle-infrared EOS-MODIS data over land surfaces: Background, operational algorithm and validation. *J. Geophys. Res.* **1997**, *102*, 17131–17141. [CrossRef]

49. Vermote, E.F.; Kotchenova, S. Atmospheric correction for the monitoring of land surfaces. *J. Geophys. Res.* **2008**, *113*. [CrossRef]

50. Alvain, S.; Moulin, C.; Dandonneau, Y.; Bréon, F.M. Remote sensing of phytoplankton groups in case 1 waters from global SeaWiFS imagery. *Deep Sea Res. I* **2005**, *52*, 1989–2004. [CrossRef]

51. Le, C.; Hu, C.; English, D.; Cannizzaro, J.; Chen, Z.; Feng, L.; Boler, R.; Kovach, C. Towards a long-term chlorophyll-α data record in a turbid estuary using MODIS observations. *Prog. Oceanogr.* **2012**, *109*, 90–103. [CrossRef]

52. Bolund, P.; Hunhammar, S. Ecosystem services in urban areas. *Ecol. Econ.* **1999**, *29*, 293–301. [CrossRef]

53. Chorus, I.; Bartram, J. *Toxic Cyanobacteria in Water: A Guide to Their Public Health Consequences, Monitoring and Management*; UNESCO/WHO/UNEP: London, UK, 1999.

54. World Health Organization. *Guidelines for Safe Recreational Water Environments, Volume I: Coastal and Fresh Waters*; World Health Organization: Geneva, Switzerland, 2003.

International Journal of
*Geo-Information*

MDPI

*Article*

# Impacts of Scale on Geographic Analysis of Health Data: An Example of Obesity Prevalence

Jay Lee [1,2,*], Mohammad Alnasrallah [1], David Wong [3,4], Heather Beaird [5,6] and Everett Logue [7]

1   Department of Geography, Kent State University, Kent, OH 44242, USA; malnasra@kent.edu
2   College of Environment and Planning, Henan University, Kaifeng 475001, China
3   Department of Geography and GeoInformation Science, George Mason University, Fairfax, VA 22030, USA; dwong2@gmu.edu
4   Department of Geography, University of Hong Kong, Hong Kong
5   Office of Epidemiology and Biostatistics, Summit County Public Health, Akron, OH 44313, USA; hbeaird@schd.org
6   College of Public Health, Kent State University, Kent, OH 44242, USA; hbeaird@kent.edu
7   Family Medicine Research Center, Summa Health System, Akron, OH 44309, USA; LogueE@summahealth.org
*   Author to whom correspondence should be addressed; jlee@kent.edu; Tel.: +1-330-672-3222.

External Editors: Fazlay Faruque and Wolfgang Kainz

Received: 23 June 2014; in revised form: 24 September 2014; Accepted: 28 September 2014; Published: 24 October 2014

**Abstract:** The prevalence of obesity has increased dramatically in recent decades. It is an important public health issue as it causes many other chronic health conditions, such as hypertension, cardiovascular diseases, and type II diabetics. Obesity affects life expectancy and even the quality of lives. Eventually, it increases social costs in many ways due to increasing costs of health care and workplace absenteeism. Using the spatial patterns of obesity prevalence as an example; we show how different geographic units can reveal different degrees of detail in results of analysis. We used both census tracts and census block groups as units of geographic analysis. In addition; to reveal how different geographic scales may impact on the analytic results; we applied geographically weighted regression to model the relationships between obesity rates (dependent variable) and three independent variables; including education attainment; unemployment rates; and median family income. Though not including an exhaustive list of explanatory variables; this regression model provides an example for revealing the impacts of geographic scales on analysis of health data. With obesity data based on reported heights and weights on driver's licenses in Summit County, Ohio, we demonstrated that geographically weighted regression reveals varying spatial trends between dependent and independent variables that conventional regression models such as ordinary least squares regression cannot. Most importantly, analyses carried out with different geographic scales do show very different results. With these findings, we suggest that, while possible, smaller geographic units be used to allow better understanding of the studies phenomena.

**Keywords:** obesity prevalence; geographic scales; geographically weighted regression

---

## 1. Introduction and Problem Statements

Geospatial analyses of health data are often carried out using census tracts as the geographic unit of analysis. This may have been largely due to two reasons. First, health data used to be released only at aggregated levels because of the confidentiality of patient data. Second, socioeconomic data from governmental sources are not available at more detail level than census tracts such as census blocks. Consequently, census tracts seem to have become the *de facto* unit of analysis for most studies in geography of health.

With the proliferation of the Internet, health data have become more accessible and are now being generated in larger volumes than before. This leads to a need to assess if analyzing health data at the scale of census tracts is sufficient and if such unit of analysis fails to reveal geographic details that we should have noticed. To that end, we report in this paper our analysis of obesity prevalence in Summit County, Ohio, using both census tracts and census block groups as the units of analysis. We show that there is often too much generalization when census tracts are used and census block groups would have been a better choice for examining geographic disparities in obesity prevalence.

As an example for examining the impacts of different geographic scales on health studies, we chose to study the issue of geographic disparity of obesity prevalence. Geographically weighted regression models were built by using obesity prevalence as the dependent variable. Racial composition, income, education, and employment were included as explanatory variables. The list of explanatory variables was determined from the obesity literature and is by no means an exhaustive list.

The obesity prevalence data are derived from calculating body mass index (BMI kg/m$^2$) that incorporated the self-reported heights and weights on all driver license data obtained from the Ohio Bureau of Motor Vehicles for years from 2008 to 2012. It should be noted that self-reported heights and weights on driver licenses tend to become obsolete as time went on. Most license holders would simply renew their licenses without updating their heights and weights. For this reason, we chose to include only data for the license holders who were between ages of 16 to 21 when they first had their licenses issued.

Data for the explanatory variables were taken from American Community Survey 2011 from the US Census Bureau. We acknowledge that these may not be the best data to use but for the purpose of comparing analytic outcomes between those from using census tracts and those using census block groups, they should serve the purpose well. We use the regression models to explore the relationships between socio-economic characteristics of small geographic units and the geographic disparities in obesity prevalence. Again, this method is used to facilitate the comparison between using census tracts and using census block groups as units of analysis and is not suggested as the best model for explaining the variations in obesity prevalence. Finally, as relationships between dependent and independent variables may vary using data at different geographic scales, analyses may be subject to what is known as the modifiable areal unit problem (MAUP) as discussed in Wong [1] In similar way, issues of using pre-aggregated data for analysis of health geography have been discussed in Cockings and Martin [2].

It should be noted that, while Summit County, Ohio, is used here as a case study. The results from the comparisons are likely applicable to many other locations in the US because the demographic profile and the socio-economic profile in the study area are very close to those of the national averages.

The prevalence of obesity among adults and children in the United States has increased dramatically in recent decades (e.g., [3–8]). Obesity is a public health issue as it often causes many other chronic health conditions, such as, hypertension, cardiovascular disease, and type II diabetes (e.g., [4,9–13]). Obesity affects life expectancy, quality of lives, and, eventually, it increases social costs in many ways due to increasing costs of health care, and workplace absenteeism, or presenteeism.

The basic cause of obesity is the imbalance between the amount of energy taken through eating and drinking, and the amount of energy expended through metabolism and physical activity [14–19]. To offset excessive energy intake, increased physical activity is encouraged as a way to keep energy in balance. However, energy imbalances appear to be facilitated by the characteristics of physical, social, and economic environments.

As reviewed in Sobal and Stunkard [20], a strong inverse relationship between the geography of socioeconomic status and the distribution of obesity exists, though slight variation was observed between developing and developed societies. This trend was confirmed by Zhang and Wang [21] from their study of the trends in the association between obesity and socioeconomic status in US adults from 1971 to 2000. McLaren [22] also concluded from reviewing 333 published studies that obesity was found to be related to most widely used SES variables, such as education, occupation, and income.

*ISPRS Int. J. Geo-Inf.* **2014**, *3*, 1198–1210

## 2. Data

In order to examine how the distribution of obese population may be related to area-specific socio-economic characteristics, we assembled our database from a number of sources:

a.    Derived BMI data—data from a five-year cycle of all holders of driver's licenses in Summit County, Ohio was obtained from Ohio Bureau of Motor Vehicles (OBMV) for 2008–2012 for public health purposes. Drivers in Ohio need to renew their licenses once every five years. By including data (age, height, weight, and home address) of all adults (16 years and older) in a five-year cycle, we basically captured everyone who had a driver's license in the county during the study period. It should be noted that this data set does not include derived BMI for population age 15 and below or those who do not hold driver's licenses. Over 480,000 addresses and associated data were geocoded to latitude/longitude coordinates. BMI was calculated for each record. Those records with BMI equal to and over 30 are selected and included in the dataset of obese population as this study focuses only on the distribution of obese population. Since self-reported heights are typically biased upward (≈1 inch) while self-reported weights are biased downward (≈10 lbs) in large surveys such as those reported by Ossiander *et al.* [23], the BMI's from the OBMV data may underestimate the true prevalence of obesity in Summit County. However, we have no reason to expect that the bias is large or strongly associated with socio-economic status (SES). For this reason, we included in this study only records of license holders who were between 16 and 21 of age at the time when their licenses were first issued. This, of course, still assumes that the self-reported weights and heights are still subject to the same potential bias as stated earlier.
b.    Socio-economic Data—we extracted the five-year data (2007–2011) from the American Community Survey to form a data set that contains both census tract and census block group data, including population counts, population counts with college or higher education attainment, median family income, unemployment, and percentages of white population.
c.    Census tract and census block group boundary files from the 2010 TIGER/Line files by the US Census Bureau.

## 3. Analysis and Results

### 3.1. Spatial Distribution of Obese Population and Geographic Scales

After residential addresses of obese adult population were geocoded (*i.e.*, BMI $\geq$ 30), they were used to calculate obesity rates, defined as the number of obese people per 1000 population, by census block groups and by census tracts. The two maps in Figure 1 provide an overview of the geography of obesity in Summit County, Ohio. Overall patterns from both maps show that higher obesity prevalence levels are observed in and around the City of Akron, the most highly urbanized portion of the county in the central part of the county. However, it should be noted that the spatial distribution of obesity ratios by census block groups provides a much higher level of geographical detail and differences in the results between the two geographical scale levels are clearly recognizable.

As shown in Figure 1a,b, in numerous parts of the county, block groups with very different obesity prevalence levels were generalized when adjacent block groups were aggregated into tracts. For example, in the northern most part of the county, it is clear that greater details of different levels of obesity prevalence are shown by block groups but generalized into a less detailed pattern by tracts. Similar generalization can be observed in other parts of the county.

Both scales are consistent in showing that the city center has very low rates. The low rates at both scales are attributable to the fact that the city center has the youngest population. The center was surrounded by areas with relatively high obesity rates, particularly to its east and west, and to a lesser extent to the south. Although many block groups had relatively high rates, they did not fill the areas surrounding the center continuously to form contiguous patches, and some high rate block groups were relatively spread outside, including some to the southwestern corner of the county. However, at the tract level, tracts with high rates were relatively contiguous, mainly because the block group rates were averaged or smoothed over larger areas (tracts). Thus, spikes of high values for block groups were lumped with neighboring units of lower levels, generating a smoother value surface over the region, and thus values are more similar over space (*i.e.*, larger positive spatial autocorrelation). This spatial smoothing process was explained in great detail in Wong [1].

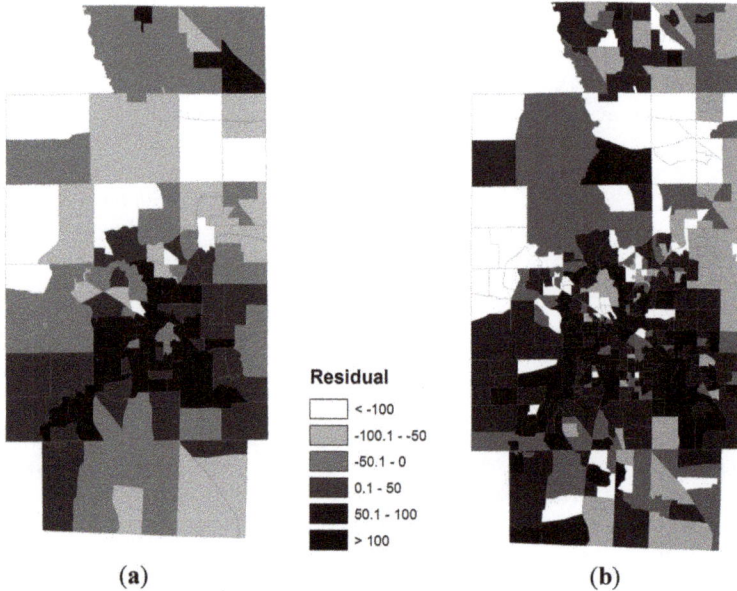

**Figure 1.** Obesity rates in Summit County, Ohio. (a) Census Tracts; (b) Census Block Groups.

### 3.2. Spatial Relationships between Obese Population and SES Attributes

To examine the socio-economic and geographical disparities of the obese population, we analyzed the spatial relationships between obesity ratios and a set of carefully selected socio-economic (SES) attributes, using both census tracts and census block groups. As suggested in *Geographies of Obesity* [24], the socio-economic attributes that may influence obesity ratio include population density, racial composition, educational attainment, income level, employment level, and other factors. Based on these, we have assembled data from the 2011 American Community Survey (US Census Bureau) for both census tracts and census block groups with the following variables for our analysis:

- Population density (POPDEN)
- Percent white population (RWHITE)
- Median family income (MEDINC)
- Percent with bachelor degree or higher (RGEBA)
- Percent unemployed (RUNEMP)

Using these areal attributes as explanatory (or independent) variables and obesity rates as the dependent variable, we first explored to what degrees the variations in obesity rates at both block groups (BGs) and tracts (TRs) levels can be explained by each of the independent variables. The results showed that only three variables are statistically significant in explaining the variation of obesity rates at both geographical levels. These variables are education (RGEBA), income (MEDINC), and unemployment (RUNEMP), as shown in Table 1. Lower educational attainment, lower income level, and higher unemployment ratios appear to be important in influencing the geographic patterns of obesity prevalence. It is also worth noting that the race variable was not significant. The adjusted-$R^2$ values listed in Table 1 indicate that these regression models are relatively weak. However, it appears that these SES variables can explain the variation in obesity rates better at the tract level than at the block group level.

**Table 1.** Regression models with the highest adjusted-$R^2$ values.

|  | Adj-$R^2$ | Regression Model |
|---|---|---|
| **Bgroups** | 0.40 | −RGEBA |
|  | 0.41 | −MEDINC |
|  | 0.36 | +RUNEMP |
| **Tracts** | 0.66 | −RGEBA |
|  | 0.65 | −MEDINC |
|  | 0.62 | +RUNEMP |

Higher correlation coefficients are expected for larger areal units (TRs *vs.* BGs), as this is part of the scale effect under the MAUP, and has been well documented and explained [25]. In short, more aggregated data have less variation and smaller variance (and standard deviation). Lower in variance (and standard deviation) will partly raise the correlation. Even at the TRs level, the $R^2$ values are not strong. One possible reason for low explanatory power of a regression model is the presence of spatial heterogeneity. While the model may have captured the pertinent variables to explain the outcomes, the relationships between the outcome and explanatory variables may vary across different observations. Such variation often follows certain geographical patterns. To address this issue, we used geographically weighted regression (GWR) [26,27] with the three explanatory variables at both the block groups and tracts levels. The results are listed in Table 2, together with results from models of ordinary least squares regression (OLS) with the same dependent and independent variables. We used ArcGIS 10.1 [28] to perform the calculations for GWR models.

**Table 2.** Summary output from geographically weighted regression (GWR) and ordinary least squares regression (OLS).

| Obesity_Ratio = Function ( RGEBA, MEDINC, RUNEMP) | | | |
|---|---|---|---|
| GWR | $R^2$ | Adjusted-$R^2$ | AICc |
| Block Groups | 0.4937 | 0.4650 | 5101.32 |
| Tracts | 0.7301 | 0.7070 | 1395.61 |
| OLS | $R^2$ | Adjusted-$R^2$ | AICc |
| Block Groups | 0.4415 | 0.4378 | 5114.80 |
| Tracts | 0.6968 | 0.6899 | 1400.02 |

From Table 2, it can be seen that the overall adjusted-$R^2$ value is higher at tracts level than at block groups level (1.9 fold). Again, the larger $R^2$ value at the tract level is expected due to scale effect as in the case of ordinary regressions. In addition, AICc values are lower at the tracts level than at the block groups level. This is true for both GWR and OLS, with only minor differences in adjusted-$R^2$ and in AICc. The performance statistics of these two models suggest that the OLS model is reasonably competent as compared to the local model using GWR because the AICc values of the OLS model are smaller than that of the GWR model. However, we will demonstrate below that despite the guidance of these model statistics favors the global OLS model, the local model has tremendous values in revealing pertinent relationships that OLS models do not reveal.

GWR essentially uses a pre-defined function to determine the level of influence that neighboring units have on each geographic unit in the regression model. For example, for census block group, $b_i$, a pre-defined function may be based on the *distance decay* concept so that block groups located farther away from $b_i$ are weighted less in the regression outcomes than the immediate neighboring block groups of $b_i$. The pre-defined function can be adjusted to reflect particular phenomena based on their spatial patterns.

Normally, the pre-defined function is applied to all geographical units. When this is the case, it is said to be using a *fixed* kernel. An option in using GWR to analyze spatial relationships is to vary the pre-defined function according to the density of data points locally. In areas where the data are spatially denser, the distance decay can be structured to reflect that in areas where the data are spatially less dense. When using the varying distance functions, it is said to be using *adaptive* kernels. In this study, we used adaptive kernel approach in our GWR models to reflect the uneven geographic distribution of the model variables.

Below in Figure 2, the distribution of residuals, *i.e.,* the differences between actual obesity rates and the predicted obesity rates by the GWR models, shows no spatial autocorrelation in either TRs or BGs. Global Moran's Index values, a widely used index for measuring spatial autocorrelation, is $-0.016$ (Z-score = $-0.3737$, Prob = 0.7086) for TRs and is 0.004 (Z-score = $-0.1667$, Prob = 0.8675) for BGs, both are not statistically significant at $\alpha = 0.025$ level. The map by census tracts shows a more generalized pattern than that by census block groups. On the map by block groups, we can easily identify areas where such residuals are larger or smaller with much detail. The different levels of details as displayed by tracts and block groups suggest that smaller geographic units may be better for modeling SES and area disparities in health. Some small areas of concern may be hidden at the tract level, but are exposed at the block group level.

From the geographically weighted regression model, it is possible to observe how a particular explanatory variable influences obesity rates more or less across the study area. This is done by mapping the regression coefficients of the explanatory variables. Figure 3 shows the distribution of coefficient values for unemployment ratios in the model. It appears that the northern parts of the county experienced increased obesity rates with increased unemployment ratios where the southern and southeastern parts of the county shows the opposite trends. Again, results from using block groups do show more spatial details than what tracts reveal. However, an important aspect of these results is that unemployment and obesity levels have opposite relationships in different parts of the region (the coefficient ranges from $-0.2$ to 0.4), a situation that is difficult to explain, but cannot be revealed by the global regression model.

Also showing the spatial patterns of coefficient values, Figure 4 suggests that educational attainment (percent of population with bachelor degrees or higher) has a stronger impact on lowering obesity rates in the northern parts than other parts of the county. This trend is better described with block groups than with census tracts because it is much generalized in the tracts. In the City of Akron, educational attainment makes less impact on obesity rates than in the northern part of the county.

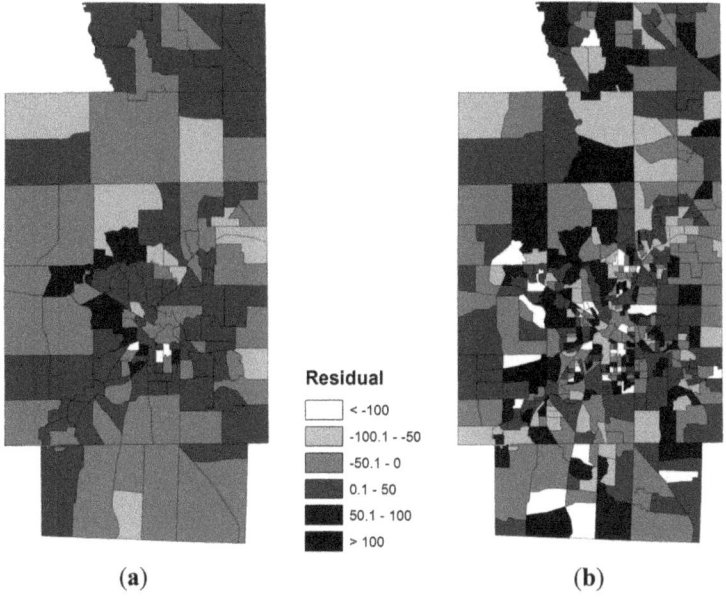

**Figure 2.** Spatial patterns of residuals from geographically weighted regression models, ObRates = *function* (RGEBA, MEDINC, RUNEMP). (**a**) Census Tracts, (**b**) Census Block Groups.

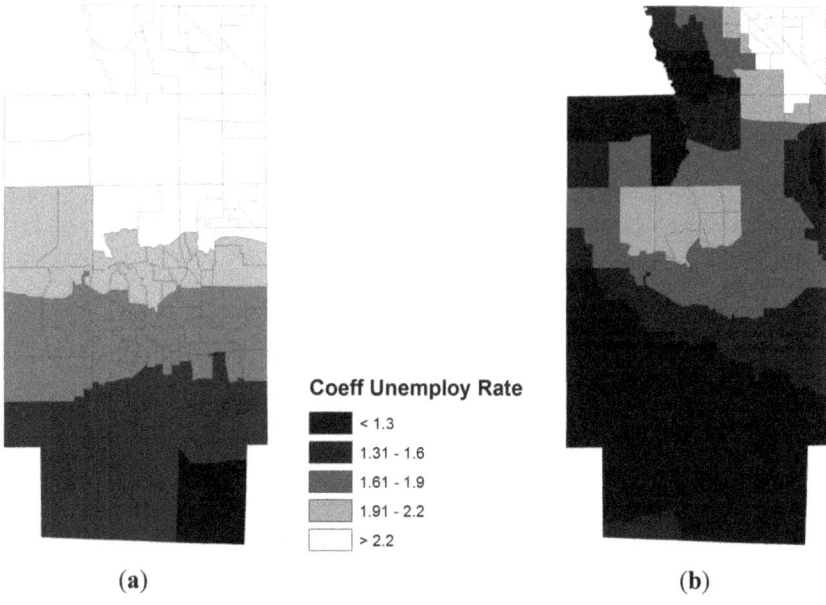

**Figure 3.** Spatial patterns of regression coefficients for unemployment ratios. (**a**) Census Tracts; (**b**) Census Block Groups.

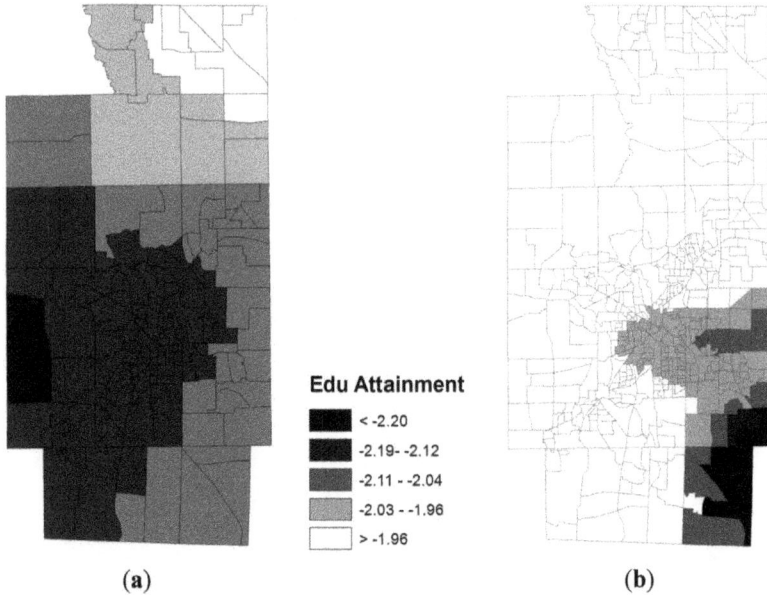

**Figure 4.** Spatial patterns of regression coefficients for educational attainment. (**a**) Census Tracts; (**b**) Census Block Groups.

Again in Figure 5, which shows regression coefficients for median family income in the GWR model, tracts also generalize the spatial pattern of how median family income influences obesity rates in Summit County. With block groups, the different levels of impacts on obesity rates by median family income are shown by circular rings that center at the City of Akron—from a positive influence of increasing median family income causing slight increases of obesity rates to a negative influence of increasing median family income causing reductions in obesity rates. Comparing what are shown by tracts and by block groups, the influences by median family income on obesity rates do show significantly different patterns on the western parts of the county. In addition, similar to the unemployment variable, the coefficient value ranges from −0.2 to 0.1, indicating that the direction of the relationship is not uniform across the region. In other words, lower income level is related to lower obesity rate in some areas (center and the east), but is related to higher obesity rate in other areas (north and west).

Overall, our analysis showed that obesity rates are indeed affected by education attainment, income level, and unemployment level. While such relationships are all statistically significant for the three SES variables included in GWR models, it is important to explore in more spatial details to appreciate where inside the county we can expect such relationships to be stronger or weaker. Thus, when making policies on how to promote health and how to allocate funding to different areas in the county, for example, at neighborhood level, geographic disparities in health can be incorporated for more effective outcomes.

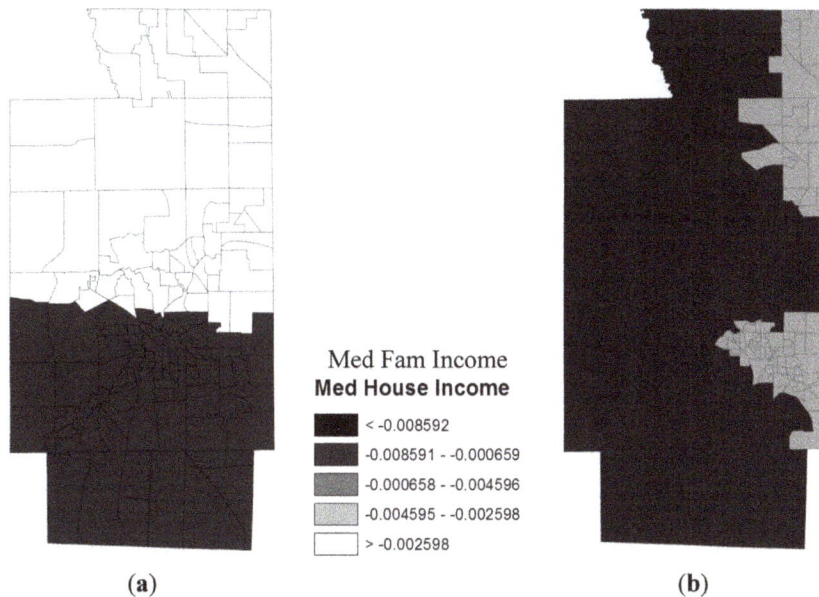

**Figure 5.** Spatial patterns of regression coefficients for median family income ($1000's). (a) Census Tracts; (b) Census Block Groups

## 4. Discussion and Concluding Remarks

We have presented in this paper our analysis of obesity rates in terms of their spatial patterns and their relationships to a set of selected socio-economic variables. Similar analytical procedures were repeated for census tracts and census block groups to show that geographic resolutions do indeed matter in such analysis.

While individual records for adults age 16 and 21 in Summit County, Ohio, as obtained from Ohio Bureau of Motor Vehicles, were used in our study, it should be noted that this is not a 100% coverage of all adult population in Summit County—this data set does not include those who chose not to acquire driver's licenses and those who failed to renew licenses. Furthermore, it is possible that heights and weights obtained from self-reporting through driver's license registrations are not accurate. For example, a person's height and weight at age 16 when first acquiring his/her driver's license may be lower than his/her height and weight by age 20 before having to renew the license. This is a well-documented phenomenon (for example, see [23]). However, BMI data as derived from heights and weights reported to the Bureau of Motor Vehicles are probably the best and the most complete data we can obtain. If more precise analysis is in order, adjustments should be made to correct such under-reported bias.

Geographic resolutions do make a difference. In general, the higher the resolution, the more details are revealed in the results of analysis. Analysis with data at lower geographic resolution may run into the risk of obscuring potentially meaningful and informative processes operational only at the finer scale. To that end, please see Lam [29] for a discussion on different types of scale and their effects on geographic studies. As a general rule of thumb, higher resolution analyses are preferred. Unfortunately, the geographic resolution of analysis is often dictated by the availability of supporting data. Although, in this study, data at the block group level were available and used, and these data are of higher resolution than the corresponding census tract data, we need to also take into account the quality of data in addition to the desirable levels of scale or resolution. If data at different geographic resolution are of similar quality, it would be preferred to use those with more geographic details. It should also be noted here that, as more micro data (e.g., individual addresses or GPS coordinates, *etc.*) are increasingly available, we argue that analyses should be performed at the highest geographic resolution whenever possible and when the supporting data allow.

In our specific case, and probably our situation is also applicable to many studies in social sciences and public health, we have to used ACS data, the only major source of data in the U.S. after 2000 census in order to obtain SES information of the social environment in which the subjects resided. An important aspect of ACS data is the survey nature such that estimates, especially for smaller geographical units, tend to be unreliable, often with relatively large margin of error [30].

As we prefer to conduct analysis with data of higher geographic resolution, and therefore using block group data is preferable to reveal detailed geographical patterns, ACS data at block group level have substantially larger error than their corresponding tract level data. Just take the median family income variable as an example, the minimum, maximum, and average coefficient of variation (*CV*) of the variable are reported in Table 3 below. Clearly, the ACS estimates at the tract level are much more reliable than those at the block group level. In fact, some of the estimates at the block group level have their 90% margin of errors larger than the estimates. On the other hand, the quality of tract level estimates is not ideal. Nonetheless, these tract estimates are more reliable. Thus, from the data quality perspective, the tract level analysis we conducted and reported here probably offer results with a higher level of confidence. This higher confidence level, unfortunately, has to be trade-off with a lower geographical resolution in the analysis results.

**Table 3.** Summary of statistics of the coefficient of variation (CV) for the variable median family income from ACS at the census block groups and tracts levels.

| Statistics for CV | Block Group Level | Tract Level |
|---|---|---|
| Minimum | 0.0137 | 0.0250 |
| Maximum | 2.2572 | 0.8192 |
| Average | 0.2703 | 0.1305 |

Many obesity studies adopted census tracts as the *de facto* geographic unit of analysis. This may be due to the obvious reasons of data availability and limits to computational resources. We argue that census tracts may generalize spatial patterns too much and that census block groups or smaller geographic units should be used whenever possible. Assuming equal in data quality, analyzing geographies of obesity at a finer geographic scale enables better decisions when formulating policies to promote health for areas with health disparities.

The use of GWR also reveals new details in terms of spatial trends of how independent variables are associated with dependent variables. These spatial trends cannot be uncovered by conventional global regression models, such as ordinary least squares regression that provides only global trends of the relationships between dependent and independent variables. For example, the varying spatial trends of how unemployment ratios impact obesity prevalence as shown in Figure 3 would never be discovered using only conventional regression modeling approaches.

**Author Contributions:** Jay Lee contributes to the work reported in this paper with the formulation of research questions, design of analytical methods, and the writing of initial drafts. Mohammad Alnasrallah is responsible for data management and analysis and the creation of maps in this paper. David Wong extends this paper by adding the discussion on modifiable area unit problems and data quality issues. Heather Beaird provides access to data used in this research and advises the interpretation of analytic outcomes. Everett Logue assists this project with the design of the conceptual framework used in the analysis.

**Conflicts of Interest:** The authors declare no conflict of interest.

### References

1. Wong, D. The modifiable areal unit problem (MAUP). In *The SAGE Handbook of Spatial Analysis*; Fotheringham, A.S., Rogerson, P.A., Eds.; Sage: London, UK, 2009; pp. 105–123.

2.  Cockings, S.; Martin, D. Zone design for environment and health studies using pre-aggregated data. *Soc. Sci. Med.* **2005**, *60*, 2729–2742. [PubMed]

3.  Anderson, B.; Rafferty, A.P.; Lyon-Callo, S.; Fussman, C.; Imes, G. Fast-food consumption and obesity among Michigan adults. *Prev. Chronic Dis.* **2011**, *8*, A71. [PubMed]

4.  Fine, L.J.; Philogene, G.S.; Gramling, R.; Coups, E.J.; Sinha, S. Prevalence of multiple chronic disease risk factors: 2001 National Health Interview Survey. *Am. J. Prev. Med.* **2004**, *27*, 18–24. [PubMed]

5.  Flegal, K.M.; Carroll, M.D.; Ogden, C.; Johnson, C.L. Prevalence and tends in obesity among US adults, 1999–2000. *J. Am. Med. Assoc.* **2002**, *288*, 1723–1727.

6.  Hedley, A.A.; Ogden, C.L.; Johnson, C.L.; Carroll, M.D.; Curtin, L.R.; Flegal, K.M. Prevalence of overweight and obesity among US children, adolescents, and adults, 1999–2002. *J. Am. Med. Assoc.* **2004**, *291*, 2847–2850.

7.  Ogden, C.L.; Flegal, K.M.; Carroll, M.D.; Johnson, C.L. Prevalence and trends in overweight among US children and adolescents, 1999–2000. *J. Am. Med. Assoc.* **2002**, *288*, 1728–1732.

8.  Ogden, C.L.; Carroll, M.D.; Kit, B.K.; Flegal, K.M. Prevalence of obesity and trends in body mass index among US children and adolescents, 1999–2010. *J. Am. Med. Assoc.* **2012**, *307*, 483–490. [CrossRef]

9.  Flynn, M.A.T.; McNeil, D.A.; Maloff, B.; Mutasingwa, D.; Wu, M.; Ford, C.; Tough, S.C. Reducing obesity and related chronic disease risk in children and youth: A synthesis of evidence with "best practice" recommendations. *Obes. Rev.* **2006**, *7*, 7–66. [CrossRef] [PubMed]

10. Must, A.; Spadano, J.; Coakley, E.H.; Field, A.E.; Colditz, G.; Dietz, W.H. The disease burden associated with overweight and obesity. *J. Am. Med. Assoc.* **1999**, *282*, 1523–1529.

11. Rippe, J.M.; Crossley, S.; Ringer, R. Obesity as a chronic disease: Modern medical and lifestyle management. *J. Am. Diet. Assoc.* **1998**, *98*, S9–S15. [PubMed]

12. Wang, Y.; Mi, J.; Shan, X.-Y.; Wang, Q.J.; Ge, K.-Y. Is China facing an obesity epidemic and the consequences? The trends in obesity and chronic disease in China. *Int. J. Obes.* **2007**, *31*, 177–188.

13. World Health Organization. Obesity: Preventing and Managing the Global Epidemic. Available online: http://libdoc.who.int/trs/WHO_TRS_984.pdf (accessed on 24 November 2013).

14. Eckel, R.H.; Krauss, R.M. American Heart Association call to action: Obesity as a major risk factor for coronary heart disease. *Circulation* **1998**, *98*, 2099–2100. [PubMed]

15. Martinez, J.A. Body-weight regulation: Causes of obesity. *Proc. Nutr. Soc.* **2000**, *59*, 337–345. [CrossRef] [PubMed]

16. Obe, C.W. Causes of obesity. In *Obesity and Weight Management in Primary Care*; Waine, C., Ed.; Blackwell Science: Oxford, UK, 2008; p. 118.

17. Sonya, A.G.; Mensinger, G.; Huang, S.H.; Kumanyika, S.K.; Stettler, N. Fast-food marketing and children's fast-food consumption: Exploring parents' influences in an ethnically diverse sample. *J. Public Policy Mark.* **2007**, *26*, 221–235.

18. Wilding, J. Causes of obesity. *Pract. Diabetes Int.* **2001**, *18*. [CrossRef]

19. Wright, S.M.; Aronne, L.J. Causes of obesity. *Abdom. Imaging* **2012**, *37*, 730–732. [CrossRef] [PubMed]

20. Sobal, J.; Stunkard, A.J. Socioeconomic status and obesity: A review of literature. *Psychol. Bull.* **1989**, *105*, 260–275. [PubMed]

21. Zhang, Q.; Wang, Y. Trends in the association between obesity and socioeconomic status in US adults: 1971–2000. *Obes. Res.* **2004**, *12*, 1622–1632. [PubMed]

22. McLaren, L. Socioeconomic status and obesity. *Epidemiol. Rev.* **2007**, *29*, 29–48. [CrossRef] [PubMed]

23. Ossiander, E.M.; Emanuel, I.; O'Brian, W.; Malone, K. Driver's license as a source of data on height and weight. *Econ. Hum. Biol.* **2004**, *2*, 219–227. [CrossRef] [PubMed]

24. Pearce, J.; Witten, K. *Geographies of Obesity: Environmental Understandings of the Obesity Epidemic*; Ashgate Publishing Ltd.: Burlinton, VT, USA, 2010; p. 331.

25. Fotheringham, A.S.; Wong, D.W.S. The modifiable areal unit problem in multivariate statistical analysis. *Environ. Plan. A* **1991**, *23*, 1025–1044. [CrossRef]

26. Brunsdon, C.; Fotheringham, A.S.; Charlton, M. Geographically weighted regression. *J. R. Stat. Soc. Ser. D* **1998**, *47*, 431–443. [CrossRef]

27. Fotheringham, A.S.; Brunsdon, C.; Charlton, M. *Geographically Weighted Regression: The Analysis of Spatially Varying Relationships*; Wiley & Sons: New York, NY, USA, 2002.

28. ESRI. Available online: http://www.esri.com (accessed on 10 October 2014).

29. Lam, N.S.-N. Fractals and scale in environmental assessment and monitoring. In *Scale and Geographic Inquiry: Nature, Society and Method*; Sheppard, E., McMaster, R.B., Eds.; John Wiley and Sons: New York, NY, USA, 2008; pp. 23–40.
30. Sun, M.; Wong, D.W.S. Incorporating data quality information in mapping the American Community Survey data. *Cartogr. Geogr. Inf. Sci.* **2010**, *37*, 285–300.

isprs International Journal of
*Geo-Information*

MDPI

*Article*

# Geographical Variation of Incidence of Chronic Obstructive Pulmonary Disease in Manitoba, Canada

Mahmoud Torabi * and Katie Galloway

Department of Community Health Sciences, University of Manitoba, 750 Bannatyne Ave., Winnipeg, MB R3E 0W3, Canada; gallowa3@cc.umanitoba.ca

* Author to whom correspondence should be addressed; Mahmoud.Torabi@med.umanitoba.ca; Tel.: +1-204-272-3136; Fax: +1-204-789-3905.

Received: 3 March 2014; in revised form: 15 July 2014; Accepted: 21 July 2014; Published: 29 July 2014

**Abstract:** We aimed to study the geographic variation in the incidence of COPD. We used health survey data (weighted to the population level) to identify 56,944 cases of COPD in Manitoba, Canada from 2001 to 2010. We used five cluster detection procedures, circular spatial scan statistic (CSS), flexible spatial scan statistic (FSS), Bayesian disease mapping (BYM), maximum likelihood estimation (MLE), and local indicator of spatial association (LISA). Our results showed that there are some regions in southern Manitoba that are potential clusters of COPD cases. The FSS method identified more regions than the CSS and LISA methods and the BYM and MLE methods identified similar regions as potential clusters. Most of the regions identified by the MLE and BYM methods were also identified by the FSS method and most of the regions identified by the CSS method were also identified by most of the other methods. The CSS, FSS and LISA methods identify potential clusters but are not able to control for confounders at the same time. However, the BYM and MLE methods can simultaneously identify potential clusters and control for possible confounders. Overall, we recommend using the BYM and MLE methods for cluster detection in areas with similar population and structure of regions as those in Manitoba.

**Keywords:** bayesian computation; chronic obstructive pulmonary disease; geographic epidemiology; prediction; random effects; spatial cluster detection

---

## 1. Introduction

Chronic obstructive pulmonary disease (COPD) is a lung disease defined by continuous airflow limitation caused by small airway disease (obstructive bronchiolitis) and parenchymal destruction (emphysema). The small airways narrow in response to chronic inflammation. As well, inflammatory processes cause the deterioration of the lung parenchyma, which leads to a decrease in the elastic recoil of the lung. As a result of these changes, the airways have a decreased ability to remain open during expiration [1]. The biggest and most widely known risk factor of COPD is cigarette smoking [2]. Other risk factors of COPD include occupational or environmental exposure to dust and hazardous gases, for example when burning biomass fuel [3]. A family history (*i.e.*, genetics), low socioeconomic status, poor nutrition, asthma, and recurrent lung infections can also be risk factors for COPD [1,4]. Therefore, COPD can be the result of a gene-environment interaction [1].

The impact of COPD is often underestimated by health authorities and government officials [5]. In Canada, one of the most overlooked chronic conditions is COPD. Patients suffering from a degenerative lung disease are often misdiagnosed as having bronchitis, a cough or a respiratory tract infection [6]. In 2008, COPD was the leading cause of hospitalizations in Canada. As well, 18% of COPD patients were readmitted to a hospital once within the year and 14% were re-admitted twice within the year. These readmission rates were higher than any other chronic illnesses [6,7]. According to a Canadian article [8], for severe COPD exacerbations or attacks, the average length of a hospital visit was 10 days with an estimated cost of $10,000. Within a single year, the estimated cost of moderate and severe COPD exacerbations exceeds $730 million. This number is expected to nearly double by 2015 [8].

There are various treatments for COPD including antibiotics and chest physiotherapy. However, early detection of COPD is crucial for a positive outcome [9]. Therefore, it is important to identify trends in COPD incidence that may suggest further epidemiological studies to identify risk factors and identify any changes in important factors. Trends may occur over a region and the focus of our paper is to examine geographical variation in the number of people diagnosed as having COPD during 2001 to 2010 in the province of Manitoba, Canada.

A spatial cluster is defined as a limited area within the entire study region which has a high proportion of disease cases [10]. Possible factors related to diseases may be determined by discovering disease clusters which may lead to an improved understanding of etiology. In fact, the identification of clusters may lead to further analyses to study how exposures and disease interventions are connected [11].

Spatial cluster detection methods can be classified into two statistical approaches, a focused approach or a non-focused (general) approach. The methodology of focused cluster detection approaches is to locate regions with an excess number of disease cases in an area near a possible cause (*i.e.*, a toxic waste site) [12,13]. On the other hand, non-focused cluster detection methods typically use various ways in order to discover areas with a high number of disease cases in the entire study region [14–16]. The circular spatial scan statistic (CSS) [17], flexible spatial scan statistic (FSS) [18], and Bayesian disease mapping (BYM) [14] are all considered to be focused cluster detection methods, whereas, the Besag and Newell (BN) [19,20] test and the maximizing excess event test (MEET) [21] are classified as non-focused cluster detection procedures. Non-focused tests are used to detect potential clusters in the study area, while focused tests are used to test the null hypothesis of no spatial cluster against the alternative hypothesis that a spatial cluster exists. In other words, the purpose of focused tests (CSS, FSS, BYM) is to find possible clusters in an area of interest and the aim of non-focused tests is to discover any significant cluster without determining a specific area of interest. These approaches were compared by analyzing childhood cancer data in the province of Alberta, Canada [22]. Recently, a frequentist approach based on the maximum likelihood estimation (MLE), via data cloning (DC) [23,24], was also proposed to obtain possible clusters [25] in an area of interest. Another cluster detection method is the local indicator of spatial association (LISA) [26]. This method is simple and easy to implement.

This paper is based on the focused cluster detection methods. In particular, the aforementioned focused approaches (CSS, FSS, BYM, MLE, and LISA) are used to analyze a real dataset of COPD cases in the province of Manitoba, Canada, from 2001 to 2010.

## 2. Methods

### 2.1. Study Subjects

This study was based on the Canadian Community Health Survey (CCHS) [27] from Statistics Canada. The CCHS is a cross-sectional survey, which gathers information from the Canadian population regarding health status, health care utilization and health determinants. The CCHS collects health related data from individuals aged twelve and older in order to provide reliable estimates at the health region level [27]. The information from the CCHS used in this study was the number of COPD cases in the province of Manitoba, Canada, from 2001 to 2010. Eleven Regional Health Authorities, which are further divided into 67 Regional Health Authority Districts (RHADs) are in charge of delivering health care services to individuals in Manitoba. The RHADs are the geographic units used in our models and all of the data used in the study are related to these RHADs which are labeled 1, 2, . . . , 67 for simplicity. As well, a population-based centroid was provided for each RHAD, however, these centroids were not necessarily geographic centres. Since the data used in the study was from a survey, appropriate weights (see Section 2.2 for more details) established by Statistics Canada [27] were applied to the data, which was then aggregated over the study period from 2001 to 2010.

The population was stable in Manitoba from approximately 1.15 million people in 2001 to 1.20 million people in 2010. Region 38 had the smallest average population size of 920 people while region 62 had the largest average population size of 91,633 people. The mean and median population sizes across the regions were 17,471 and 9466, respectively. The total number of COPD cases in Manitoba was 56,944 with a mean of 850 and median of 504 cases. These observations are based on the weighted results of COPD cases across the 67 regions.

The observed number of COPD cases and the expected number of COPD cases as well as the population size of each region are important requirements for focused spatial cluster detection methods. Adjustments may be made when the expected number of cases varies by different factors such as year, age, and gender. The expected number of disease cases was then adjusted by year (1–10), age group ((0–5), (6–20), (21–40), (41, 88), (89+)) and gender (male, female). A review of the CSS, FSS, BYM, MLE, and LISA spatial cluster detection methods is given in the Appendix A.

The five focused spatial cluster detection procedures (CSS, FSS, BYM, MLE, and LISA) have different assumptions. Although the CSS, FSS, and LISA approaches are distribution free, it is assumed that the number of disease cases follow a Poisson distribution in the BYM and MLE methods. As well, while the number of regions to be included in the cluster must be specified for the CSS and FSS methods, this is not a requirement for the BYM and MLE approaches. For the model-based cluster identification methods (BYM and MLE), if the model does not fit the data well, the result can be misleading. So, the *deviance residual* [28] should also be checked. While the expected number of disease cases or the population of each region is required for the above methods, they are not a requirement for the LISA method.

The University of Manitoba's Research Data Centre approved the study, and Statistics Canada approved administrative data access. ArcGIS version 10.0 (Environmental Systems Research Institute, Redlands, CA, USA) was used to produce choropleth maps of risks.

### 2.2. Weighting Process

The weighting was completed by Statistics Canada using a detailed weighting process [27]. A brief summary of this procedure is given here. First, the weighting depends on the sampling method (area frame *vs.* telephone frame) used in each region. In the area frame an initial weight is assigned based on the Labour Force Survey (LFS). Out-of-scope units (*i.e.,* Dwellings that are under construction, vacant, seasonal or secondary and institutions) are removed from the sample. As well, sub-clusters (*i.e.,* Sub-sampling within a selected dwelling), larger sample sizes and non-response units are adjusted for in the weighting process. In the telephone frame (the survey is conducted by telephone) an initial weight is assigned as the probability that phone number will be selected, which depends on the number of units sampled and the number of units available to be sampled. In this method, samples are drawn every two months therefore, an adjustment factor is applied to reduce the weights of each two-month sample so that the total sample is representative of the population only once. Similar to the area frame method, out-of-scope numbers (*i.e.,* Businesses, institutions, out-of-scope dwellings or numbers that are not in service) are removed from the sample. Also, non-response units and dwellings with multiple phone numbers are adjusted for in the weighting process [27].

The weights common to the area frames and telephone frames need to be integrated using an adjustment factor $\alpha$ ($0 < \alpha < 1$). Then a person-level weight is created by taking the inverse of the probability a person in the selected dwelling will be selected, which depends on the number of people in the household and the ages of those people. After the appropriate adjustment is made, a "winsorization" trimming method is used to decrease any extreme weights that occur. Finally, a calibration approach is used to ensure the weights are representative of the population estimates for the different age groups and genders in each health region [27].

*2.3. Specific Hypotheses*

We specify the alternative hypotheses for the CSS, FSS, BYM, and MLE approaches. We consider multiple alternatives that are tested separately. Further, let $RR_i$ indicate the relative risk for the $i$-th region within a cluster when compared with the region outside a cluster; the latter has $RR_i = 1$. For example for cluster $X$ , the $RR_i$ is given by

$$RR_i = \begin{cases} 3 & i \in X \\ 1 & otherwise. \end{cases}$$

## 3. Results

The results of five different cluster detection techniques when applied to a COPD dataset in the province of Manitoba, Canada, from 2001 to 2010 are shown and compared in this section.

Based on the 67 regions, four different clusters were tested: (1) a case of no clusters (called A); (2) seven regions from the northern part of the province (called B); (3) seven regions from south-central part of the province (called C); and (4) 12 regions which comprise the Winnipeg region (called D). For A, no region was specified as a potential cluster. Moreover, the regions belonging to clusters B, C, and D are: B = {31, 33, 34, 36, 38, 40, 41}, C = {27, 28, 29, 30, 50, 51, 52}, and D = {56, 57, 58, 59, 60, 61, 62, 63, 64, 65, 66, 67}. Since the LISA method does not depend on the expected number of observations, it could only be applied to cluster A as the other clusters require the adjustment of the expected number of disease cases for those regions inside the specified cluster.

The areas that are statistically significant (potential clusters) are shown for each cluster and each method separately (Figures 1–4). The summary of cluster A, no region specified as a potential cluster, is presented in Table 1. For the CSS and FSS procedures, the regions that are most likely to constitute a disease cluster are presented, as well as the regions that are second and third most likely to be considered as a cluster. For the BYM and MLE methods, a region is considered (and ranked) to be a significant cluster if the lower limit of the credible/prediction interval follows the specified criteria. For example, in the BYM method region 10 is most likely to be considered as a cluster and region 61 is least likely to be considered as a cluster under the criteria that the lower bound of RR is greater than one. For the LISA method, a region is determined (and ranked) to be significant if the p-value is less than 0.1.

The FSS method identified more regions as potential clusters than the CSS approach for cluster A, although, the regions with potential clusters that were detected by the CSS method were also identified by the FSS approach. The CSS approach detected regions {10, 43, 45, 61, 62} as potential clusters, and the FSS method identified the same regions as the CSS method, as well as regions {1, 11, 12, 13, 14, 20, 21, 27, 46, 50, 51, 54, 56, 60, 64, 65}. The BYM and MLE methods identified regions {1, 3, 6, 10, 11, 12, 20, 21, 24, 27, 43, 45, 50, 54, 61, 62, 64, 65} as possible clusters. The only difference between the results of these two procedures was the order of significance for the potential clusters. As well, most of the regions identified using these two approaches were also identified by the FSS approach and the regions identified by the CSS technique were also detected by the BYM and MLE approaches. Note that by evaluating the criterion of the RR values from greater than 1 to 1.5 or even 2, the number of potential clusters decreases (Table 1). Based on the deviance residual plots for both the BYM and MLE methods, we found that there is no serious lack of fit in the model. The LISA method found regions {2, 7, 16, 24, 43, 56, 57, 58, 60, 62, 64, 67} to be possible clusters of COPD. This approach identified some different regions to be potential clusters as compared to the other methods.

For the case of cluster B, none of the methods were able to detect all the regions in cluster B as a potential cluster. However, the CSS method identified regions 10, 43, and 62 as a potential cluster while the FSS method detected the same regions as the CSS method in addition to regions {11, 12, 13, 14, 20, 21, 27, 31, 45, 46, 50, 51, 54}. The BYM approach could identify regions {1, 3, 6, 10, 11, 12, 20, 21, 24, 27, 31, 43, 45, 50, 54, 61, 62, 64, 65} as potential clusters. The MLE method was also able to identify the same regions as the BYM method in addition to region 19.

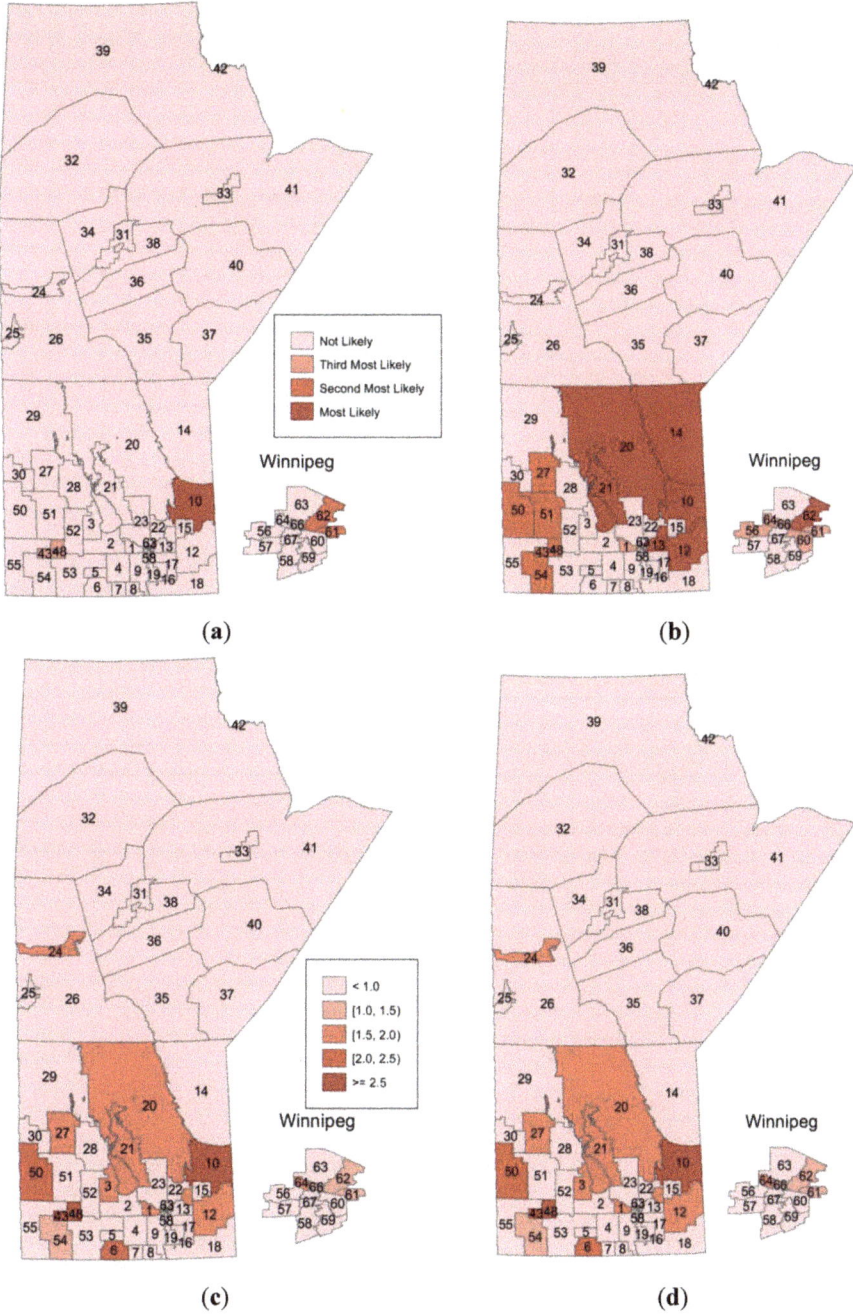

**Figure 1.** *Cont.*

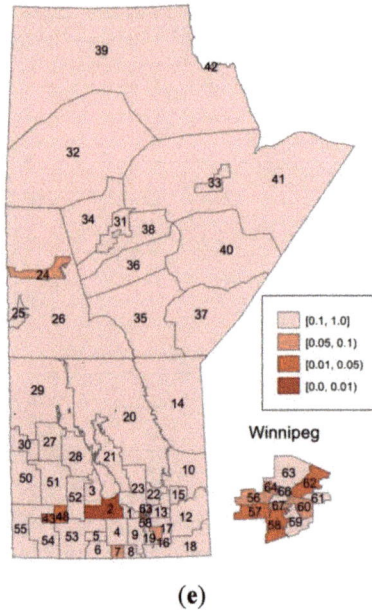

(e)

**Figure 1.** The order of most likely clusters of COPD for the CSS, FSS, and LISA (based on the *p*-value) methods, and the special effects of the regional COPD risks for the BYM and MLE methods; in the case of cluster A. Major urban centre (Winnipeg region) is incorporated as an inset. (**a**) CSS; (**b**) FSS; (**c**) BYM; (**d**) MLE; (**e**) LISA.

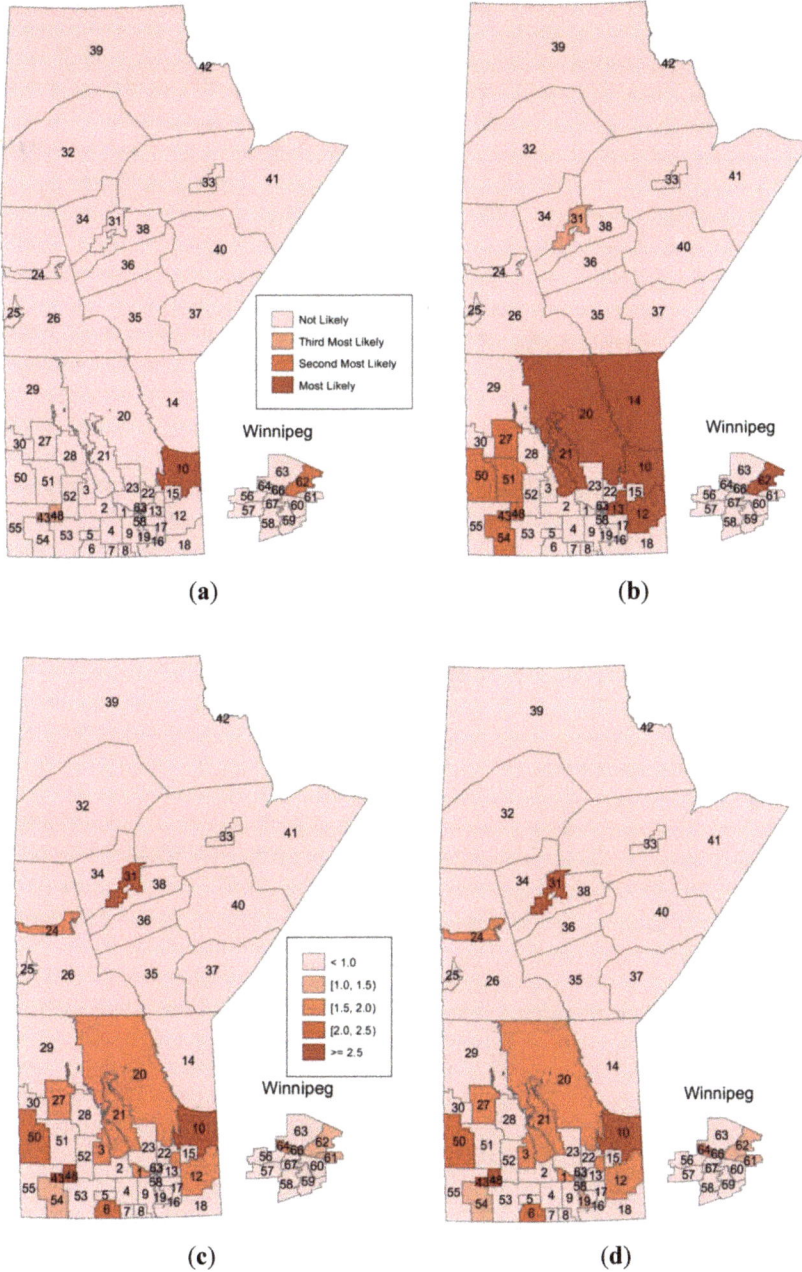

**Figure 2.** The order of most likely clusters of COPD for the CSS and FSS methods, and the special effects of the regional COPD risks for the BYM and MLE methods; in the case of cluster B. Major urban centre (Winnipeg region) is incorporated as an inset. (**a**) CSS; (**b**) FSS; (**c**) BYM; (**d**) MLE.

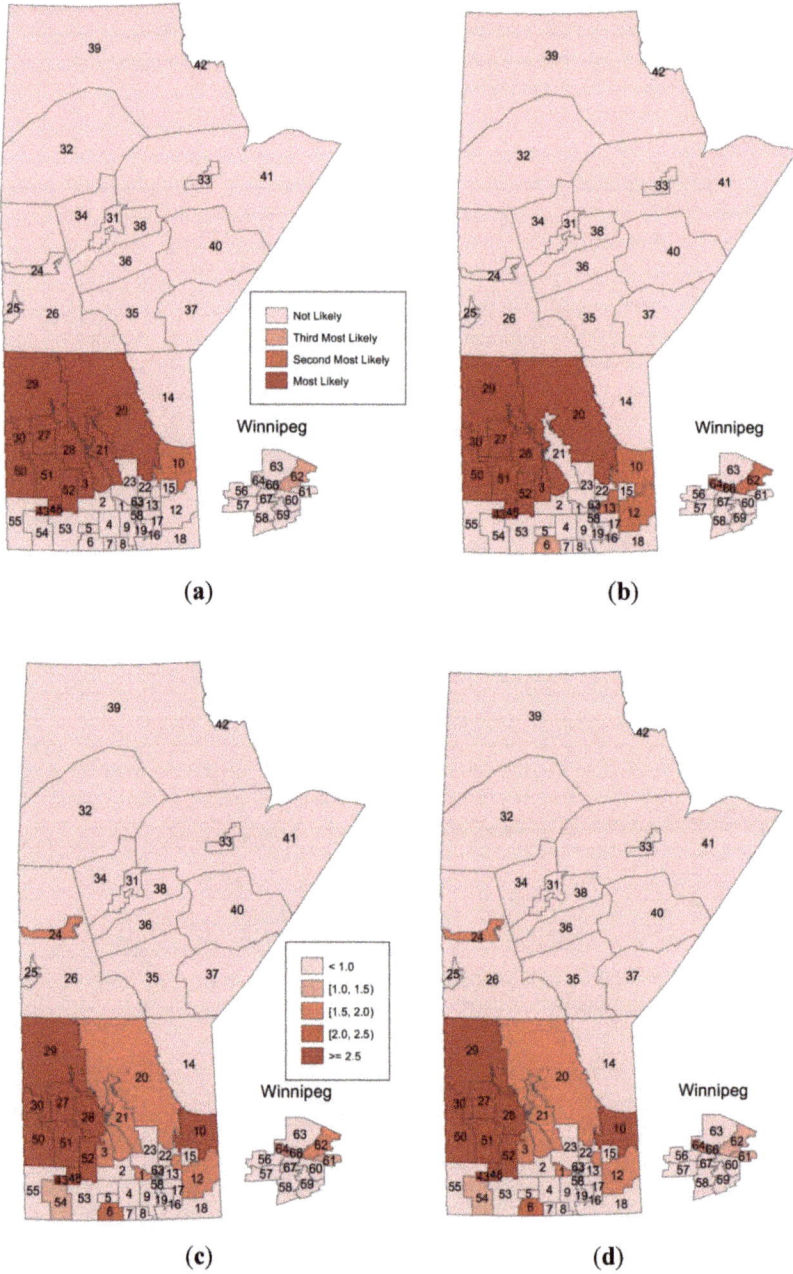

**Figure 3.** The order of most likely clusters of COPD for the CSS and FSS methods, and the special effects of the regional COPD risks for the BYM and MLE methods; in the case of cluster C. Major urban centre (Winnipeg region) is incorporated as an inset. (**a**) CSS; (**b**) FSS; (**c**) BYM; (**d**) MLE.

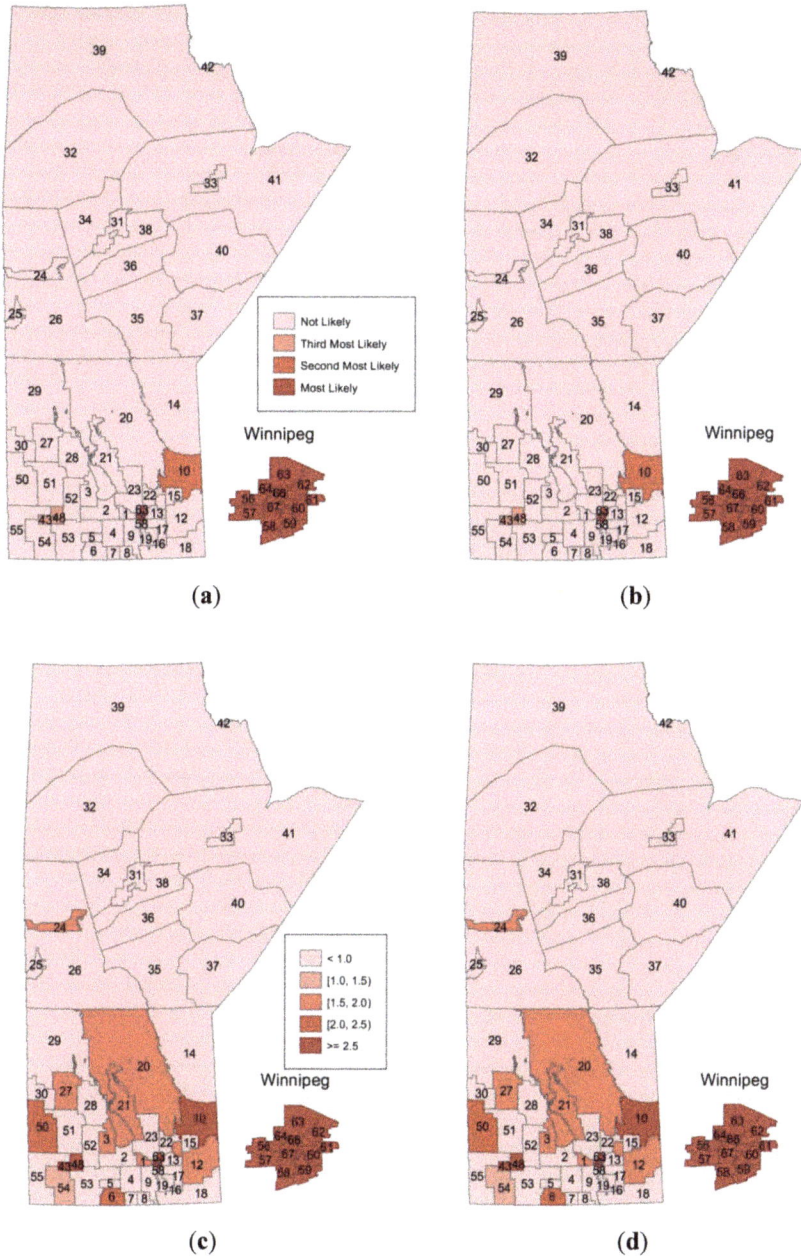

**Figure 4.** The order of most likely clusters of COPD for the CSS and FSS methods, and the special effects of the regional COPD risks for the BYM and MLE methods; in the case of cluster D. Major urban centre (Winnipeg region) is incorporated as an inset. (**a**) CSS; (**b**) FSS; (**c**) BYM; (**d**) MLE.

**Table 1.** The order of significant regions for the LISA, CSS, FSS, BYM, and MLE methods for cluster A.

| Region | Method | | | | | | | | |
|--------|--------|--------|--------|--------|--------|--------|--------|--------|--------|
| | LISA | CSS | FSS | RR > 1.0 | | RR > 1.5 | | RR > 2.0 | |
| | | | | BYM | MLE | BYM | MLE | BYM | MLE |
| 1 | - | - | 3 | 13 | 11 | - | - | - | - |
| 2 | 1 | - | - | - | - | - | - | - | - |
| 3 | - | - | - | 11 | 13 | - | - | - | - |
| 6 | - | - | - | 5 | 5 | 5 | 5 | - | - |
| 7 | 12 | - | - | - | - | - | - | - | - |
| 10 | - | 1 | 1 | 1 | 1 | 1 | 1 | 1 | 1 |
| 11 | - | - | 1 | 15 | 15 | - | - | - | - |
| 12 | - | - | 1 | 10 | 10 | - | - | - | - |
| 13 | - | - | 1 | - | - | - | - | - | - |
| 14 | - | - | 1 | - | - | - | - | - | - |
| 16 | 9 | - | - | - | - | - | - | - | - |
| 20 | - | - | 1 | 4 | 4 | 4 | 4 | - | - |
| 21 | - | - | 1 | 14 | 14 | - | - | - | - |
| 24 | 7 | - | - | 6 | 6 | 6 | 6 | - | - |
| 27 | - | - | 2 | 12 | 12 | - | - | - | - |
| 43 | 2 | 3 | 2 | 2 | 2 | 2 | 2 | 2 | 2 |
| 45 | - | 3 | 2 | 7 | 8 | - | - | - | - |
| 46 | - | - | 2 | - | - | - | - | - | - |
| 50 | - | - | 2 | 3 | 3 | 3 | 3 | - | - |
| 51 | - | - | 2 | - | - | - | - | - | - |
| 54 | - | - | 2 | 17 | 17 | - | - | - | - |
| 56 | 6 | - | 3 | - | - | - | - | - | - |
| 57 | 4 | - | - | - | - | - | - | - | - |
| 58 | 5 | - | - | - | - | - | - | - | - |
| 60 | 8 | - | 3 | - | - | - | - | - | - |
| 61 | - | 2 | 3 | 18 | 18 | - | - | - | - |
| 62 | 3 | 2 | 1 | 8 | 7 | - | - | - | - |
| 64 | 11 | - | 3 | 9 | 9 | - | - | - | - |
| 65 | - | - | 3 | 16 | 16 | - | - | - | - |
| 67 | 10 | - | - | - | - | - | - | - | - |

For cluster C, all four methods were able to detect all the regions of cluster C as a potential cluster. Furthermore, the CSS method also identified regions {3, 10, 20, 21, 43, 45, 49, 62} as potential clusters while the FSS method detected those regions identified by the CSS method in addition to regions {6, 11, 12, 13, 64, 65}. Both the BYM and MLE methods identified regions {1, 3, 6, 10, 11, 12, 20, 21, 24, 43, 45, 54, 61, 62, 64, 65} in addition to the regions in cluster C as potential clusters.

For cluster D, all four methods detected the regions belonging to the D cluster as a potential cluster. In addition to the regions in Winnipeg (cluster D), the BYM and MLE approaches were also able to detect some neighbours of Winnipeg (14 regions) as potential clusters. However, the CSS and FSS methods only detected two regions, 10 and 43 as a potential cluster in addition to cluster D.

## 4. Discussion and Conclusions

We used five popular approaches in spatial epidemiology to identify potential clusters of COPD cases in the Canadian province of Manitoba, Canada. These five methods have been used extensively in the literature and are relatively comprehensive. These methods use different approaches (non-parametric to parametric) to test for significant clusters.

We considered four different alternative hypotheses to compare the results of the CSS, FSS, BYM, and MLE methods. All four methods did a good job of identifying potential clusters for dense populations (clusters C and D) but not for a dispersed population (cluster B). In general, the CSS method identified a lower number of regions combined as a potential cluster compared to the FSS method, due to the non-circular shape of some regions in the province of Manitoba. A disadvantage of the LISA method is that the results do not depend on the expected number of cases or the population in each region. This is concerning since regions with high populations will likely have higher observed numbers of disease, however, this is not taken into account when using the LISA method. Therefore, the LISA method could only be applied to cluster A as cluster B, C, and D require the expected numbers to be adjusted for the respective regions in a cluster. This may explain why the LISA approach identified some different regions as potential clusters when compared with the other procedures.

A region was identified as a potential cluster if the lower bound of credible/prediction interval of the estimated relative risk was larger than one for the BYM and MLE approaches. Different decision rules may be defined where the estimated relative risk (in terms of the credible/prediction interval) would be larger or smaller than one [29]. One could also consider the exceedance probability $Pr(RR_i > b) > c$, where $b$ can be 1, 2 or 3 and $c$ may be a large value such as 0.90 [30]. For the LISA method, a region was defined to be significant if the associated $p$-value was less than 0.1. However, different decision rules could be used where the level of significance is smaller than 0.1.

Here, three important factors, age, gender and year were used to adjust the expected number of COPD cases in the province of Manitoba. Unlike the CSS, FSS, and LISA methods, we can extend the model A2–A3 in the Appendix A, for both the BYM and MLE methods, to include other covariates directly, which may be required for some applications.

We also note that the methods have different settings and assumptions, which motivate our comparison. User-chosen settings are part of all cluster tests and different choices could lead to different results. All five methods have been proposed for local clusters. Under the null hypothesis, the number of COPD cases follows a Poisson distribution for the BYM and MLE methods, while the test statistic for the CSS and FSS methods has an asymptotically $\chi^2$distribution and the LISA method uses an empirical distribution. These features motivated us to consider these important methods and apply them to our COPD cases.

As limitations of this study, we assumed that our COPD cases are rare cases to be able to use the Poisson model in the BYM and MLE methods. Also, we used survey data (weighted to the population level) in our study. Strengths of the study include the evaluation of multiple cluster detection methods.

Overall, the potential clusters of COPD were located in the southern part of the province with the exception of region 24, which was identified by the BYM and MLE methods (cluster A). According to findings from Fransoo *et al.* [31], which are based on the Community Health Survey [27], binge drinking and smoking levels are higher than the Manitoba average in the south-central part of the province. As well, there are a higher percentage of people who consume low levels of fruits and vegetables in this region, although these differences are not statistically significant. Obesity levels are significantly higher than the Manitoba average in this region as well. In the south-eastern part of the province binge drinking, exposure to second hand smoke, and overweight and obesity levels are higher than the Manitoba average although the results are not statistically significant [31]. These are some possible health determinants that may explain the clusters of COPD in these regions. We found that the BYM and MLE methods are the best approaches in terms of identifying potential clusters and controlling for possible confounders (if any). These results may represent real increases in COPD or may be due to unmeasured covariates that need to be adjusted for in the model. Further investigation is needed to examine these findings, and also to explore the cause of these increases.

**Acknowledgments:** We would like to thank two referees for constructive comments and suggestions, which led to an improved version of the manuscript. This work was supported by grants from the Manitoba Health Research Council (MHRC) and the Natural Sciences and Engineering Research Council of Canada (NSERC) to M. Torabi. Disclaimer: The interpretations, conclusions and opinions expressed in this paper are those of the authors and do not necessarily reflect the position of Statistics Canada. This study is based in part on data provided by the Research Data Centre at the University of Manitoba. The interpretation and conclusions contained herein are those of the researchers and do not necessarily represent the views of Statistics Canada.

**Author Contributions:** The authors were both involved in the study concept and design, acquisition of data, interpretation of data, and the writing and critical revision of the manuscript for important intellectual content.

**Conflicts of Interest:** The authors declare no conflict of interest.

## Appendix A

*Circular Spatial Scan Statistic (CSS)*

The spatial scan statistic is widely used in the field of epidemiology for a variety of purposes [32]. A circular window $S$ is set on each region by the circular spatial scan statistic with the radius of the circle ranging from zero to a pre-specified maximum distance $d$ or a pre-determined maximum number of regions $J$ to be considered in the cluster. The window made up of the $(j-1)$-th nearest neighbours to region $i$ is denoted by $S_{i:j}(j = 1, \ldots, J)$. As well, $S_1 = \{S_{i:j}; i = 1, \ldots, m; j = 1, \ldots, J\}$ denotes the set of all windows to be scanned by the circular scan statistic. A likelihood ratio statistic based on the number of observed and expected cases inside and outside the circle is computed for each circle. Now, $L_0$ denotes the likelihood under the null hypothesis and $L_i(i = 1, \ldots, m)$ represents the likelihood under the alternative hypothesis. The null hypothesis states that there is no cluster in region $i$ and is tested against the alternative hypothesis that a cluster exists in region $i$ based on its $j$-th nearest neighbours. The likelihood ratio statistic is given by

$$\max_i \frac{L_i}{L_0} = \left(\frac{C_i}{E_i}\right)^{C_i} \left(\frac{N - C_i}{N - E_i}\right)^{N - C_i} I(C_i > E_i) \tag{A1}$$

where $C_i$ denotes the observed number of cases inside a circle and $E_i$ represents the expected number of cases inside a circle. Also, $(N - C_i)$ and $(N - E_i)$ denote the observed and expected number of cases outside the circle, respectively. The indicator function $I(C_i > E_i)$ is equal to 1 when $C_i > E_i$ and 0 otherwise. Circles with high likelihood ratios are identified as possible clusters [17].

Using SaTScan [33] or FleXScan [34] software, the CSS method can be applied. Generally, $J$ is chosen to include at most 50% of the population at risk, however, we used the FleXScan default which is a maximum spatial cluster size of $J = 15$. In order for the region to be part of the circle, the region centroid had to be included in the radius of the circle.

*Flexible Spatial Scan Statistic (FSS)*

The flexible spatial scan statistic works in the same manner as the circular spatial scan statistic but now the shape of the potential cluster is flexible while still being restricted to a small neighbourhood of each region. By connecting adjacent regions, the flexible scan statistic places an irregularly shaped window $S$ on each region. For any region $i$, the set of irregularly shaped windows of length $j$, containing $j$ connected regions including region $i$, can vary from 1 to the pre-determined maximum $J$, where $J$ is the maximum length of a cluster. Moreover, to prevent unlikely cluster shapes, the joined regions are restricted to the subsets of the set of regions $i$ and $(J - 1)$ -th nearest neighbours of region $i$. The set of all windows to be scanned by the flexible spatial scan statistic is then $S_2 = \{S_{i:j(k)}; i = 1, \ldots, m; j = 1, \ldots, J; k = 1, \ldots, k_{ij}\}$. The circular spatial scan statistic examines $J$ circles for any region $i$ and the flexible spatial scan statistic examines $J$ circles plus all the sets of connected regions whose centroids are found within the $J$-th largest concentric circle. Subsequently, the size of $S_2$ is much larger than $S_1$ which is at most $mJ$. The test statistic used in the FSS method under the Poisson assumption is based on the likelihood ratio test given in Equation (A1). Now, the circle defined in Equation (A1) refers to $S_2$ instead of $S_1$. Using the FSS method, circles with high likelihood ratio values are considered to be potential regions of disease clusters [18]. The FSS method can also be applied using the FleXScan software [34] with the FleXScan default which is $J = 15$.

*Bayesian Disease Mapping (BYM)*

Identifying clusters can also be done through a Bayesian framework using Markov chain Monte Carlo (MCMC) methods [14,15,35,36]. First used by Besag *et al.* [14], Bayesian disease mapping (BYM) is a modeling approach consisting of two parts. In the first part of the model, it is assumed that the cases follow a Poisson distribution with an area specific parameter $\theta_i E_i$ :

$$C_i \sim Poisson(\theta_i E_i) \tag{A2}$$

where the observed and expected number of cases in region $i$ are given by $C_i$ and $E_i$, respectively. The second part of the model is achieved through

$$log(\theta_i) = \mu + \eta_i \tag{A3}$$

where the relative risk ( $RR$ ) in region $i$ is given by $\theta_i$ , $\mu$ represents the overall mean ratio over the entire region and $\eta_i$ represents spatially correlated random effects. These spatial random effects are captured using the usual CAR model. A variety of CAR models can be used by obtaining a collection of mutually compatible conditional distributions $p(\eta_i|\eta_{-i})$, $i = 1,\ldots,m$, where $\eta_{-i} = \{\eta_j : j \neq i, \ j \in \delta_i\}$ and $\delta_i$ refers to a set of neighbours for the $i$-th region [14]. We use the following general model for spatial effects $\eta_i$ :

$$\eta = (\eta_1, \ldots, \eta_m)' \sim N(0, \Sigma_\eta)$$

$$\Sigma_\eta = \sigma_\eta^2 (I_m - \lambda_\eta D)^{-1} P$$

where $P$ is a $m \times m$ diagonal matrix with elements $P_{ii} = 1/e_i$ ; $D$ is a $m \times m$ matrix with elements $D_{ij} = (e_j/e_i)^{1/2}$ if region $i$ and $j$ are adjacent and $D_{ij} = 0$ otherwise; $e_i$ is the number of regions adjacent to region $i$ ; $\sigma_\eta^2$ is the spatial dispersion parameter; $\lambda_\eta$ measures the spatial autocorrelation, $\lambda_{min} \leq \lambda_\eta \leq \lambda_{max}$ , where $\lambda_{min}^{-1}$ and $\lambda_{max}^{-1}$ are the smallest and largest eigenvalues of $P^{-1/2}DP^{1/2}$ ; and $I_m$ is an identity matrix of dimension $m$ . We refer to [37] for details of this proper CAR model. Using vague prior distributions and within the Bayesian framework (MCMC), the parameters may be estimated to produce posterior distributions for the parameters in the model [14].

A cluster is specified as a region where the estimated relative risk (in terms of the lower credible set) is significantly larger than one [38]. To apply this method, WinBUGS software [37] is used to calculate the relative risk values.

*Frequentist Approach Using Maximum Likelihood Estimation (MLE)*

The data cloning (DC) approach is a computational algorithm to obtain the MLE for hierarchical models [23,24]. This approach is based on the Bayesian computational method and is used for frequentist purposes. This method involves independently repeating the observations $C = (C_1, \ldots, C_m)'$ for $L$ different individuals. Subsequently, these individuals all have the exact same set of observations C which are represented by $\mathbf{C}^{(L)} = (\mathbf{C}, \mathbf{C}, \ldots, \mathbf{C})$ . The posterior distribution of $\alpha = (\mu, \lambda_\eta, \sigma_\eta^2)'$ conditional on the data $\mathbf{C}^{(L)}$ is then given by

$$\pi_L\left(\mathbf{ff} | \mathbf{C}^{(L)}\right) = \frac{\{L(\alpha, \mathbf{C})\}^L \pi(\alpha)}{H(\mathbf{C}^{(L)})} \tag{A4}$$

where the prior distribution on the parameter space is $\pi(\alpha)$ and $H\left(\mathbf{C}^{(L)}\right) = \int \{L(\alpha, \mathbf{C})\}^L \pi(\alpha)d\alpha$ is the normalizing constant. Also, $\{L(\alpha, \mathbf{C})\}^L$ represents the likelihood for $L$ copies of the original data. As shown by Lele *et al.* [23,24], when $L$ is large enough, $\pi_L(\alpha|\mathbf{C}^{(L)})$ converges to a multivariate Normal distribution with the mean given by the MLE of the model parameters and variance-covariance matrix equal to $1/L$ times the inverse of the Fisher information matrix for the MLE. Hence, the sample mean vector of the generated random numbers from Equation (A4) acts as an estimate of the MLE and an estimate of the asymptotic variance-covariance matrix for the MLE $\hat{\alpha}$ is given by $L$ times the sample variance-covariance matrix of the generated random numbers from Equation (A4). Lele *et al.* [24] also provided various tests to determine the adequate number of clones $L$ .

*Prediction of Relative Risk:*

The prediction of relative risk (random effects) can be fairly problematic, especially in the frequentist framework. One approach to estimate **r** using the data is to use $\pi(R{=}r|C,\hat{\alpha})$ where $R = (RR_1,\ldots,RR_m)'$. However, the variability introduced by the model parameters estimate is not captured in this approach. In the literature [39], it has been suggested to use the following density in order to take into consideration the variation of the estimator,

$$\pi(\mathbf{r}|\mathbf{y}) = \frac{\int f(\mathbf{C}|\mathbf{r},\alpha_1)g(\mathbf{r}|\alpha_2)\phi(\alpha,\hat{\alpha},I^{-1}(\hat{\alpha}))d\alpha}{H(\mathbf{C})} \tag{A5}$$

where $\alpha_1 = \mu$, $\alpha_2 = (\lambda_\eta,\sigma_\eta^2)'$, $f(\cdot)$ and $g(\cdot)$ are Poisson and Normal distributions, respectively, and $\phi(\cdot, \xi, \Sigma)$ denotes a multivariate Normal density with mean $\xi$ and variance-covariance $\Sigma$. In this paper, the prediction of the **r** was found using Equation (A5) through MCMC sampling. A disease cluster is defined as a region where the estimated relative risk (in terms of the lower prediction interval) is significantly larger than one. The dclone package [40] is utilized within the R software [41] in order to calculate the relative risk values.

*Local Indicator of Spatial Association (LISA)*

Another method for identifying spatial clusters is a local indicator of spatial association (LISA) statistic [26]. In general, for an observation, $y_i$ in the $i^{th}$ region, the LISA statistic is given by

$$L_i = f\left(y_i, y_{J_i}\right)$$

where $f$ is a function and the values observed in the $J^{th}$ neighbourhood of region $i$ are given by $y_{J_i}$. In order to determine the statistical significance of the spatial association at region $i$, the following must be satisfied

$$\Pr(L_i > \delta_i) \le \alpha_i$$

where a critical value is given by $\delta_i$ and $\alpha_i$ is a given level of significance. Another condition of a LISA statistic is the total of all LISA statistics in a region must be proportional to a global indicator of spatial association. In other words,

$$\sum_i L_i = \gamma\Lambda$$

where $\Lambda$ is an indicator of the global indicator with a scale factor defined by $\gamma$. In order to test whether there is statistically significant spatial association over all the regions, the following statement must be true [26]

$$\Pr(\Lambda > \delta) \le \alpha.$$

A general LISA statistic may be used to test the null hypothesis of no spatial association against the alternative hypothesis that spatial clustering exists across a region. However, the distribution of the general LISA may be hard to find. For this reason, conditional randomization or a permutation approach is used to find an empirical distribution. The randomization is done by holding the observed value ($y_i$) in region $i$ constant and the remaining observed values across the entire study region are randomly permuted and the value of $L_i$ is computed. This is done for each region in the study area. The result is an empirical distribution function, which expresses the extent to which each observation is considered to be extreme in comparison with the other observed values [26].

The LISA method is usually a simple method to apply, however, it is complicated by the fact that the LISA statistics for individual regions may be correlated. For example, when regions $i$ and $k$ are neighbours or have common elements in their neighbourhood sets, the corresponding LISA statistics, $L_i$ and $L_k$ will be correlated. Typically, it is extremely hard to derive the marginal distributions of each statistic and therefore, the significance levels must be approximated by Bonferroni inequalities or the method outlined by Sidák [42]. Using Bonferroni inequalities, the individual significance levels ($\alpha_i$) are set to $\alpha/m$ and using Sidák's method, they are equal to $1 - (1 - \alpha)^{1/m}$ , where the overall significance level is set to $\alpha$ and there are $m$ comparisons. It has been suggested that $m$ is taken to be the number of observations $n$ . However, this may result in bounds that are too conservative and in fact very few observations may be deemed to be significant clusters [26]. Further investigation is being conducted to determine the best value for $m$ . In our study this method is implemented in R [41] using the ncf package [43].

## References

1.  Global Initiative for Chronic Obstructive Lung Disease (GOLD). Global Strategy for the Diagnosis, Management and Prevention of COPD. (updated 2013). Available online: http://www.goldcopd.org (accessed on 3 July 2013).
2.  Eisner, M.D.; Anthonisen, N.; Coultas, D.; Kuenzli, N.; Perez-Padilla, R.; Postma, D.; Romieu, I.; Silverman, E.K.; Balmes, J.R. An official American Thoracic Society public policy statement: Novel risk factors and the global burden on chronic obstructive pulmonary disease. *Amer. J. Respir. Crit. Care Med.* **2010**, *182*, 693–718. [CrossRef]
3.  Sezer, H.; Akkurt, I.; Guler, N.; Marakoğlu, K.; Berk, S. A case-control study on the effect of exposure to different substances on the development of COPD. *Ann. Epidemiol.* **2006**, *16*, 59–62. [CrossRef] [PubMed]
4.  Burt, L.; Corbridge, S. COPD exacerbations. *Amer. J. Nurs.* **2013**, *113*, 34–43. [CrossRef]
5.  Lamprecht, B.; McBurnie, M.A.; Vollmer, W.M.; Gudmundsson, G.; Welte, T.; Nizankowska-Mogilnicka, E.; Studnicka, M.; Bateman, E.; Anto, J.M.; Burney, P.; *et al.* COPD in never smokers: Results from the population-based burden of obstructive lung disease study. *Chest* **2011**, *139*, 752–763. [CrossRef] [PubMed]
6.  Canadian Thoracic Society. *The Human and Economic Burden of COPD: A Leading Cause of Hospital Admission in Canada*; Canadian Thoracic Society: Ottawa, ON, Canada, 2010.
7.  Canadian Institute of Health Information. Health Indicators 2008. Available online: https://secure.cihi.ca/ free_products/HealthIndicators2008_ENGweb.pdf (accessed on 3 July 2013).
8.  Mittmann, N.; Kuramoto, L.; Seung, S.J.; Haddon, J.M.; Bradley-Kennedy, C.; FitzGerald, J.M. The cost of moderate and severe COPD exacerbations to the Canadian healthcare system. *Respir. Med.* **2008**, *102*, 413–421. [CrossRef] [PubMed]
9.  Scheinfeld, M.H.; Maniatis, T.; Gurell, D. COPD? *Amer. J. Med.* **2006**, *119*, 839–842. [CrossRef]
10. Lawson, A.B. *Statistical Methods in Spatial Epidemiology*, 2nd ed.; John Wiley & Sons, Ltd.: London, UK, 2006.
11. Jennings, J.M.; Curriero, F.C.; Celentano, D.; Ellen, J.M. Geographic identification of high gonorrhea transmission areas in Baltimore, Maryland. *Amer. J. Epidemiol.* **2005**, *161*, 73–80. [CrossRef]
12. Elliott, P.; Briggs, D.; Morris, S.; de Hoogh, C.; Hurt, C.; Jensen, T.K.; Maitland, I.; Richardson, S.; Wakefield, J.; Jarup, L. Risk of adverse birth outcomes in populations living near landfill sites. *Brit. Med. J.* **2001**, *323*, 363–368. [CrossRef] [PubMed]
13. Lawson, A.B.; Biggeri, A.; Williams, F.L.R. A review of modeling approaches in health risk assessment around putative sources. In *Disease Mapping and Risk Assessment for Public Health*; Lawson, A.B., Biggeri, A., Böhning, D., Lesaffre, E., Viel, J., Bertollini, R., Eds.; Wiley: New York, NY, USA, 1999; pp. 231–245.
14. Besag, J.E.; York, J.C.; Mollìe, A. Bayesian image restoration with two applications in spatial statistics (with discussion). *Ann. Inst. Stat. Math.* **1991**, *43*, 1–59. [CrossRef]
15. Clayton, D.; Bernardinelli, L. Bayesian methods for mapping disease risk. In *Geographical and Environmental Epidemiology: Methods for Small-Area Studies*; Elliott, P., Cuzick, J., English, D., Stern, R., Eds.; Oxford University Press: Oxford, UK, 1996; pp. 205–220.
16. Clayton, D.; Kaldor, J. Empirical Bayes estimates of age-standardized relative risks for use in disease mapping. *Biometrics* **1987**, *43*, 671–681. [CrossRef] [PubMed]
17. Kulldorff, M. A spatial scan statistics. *Commun. Statist. A—Theor. Method.* **1997**, *26*, 1481–1496.
18. Tango, T.; Takahashi, K. A flexibly shaped spatial scan statistic for detecting clusters. *Int. J. Health Geogr.* **2005**, *4*, 1–15. [CrossRef] [PubMed]
19. Besag, J.E.; Newell, J. The detection of clusters in rare diseases. *J. R. Stati. Soc. Ser. A* **1991**, *154*, 143–155. [CrossRef]

*ISPRS Int. J. Geo-Inf.* **2014**, *3*, 1039–1057

20. Torabi, M.; Rosychuk, R.J. Spatial event cluster detection using an approximate normal distribution. *Int. J. Health Geogr.* **2008**, *7*, 1–22. [CrossRef] [PubMed]

21. Tango, T. A test for spatial disease clustering adjusted for multiple testing. *Stat. Med.* **2000**, *19*, 191–204. [CrossRef] [PubMed]

22. Torabi, M.; Rosychuk, R.J. An examination of five spatial disease clustering methodologies for the identification of childhood cancer clusters in Alberta, Canada. *Spat. Spatiotemporal Epidemiol.* **2011**, *2*, 321–330. [CrossRef] [PubMed]

23. Lele, S.R.; Dennis, B.; Lutscher, F. Data cloning: Easy maximum likelihood estimation for complex ecological models using Bayesian Markov chain Monte Carlo methods. *Ecol. Lett.* **2007**, *10*, 551–563. [CrossRef] [PubMed]

24. Lele, S.R.; Nadeem, K.; Schmuland, B. Estimability and likelihood inference for generalized linear mixed models using data cloning. *J. Am. Stat. Assoc.* **2010**, *105*, 1617–1625. [CrossRef]

25. Torabi, M. Spatial disease cluster detection: An application to childhood asthma in Manitoba, Canada. *J. Biom. Biostat.* **2012**. [CrossRef]

26. Anselin, L. Local indicators of spatial association—LISA. *Geogr. Anal.* **1995**, *2*, 93–115.

27. Statistics Canada. *Canadian Community Health Survey User Guide (2001–2010)*; Statistics Canada: Ottawa, ON, Canada, 2010.

28. McCullagh, P.; Nelder, J.A. *Generalized Linear Models*, 2nd ed.; Chapman and Hall: London, UK, 1989.

29. Richardson, S.; Thomson, A.; Best, N.; Elliott, P. Interpreting posterior risk estimates in disease-mapping studies. *Environ. Health Perspect.* **2004**, *112*, 1016–1025. [CrossRef] [PubMed]

30. Banerjee, S.; Gelfand, A.E.; Carlin, B.P. *Hierarchical Modeling and Analysis for Spatial Data*; Chapman and Hall: London, UK, 2004.

31. Fransoo, R.; Martens, P.; Burland, E. *The Need to Know Team*; Prior, H., Burchill, C., Eds.; Manitoba Centre for Health Policy, Manitoba RHA Indicators Atlas: Winnipeg, MB, Canada, 2009.

32. Fukuda, Y.; Umezaki, M.; Nakamura, K.; Takano, T. Variations in social characteristics of spatial disease clusters: Examples of colon, lung and breast cancer in Japan. *Int. J. Health Geogr.* **2005**, *4*, 1–13. [CrossRef] [PubMed]

33. Kulldorff, M.; Rand, K.; Gherman, G.; Williams, G.; DeFrancesco, D. *SaTScan V2.1: Software for the Spatial and Space-Time Scan Statistics*; National Centre Institute: Bethesda, MD, USA, 1998.

34. Takahashi, K.; Yokoyama, T.; Tango, T. *FleXScan: Software for the Flexible Scan Statistic*; National Institute of Public Health: Nagoya, Japan, 2006.

35. Bernardinelli, L.; Montomoli, C. Empirical Bayes *versus* fully Bayesian analysis of geographical variation in disease risk. *Stat. Med.* **1992**, *11*, 983–1007. [CrossRef] [PubMed]

36. Gilks, W.R.; Richardson, S.; Spielhalter, D.J. *Markov Chain Monte Carlo in Practice*; Chapman and Hall/CRC: London, UK, 1995.

37. Spiegelhalter, D.; Thomas, A.; Best, N.; Lunn, D. *WinBUGS Version 1.4 User Manual*; MRC Biostatistics Unit, Institute of Public Health: London, UK, 2004.

38. Aamodt, G.; Samuelsen, S.O.; Skrondal, A. A simulated study of three methods for detecting disease clusters. *Int. J. Health Geogr.* **2006**, *5*, 1–11. [CrossRef] [PubMed]

39. Hamilton, J.D. A standard error for the estimated state vector of a state-space model. *J. Econometrics* **1986**, *33*, 387–397. [CrossRef]

40. Sólymos, P. Dclone: Data cloning in R. *R J.* **2010**, *2*, 29–37.

41. R Core Team. *R: A Language and Environment for Statistical Computing*; R Foundation for Statistical Computing: Vienna, Austria, 2013.

42. Sidák, Z. Rectangular confidence regions for the means of multivariate normal distributions. *J. Am. Stat. Assoc.* **1967**, *62*, 626–633.

43. Bjornstad, O.N. ncf: Spatial Nonparametric Covariance Functions. R Package Version 1.1-5. 21 November 2013. Available online: http://cran.r-project.org/web/packages/ncf/index.html (accessed on 24 July 2014).

isprs International Journal of
*Geo-Information*

MDPI

Communication

# Holistics 3.0 for Health

**David John Lary [1,\*], Steven Woolf [2], Fazlay Faruque [3] and James P. LePage [4]**

[1]  Hanson Center for Space Science, University of Texas at Dallas, Richardson, TX 75080, USA

[2]  Center on Society and Health, Virginia Commonwealth University, Richmond, VA 23298, USA;
    swoolf@vcu.edu

[3]  GIS & Remote Sensing Program, University of Mississippi Medical Center, MS 39216, USA;
    ffaruque@umc.edu

[4]  ACOS Research and Development, VA North Texas Health Care System, Dallas, TX 75216, USA;
    james.lepage@va.gov

\*  Author to whom correspondence should be addressed; David.Lary@utdallas.edu; Tel.: +1-972-489-2059.

Received: 21 February 2014; in revised form: 1 July 2014; Accepted: 10 July 2014; Published: 24 July 2014

**Abstract:** Human health is part of an interdependent multifaceted system. More than ever, we have increasingly large amounts of data on the body, both spatial and non-spatial, its systems, disease and our social and physical environment. These data have a geospatial component. An exciting new era is dawning where we are simultaneously collecting multiple datasets to describe many aspects of health, wellness, human activity, environment and disease. Valuable insights from these datasets can be extracted using massively multivariate computational techniques, such as machine learning, coupled with geospatial techniques. These computational tools help us to understand the topology of the data and provide insights for scientific discovery, decision support and policy formulation. This paper outlines a holistic paradigm called Holistics 3.0 for analyzing health data with a set of examples. Holistics 3.0 combines multiple big datasets set in their geospatial context describing as many areas of a problem as possible with machine learning and causality, to both learn from the data and to construct tools for data-driven decisions.

**Keywords:** geospatial; machine learning; Big Data; health; remote sensing; Holistics 3.0; data-driven decisions

## 1. Introduction

For decades, experts in public health and the social sciences have recognized the geospatial variation of populations; health is shaped by multiple factors, including healthcare, public health systems, individual behaviors (e.g., smoking) and risk factors (e.g., obesity), socioeconomic factors (e.g., income, education), the physical environment (e.g., air pollution), the social environment (e.g., social support), public policies and the "macro-structural" elements of society that shape this entire list. Linking these data to resolve public health concerns requires accounting for non-linear multi-variate relationships, as well as the spatial dimension of the data.

Decades of research have also attempted to isolate the specific factors that matter the most, not only as an academic exercise, but to help policy makers set priorities in a decision-making environment of limited resources. For example, in a specific community beset with high rates of chronic diseases, should the city council or board of supervisors give priority to hospital budgets, expanding primary care, passing laws to ban smoking indoors, addressing unemployment, strengthening schools, and so on? All are clearly important, but which one or which combination matters the most and which will give the best return on investment?

Posed with such questions, scientists have typically resorted to traditional statistical techniques, such as regression equations, to try to quantify or model the relative importance of different factors. This approach has shed insight onto some of the key factors, but also carries limitations, two of which bear mentioning here. First, these calculations often examine associations rather than causality: for example, the fact that people who have not graduated from high school have worse health does not mean that handing out diplomas will fully erase the disparity; rather, the educational level proves to be a useful proxy. Second, the variables that researchers insert into their formulas are chosen selectively based on the variables for which data are available and those that the researchers think are most important to consider. For example, a researcher forced to choose whether to adjust for poverty, voter registration or social trust will invariably choose poverty, because there is more evidence available linking poverty to adverse health outcomes.

The *a priori* selectivity in choosing the variables to consider is partly a legacy of the age-old scientific method (pose a hypothesis first and then collect data to support or refute it), but partly a practical necessity, because examining all of the data has previously often been an untenable option, especially as the volume of available data has expanded. The advent of machine learning is removing the second barrier at a time when the availability of "Big Data" is ascendant in all fields [1–6].

## 1.1. Big Data

Different people use the term Big Data in slightly different ways. However, the common idea is large datasets in terms of the volume of data (e.g., because of temporal or spatial resolution and/or coverage) and/or large in terms of the number of variables included. One of the prime differences in the use of Big Data typically relates to: exactly how big is big?

For the specific examples used in this paper, these datasets also describe geospatial variations; the number of variables ranges from tens to thousands of variables, and the number of records for each variable ranges from thousands to many millions, covering both a snapshot in time as well as the daily variation tracked for nearly two decades. This quantity of data would probably be classified as Big Data by most investigators.

## 1.2. Machine Learning

Machine learning is a valuable set of tools for empirically estimating and classifying variables of interest when we do not have a complete theoretical description of a process, but we do have useful data. Further, we often would like to use these data to provide insights and/or help make decisions.

Machine learning encompasses a very broad range of algorithms (for example, neural networks, support vector machines, Gaussian processes, decision trees, random forests, *etc.*) that can provide multi-variate, non-linear, non-parametric regression or classification based on a training dataset (*i.e.*, a set of examples to learn from) and give insight into the underlying topology of the data. This approach allows the data to speak for themselves.

Machine learning has widespread and growing applications. A few examples of its daily use include credit checking, use by Amazon and other online stores to suggest other products of potential interest to consumers, Netflix movie suggestions, remote sensing applications, various Google tools and inventory decisions made by large retailers, such as Walmart [7–9].

Machine learning has not yet made its large-scale entry into public health, but Big Data exist widely in population health, a sea of data that can be mined in healthcare (e.g., electronic medical records), public health statistics, census data on population living conditions, environmental hazards and public programs. Within the past year, articles and entire theme issues of major medical journals have called attention to the exciting opportunities that exist in applying machine learning to these data sets. Organizations have launched prizes to encourage innovations in this area [10]. The federal government has hosted four annual Health Datapalooza conferences born from efforts by the Obama administration to "liberate" health data (http://healthdatapalooza.org/about/). Many of these initiatives focus on creating decision support tools to track diseases, such as the early identification of infectious disease outbreaks or dashboards for tracking health conditions and costs, but the application of machine learning to these datasets opens a much larger horizon.

*1.3. Holistics 3.0*

We define **Holistics 3.0** as (1) bringing together multiple datasets describing as many aspects of a problem as possible, *i.e.*, holistically describing the problem with data; (2) coupling this holistic description of the problem with machine learning to build empirical decision support tools for data-driven decisions; and (3) where relevant, augment the correlations and associations exploited by machine learning to address the further issue of causality. Taken together this paradigm is called Holistics 3.0 and is illustrated schematically in Figure 1.

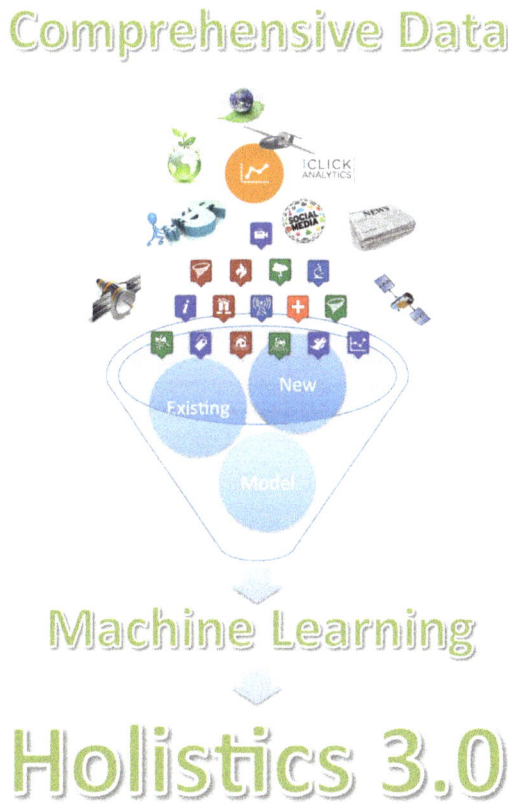

**Figure 1.** A schematic illustrating the key components of Holistics 3.0: (1) multiple geospatial datasets describing many aspects of a problem holistically; (2) use of machine learning to build empirical decision support tools; and (3) augmentation by inferences about causality. Taken together, this paradigm is called Holistics 3.0.

## 2. Mining Meaning from Data

Is unleashing a powerful computer to scan for associations across a wide net of countless variables useful? Data-driven decisions can help policymakers, from clinicians to elected officials, to identify the factors that are most likely to improve health. Variables that have not been previously considered, or even noticed, may hold the key to understanding important drivers of health outcomes that offer new solutions and new understandings of how disease complications arise in the first place. In many of our previous machine learning studies, we find that in many cases, an accurate description of the problem requires us to simultaneously describe multiple aspects of the problem. Sometimes, this encompasses just five or six variables, but sometimes forty or more. There is usually not a single "magic" variable; rather, we are typically faced with truly multi-variate problems, where many factors must simultaneously be considered, which are often also non-linear. In many cases, we do not know the functional form of the relationship (*i.e.*, they are also non-parametric), and many times, the variables may have non-Gaussian distributions.

Machine learning can help tremendously with just these types of problems, not only for identifying key variables, but also in providing empirical tools, often of remarkable fidelity. If we then add the concept of causality as described by Judea Pearl [11], we have a very powerful paradigm for data-driven decisions. Current understandings of the causal pathway for diseases have been built on the scientific method, which has a strong legacy, but also carries the limitation of relying on human minds to choose which variables (and hypotheses) are worthy of further study.

Machine learning offers an unprecedented opportunity to let the data speak for themselves, without human presupposition, and to draw attention to previously unrecognized variables that seem to have important associations with health outcomes and warrant further study under the traditional scientific method to confirm/refute the association. Machine learning thereby can augment the scientific method by offering a more "open-minded" approach to hypothesis testing and scientific lead generation that relies on the data to draw attention to intriguing questions. In this way, using machine learning is more evidence-based than relying on a few individuals to selectively pick and choose variables they "think" are important.

Apart from discovering new levers for improving health, machine learning also has the potential to quantify the relative importance of levers already known to be important. For example, claims that healthcare accounts for 10% of health outcomes are based largely on linear regression equations. Machine learning can improve on this approach by not only providing a multi-variate, non-linear, non-parametric regression, which requires no prior knowledge of functional form, but also providing an objectively ranked list of the relative importance of the variables used in the regression.

Moreover, access to multiple big datasets required to perform machine learning also provides a ready resource for policymakers who need quick access to descriptive statistics or trends for their communities or special populations. For example, the Affordable Care Act (ACA) requires hospitals to prepare community health needs assessments to maintain their nonprofit status with the Internal Revenue Service. Hospitals across the country that have little experience in studying population statistics are scrambling to find colleagues in public health or community organizations that can be contracted to produce these reports. The enormous insights available through machine learning would enable health systems to profile their communities at a level of detail not currently imaginable.

Large health systems are also forming accountable care organizations (ACOs) that, under the ACA, are required to assume responsibility for the health of the population they serve. The impetus behind ACOs is to encourage healthcare systems to identify new models of intervention, including those that involve determinants of health outside the clinic, to prevent disease, reduce complications and control costs. Most ACO leaders are currently relying on educated guesses to decide how best to invest their dollars to improve population health [12]. Decision support tools based on machine learning have the potential to arm decision-makers and policy-makers with more granular information about the health of the population, the prevalence and geography of local factors that are shaping community health and where the greatest potential return on investment might lie if confirmatory research supports a causal link.

## 3. Some Examples

Let us now examine two very different examples relevant to geospatial health that illustrate this approach.

**Table 1.** Health outcomes associated with particulate matter (PM) and ultra-fine particles (UFP) (modified from [13]).

| Health Outcomes | Short-Term Studies | | | Long-Term Studies | | |
|---|---|---|---|---|---|---|
| | PM10 | PM2.5 | UFP | PM10 | PM2.5 | UFP |
| **Mortality** | | | | | | |
| All causes | xxx | xxx | x | xx | xx | x |
| Cardiovascular | xxx | xxx | x | xx | xx | x |
| Pulmonary | xxx | xxx | x | xx | xx | x |
| **Pulmonary effects** | | | | | | |
| Lung function, e.g., PEF | xxx | xxx | xx | xxx | xxx | |
| Lung function growth | | | | xxx | xxx | |
| **Asthma and COPD exacerbation** | | | | | | |
| Acute respiratory symptoms | | xx | x | xxx | xxx | |
| Medication use | | | x | | | |
| Hospital admission | xx | xxx | x | | | |
| **Lung cancer** | | | | | | |
| Cohort | | | | xx | xx | x |
| Hospital admission | | | | xx | xx | x |
| **Cardiovascular effects** | | | | | | |
| Hospital admission | xxx | xxx | | x | x | |
| **ECG-related endpoints** | | | | | | |
| Autonomic nervous system | xxx | xxx | xx | | | |
| Myocardial substrate and vulnerability | | xx | x | | | |
| **Vascular function** | | | | | | |
| Blood pressure | xx | xxx | x | | | |
| Endothelial function | x | xx | x | | | |
| **Blood markers** | | | | | | |
| Pro inflammatory mediators | xx | xx | xx | | | |
| Coagulation blood markers | xx | xx | xx | | | |
| Diabetes | x | xx | x | | | |
| Endothelial function | x | x | xx | | | |
| **Reproduction** | | | | | | |
| Premature birth | x | x | | | | |
| Birth weight | xx | x | | | | |
| IUR/SGA | x | x | | | | |
| **Fetal growth** | | | | | | |
| Birth defects | x | | | | | |
| Infant mortality | xx | x | | | | |
| Sperm quality | x | x | | | | |
| **Neurotoxic effects** | | | | | | |
| Central nervous system | | x | xx | | | |

Legend: x, few studies ($\leq 6$); xx, many studies (7–10); xxx, large number of studies (>10).

## 3.1. Airborne Particulate Matter

With the increasing awareness of the many health impacts (Table 1) of particulate matter, ranging from general mortality to specific pulmonary, respiratory, cardiovascular, cancer and reproductive conditions, to name but a few, there is a growing and pressing need to have global daily estimates of the concentration of ground-level airborne particulate matter with a diameter of 2.5 microns or less ($PM_{2.5}$). The Holistics 3.0 paradigm can be applied to existing NASA remote sensing datasets coupled with meteorological analyses, demographic data and *in situ* observations to effectively meet this need. We have already successfully employed machine learning to estimate the daily global $PM_{2.5}$

concentration on a routine basis. The approach uses a suite of remote sensing and meteorological data products and ground-based observations of particulate matter at 8329 measurement sites in 55 countries (Figure 2) made between 1997 to the present to estimate the daily distributions of $PM_{2.5}$.

The many health impacts of $PM_{2.5}$ depend on the airborne concentration at ground level, where $PM_{2.5}$ can be inhaled (Table 1). However, as can be seen in Figure 2, the spatial coverage has many gaps, and in some countries, there are no $PM_{2.5}$ observations altogether. This is largely due to the costs involved in operating such a sensor network. Several studies have sought to overcome the lack of direct $PM_{2.5}$ observations by using remote sensing and satellite-derived aerosol optical depth (AOD) coupled with regression and/or numerical models to estimate the ground-level concentration of $PM_{2.5}$ [14–32].

Many studies have shown that the relationship between $PM_{2.5}$ and AOD is a multi-variate function of a large number of parameters, including: humidity, temperature, boundary layer height, surface pressure, population density, topography, wind speed, surface type, surface reflectivity, season, land use, normalized variance of rainfall events, size spectrum and phase of cloud particles, cloud cover, cloud optical depth, cloud top pressure and the proximity to particulate sources [17,18,20,27,29,33–51]. In some cases, such as for wind speed, the relationship is highly non-linear, and in many cases not well characterized.

The picture is further complicated by the biases present in the satellite AOD products [52–56], the difference in spatial scales of the *in situ* point $PM_{2.5}$ observations and the remote sensing data (several kilometers per pixel) and, finally, the sharp $PM_{2.5}$ gradients that can exist in and around cities, particularly in Asia.

Taken together, all of these factors naturally suggest that any successful regression must be multivariate, non-linear and non-parametric. The natural choice is therefore machine learning, an approach which excels in describing multivariate, non-linear, non-parametric problems. Machine learning, does an outstanding job of providing a new $PM_{2.5}$ product. Figure 3 shows the monthly average of our machine learning $PM_{2.5}$ product ($\mu g/m^3$) for August, 2001. The average of the observations at a given site are overlaid as color-filled circles when observations were available for at least a third of the days. Notice the good agreement between the $PM_{2.5}$ product and the observations (*i.e.*, the color fill of the circles depicting the observations is in good agreement with the background color depicting the new machine learning $PM_{2.5}$ product).

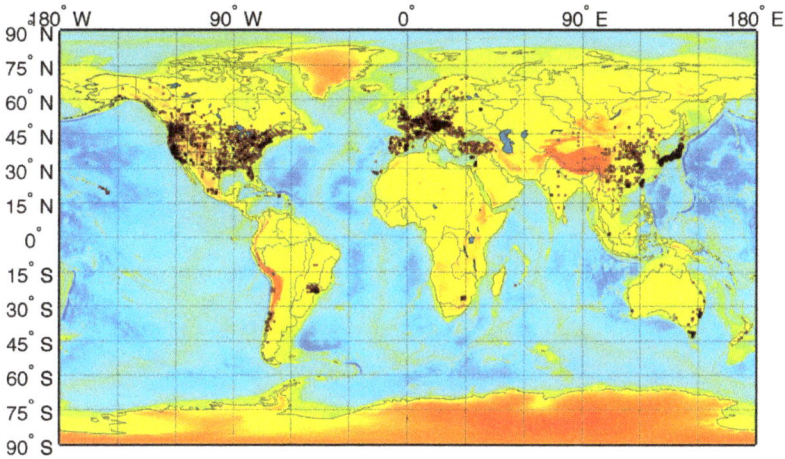

**Figure 2.** A map showing the 8329 $PM_{2.5}$ measurement site locations from 55 countries (red squares) that were used over the period 1997–present. The greatest density of sites is in North America, Europe and Asia. However, there are also southern hemisphere sites in South America, South Africa, Australia and New Zealand. The background color scale shows the global topography and bathymetry.

As would be expected in late summer, the eastern U.S. has much higher $PM_{2.5}$ concentration than the western U.S. Figure 3a is of Alaska and highlights common fire zones associated with elevated $PM_{2.5}$. Figure 3b,c show the good agreement between our product and observations. Figure 3d shows the elevated $PM_{2.5}$ with the heavily agricultural Central Valley in California, the highly populated Los Angeles Metropolitan Area, the Sonoran Desert, one of the most active dust source regions in the U.S., the Four Corners Power Plants, some of the largest coal-fired generating stations in the U.S., and the Great Salt Lake Desert.

We are in the process of combining the daily particulate data product produced using machine learning (an example is shown in Figure 3) and other environmental data products with the VA's Electronic Health Record (EHR) System to facilitate data-driven insights and decisions.

### 3.2. Life Expectancy and Socioeconomic Data from the U.S. Census

In 2012, the Center on Society and Health at Virginia Commonwealth University launched a two-year study of factors that affect life expectancy in California census tracts with high poverty rates. They calculated the life expectancy of thousands of census tracts in California, working with vital statistics supplied by the California Department of Health. The researchers were using traditional statistical techniques, such as regression equations, to study factors in these census tracts that might affect life expectancy, such as healthcare and public health, the physical and social environment, socioeconomic conditions of individuals and households and macrostructural resources. Their goal was to see if these factors seem to differ in outlier census tracts with unexpectedly high or low life expectancy given their poverty rate. The goal is to help elected officials and other policymakers identify assets that act positively to help poor communities buffer the adverse health effects of poverty. The parsimony required for traditional methods, as discussed above, required the researchers to handpicked a limited number of variables in each domain to enter into their regression equation.

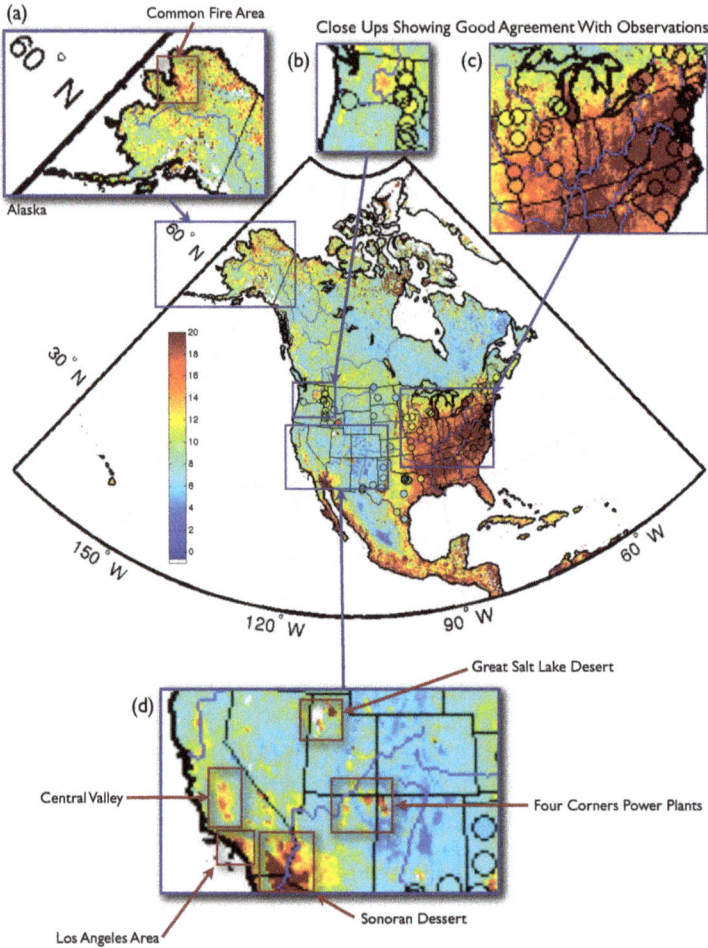

**Figure 3.** The monthly average of our prototype machine learning PM$_{2.5}$ product ($\mu$g/m$^3$) for August 2001. The average of the observations at a given site is overlaid as color-filled circles when observations were available for at least a third of the days. Notice the good agreement between the PM$_{2.5}$ product and the observations. Furthermore, as would be expected, in summer, the eastern U.S. has much higher PM$_{2.5}$ concentration than the western U.S. (**a**) Alaska highlighting common fire areas associated with elevated PM$_{2.5}$; (**b,c**) the good agreement between our product and the observations; (**d**) the elevated PM$_{2.5}$ with the heavily agricultural Central Valley in California, the highly populated Los Angeles Metro Area, the Sonoran Desert, one of the most active dust source regions in the U.S., the Four Corners Power Plants, some of the largest coal-fired generating stations in the U.S., and the Great Salt Lake Desert.

Building on this study, we conducted a parallel project to serve as a demonstration of the power of machine learning. As a proof of concept, it was intentionally limited to one geographic area (the State of California) and one domain in the researchers' model: socioeconomic conditions in individuals and households.

The 2007–2011 U.S. Census Bureau's American Community Survey provides a rich data set on socioeconomic conditions for individuals and households that are too extensive for traditional researchers to examine in total, but that are arrayed at the census tract level in a downloadable form that supercomputers can easily analyze. The total number of variables available at the census tract level is extensive: a total of 21,038 variables spread across 113 data files. Some of the variables are duplicated in more than one file, and when these duplicates are removed, 18,528 unique variables remain. Some of these variables have missing values for certain census tracts; if variables with missing values are removed, then 13,065 variables remain. Although this is a staggering number of variables that most conventional investigators would not even attempt to analyze, we easily used machine learning to design a fully non-linear, non-parametric, multi-variate fit of all 13,065 variables. We calculated the bivariate statistical significance of the individual correlation of these variables with life expectancy and found that a staggering 10,339 variables had a p-value less than 0.05. The variables in the census are not orthogonal, and aspects of the same information content are duplicated in many census variables. If the p-value threshold is progressively decreased, we find that, remarkably, seven variables have a p-value of less than $10^{-240}$.

Whether we use all 13,065 variables or the 10,339 with a *p*-value of less than 0.05, or just the seven variables with a *p*-value of less than $10^{-240}$, machine learning is able to do a good job of estimating life expectancy. Two examples of the fully non-linear, non-parametric, multi-variate estimates of life expectancy using Random Forests [57,58] are shown in Figure 4.

A few obvious take away messages were found.

First, the variables that machine learning highlighted as most important replicated classic epidemiological studies in highlighting the importance of factors with known associations with life expectancy—age, sex, race-ethnicity, income/poverty, and education—however, in addition, the machine learning approach identified some additional key factors. Among the top 50, it identified other factors that ranked very highly in importance, including: relocation (first); employment, especially in certain industries (third, seventh, eighth, 13th, 15th, 19th, 20th, 36th, 42th, 49th); occupied housing (sixth, 11th, 16th, 18th, 21st, 31st, 50th); single-parent households (sixth, 12th); language (14th); grandparents living with grandchildren (26th, 48th); and SNAP eligibility (38th). Whether these variables are proxies for socioeconomic status or reflect unique influences on life expectancy independent of classic social determinants of health requires further analysis beyond the scope of this project.

Second, perhaps because key information in the census is duplicated many times, the fit using just seven variables was almost as good as that using all 13,065 variables, or the 10,339 with a *p*-value of less than 0.05.

Third, as good as the machine learning fit was, the slope of the scatter diagram is not exactly one. This data signature indicates that additional key factors have not been considered. This is not unexpected; this study focused only on the socioeconomic dimension and not other determinants of health, such as environmental factors, like air quality, or dietary habits and access to health care. It is clear that important factors related to life expectancy that need to be simultaneously accounted for are healthcare, health behaviors, the social environment, the physical environment (including pollution) and public spending on social and health services.

## 10,339 variables with p<0.05

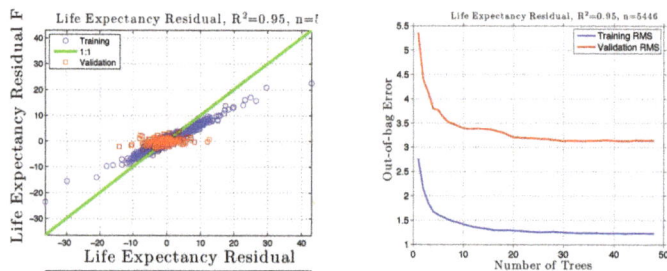

**(a)**

## 7 variables with p<10⁻²⁴⁰

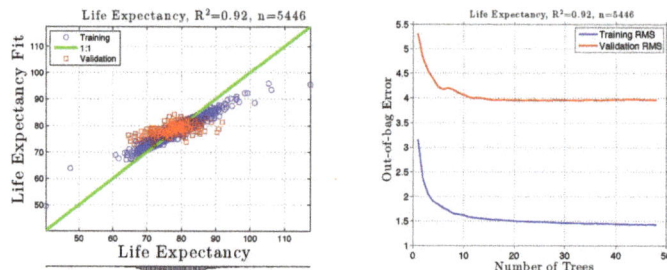

**(b)**

**Figure 4.** Two examples of the scatter diagrams for fully non-linear, non-parametric, multi-variate estimates of life expectancy: **(a)** 10,339 variables in the American Community Survey (U.S. Census Bureau) with a bivariate p-value for life expectancy of less than 0.05; **(b)** seven variables in the American Community Survey (U.S. Census Bureau) with a bivariate p-value for life expectancy of less than $10^{-240}$. Blue circles depict the training data. Red squares depict randomly selected, totally independent validation data not used in the training. The green line is the ideal 1:1 line for a perfect fit.

Therefore, we have seen that machine learning is a powerful tool for dealing with massively multi-variate systems, for letting the data speak, for highlighting key drivers and for providing objective tools that tell us when additional factors need to be considered.

### 3.3. False Positives

It is critical to recognize that correlation is not causation; further, when using machine learning with many variables, there is always the possibility of false associations. The full analysis of such possibilities are beyond the scope of this paper. However, one way to address these questions is some kind of experimental control in the data mining exercise, but a variety of other methods have been proposed.

## 4. Summary

Human health is part of an interdependent, multifaceted system, with many aspects varying geospatialy. The Holistics 3.0 paradigm that brings together data on as many aspects of a problem as possible and combines it with machine learning (and where necessary, causality) is a powerful tool for informing data-driven decisions that can incorporate and account for geospatial variations. Key in this is allowing the data to "speak for itself" and the ability to process thousands of variables simultaneously in a fully multi-variate, non-linear, non-parametric, non-Gaussian framework.

**Acknowledgments:** It is a pleasure to acknowledge The Institute for Integrative Health, the University of Texas at Dallas, DoD Telemedicine & Advanced Technology Research Center (TATRC) for Award W81XWH-11-2-0165, NASA for research funding through the award, NNX11AL18G, and Grant 20111592 from The California Endowment. The project described was also partially supported by Grant Number R21ES019713 from the National Institute of Environmental Health Sciences. The content is solely the responsibility of the authors and does not necessarily represent the official views of the National Institute of Environmental Health Sciences or the National Institutes of Health.

**Author Contributions:** David Lary contributed most of the numerical analyses and substantial portions of the text. Steven Woolf's team provided the life expectancy data, and contributed significantly to the introductory text and §3.2. Fazlay Faruque contributed Table 1 and collaborated on the $PM_{2.5}$ study. James LaPage collaborated on linking the environmental data to health outcomes at the VA.

**Conflicts of Interest:** The authors declare no conflicts of interest.

## References

1. Jacobs, A. The pathologies of big data. *Commun. ACM* **2009**, *52*, 36–44. [CrossRef]
2. Guhaniyogi, R.; Finley, A.O.; Banerjee, S.; Gelfand, A.E. Adaptive Gaussian predictive process models for large spatial datasets. *Environmetrics* **2011**, *22*, 997–1007. [CrossRef] [PubMed]
3. Finley, A.O.; Banerjee, S.; Gelfand, A.E. Bayesian dynamic modeling for large space-time datasets using Gaussian predictive processes. *J. Geogr. Syst.* **2012**, *14*, 29–47. [CrossRef]
4. European Space Agency (ESA). *Big Data from Space*; European Space Agency: Frascati, Italy, 2013.
5. Hay, S.; George, D.; Moyes, C.; Brownstein, J. Big data opportunities for global infectious disease surveillance. *PLoS Med.* **2013**, *10*, e1001413. [CrossRef] [PubMed]
6. Karimi, H.A. (Ed.) *Big Data: Techniques and Technologies in Geoinformatics*; CRC Press: Boca Raton, FL, USA, 2014; p. 312.
7. Barton, D.; Court, D. Making advanced analytics work for you. *Harv. Bus. Rev.* **2012**, *90*, 78–83. [PubMed]
8. Davenport, T.; Patil, D. Data Scientist: The Sexiest Job of the 21st Century. *Harvard Business Review*, Ocotber 2012.
9. McAfee, A.; Brynjolfsson, E. Big Data: The Management Revolution. *Harvard Business Review*, October 2012.
10. Murdoch, T.; Detsky, A. The inevitable application of big data to health care. *JAMA* **2013**, *309*, 1351–1352. [CrossRef] [PubMed]
11. Pearl, J. *Causality: Models, Reasoning and Inference*; Cambridge University Press: New York, NY, USA, 2009.
12. Noble, D.; Casalino, L. Can accountable care organizations improve population health? Should they try? *JAMA* **2013**, *11*, 119–120.
13. Ruckerl, R.; Schneider, A.; Breitner, S.; Cyrys, J.; Peters, A. Health effects of particulate air pollution: A review of epidemiological evidence. *Inhalation Toxicol.* **2011**, *23*, 555–592. [CrossRef]
14. Engel-Cox, J.A.; Hoff, R.M.; Haymet, A.D.J. Recommendations on the use of satellite remote-sensing data for urban air quality. *J. Air Waste Manag. Assoc.* **2004**, *54*, 1360–1371. [CrossRef] [PubMed]
15. Engel-Cox, J.A.; Holloman, C.H.; Coutant, B.W.; Hoff, R.M. Qualitative and quantitative evaluation of MODIS satellite sensor data for regional and urban scale air quality. *Atmos. Environ.* **2004**, *38*, 2495–2509. [CrossRef]

16. Engel-Cox, J.A.; Hoff, R.M.; Rogers, R.; Dimmick, F.; Rush, A.C.; Szykman, J.J.; Al-Saadi, J.; Chu, D.A.; Zell, E.R. Integrating lidar and satellite optical depth with ambient monitoring for 3-dimensional particulate characterization. *Atmos. Environ.* **2006**, *40*, 8056–8067. [CrossRef]

17. Liu, Y.; Sarnat, J.A.; Kilaru, A.; Jacob, D.J.; Koutrakis, P. Estimating ground-level PM2.5 in the Eastern United States using satellite remote sensing. *Environ. Sci. Technol.* **2005**, *39*, 3269–3278. [CrossRef] [PubMed]

18. Liu, Y.; Franklin, M.; Kahn, R.; Koutrakis, P. Using aerosol optical thickness to predict ground-level PM2.5 concentrations in the St. Louis area: A comparison between MISR and MODIS. *Remote Sens. Environ.* **2007**, *107*, 33–44. [CrossRef]

19. Liu, Y.; Paciorek, C.; Koutrakis, P. Estimating daily PM2.5 exposure in Massachusetts with satellite aerosol remote sensing data, meteorological, and land use information. *Epidemiology* **2008**, *19*, S116.

20. Van Donkelaar, A.; Martin, R.V.; Park, R.J. Estimating ground-level PM2.5 using aerosol optical depth determined from satellite remote sensing. *J. Geophys. Res.* **2006**, *111*. [CrossRef]

21. Van Donkelaar, A.; Martin, R.; Verduzco, C.; Brauer, M.; Kahn, R.; Levy, R.; Villeneuve, P. A hybrid approach for predicting PM2.5 exposure response. *Environ. Health Perspect.* **2010**, *118*. [CrossRef]

22. Van Donkelaar, A.; Martin, R.V.; Brauer, M.; Kahn, R.; Levy, R.; Verduzco, C.; Villeneuve, P.J. Global estimates of ambient fine particulate matter concentrations from satellite-based aerosol optical depth: Development and application. *Environ. Health Perspect.* **2010**, *118*, 847–855. [CrossRef] [PubMed]

23. Van Donkelaar, A.; Martin, R.V.; Levy, R.C.; da Silva, A.M.; Krzyzanowski, M.; Chubarova, N.E.; Semutnikova, E.; Cohen, A.J. Satellite-based estimates of ground-level fine particulate matter during extreme events: A case study of the Moscow fires in 2010. *Atmos. Environ.* **2011**, *45*, 6225–6232. [CrossRef]

24. Martin, R.V. Satellite remote sensing of surface air quality. *Atmos. Environ.* **2008**, *42*, 7823–7843. [CrossRef]

25. Hoff, R.M.; Christopher, S.A. Remote sensing of particulate pollution from space: Have we reached the promised land? *J. Air Waste Manag. Assoc.* **2009**, *59*, 645–675. [CrossRef] [PubMed]

26. Hoffmann, B.; Moebus, S.; Dragano, N.; Stang, A.; Moehlenkamp, S.; Schmermund, A.; Memmesheimer, M.; Broecker-Preuss, M.; Mann, K.; Erbel, R.; *et al.* Chronic residential exposure to particulate matter air pollution and systemic inflammatory markers. *Environ. Health Perspect.* **2009**, *117*, 1302–1308. [CrossRef] [PubMed]

27. Zhang, H.; Hoff, R.M.; Engel-Cox, J.A. The relation between Moderate Resolution Imaging Spectroradiometer (MODIS) aerosol optical depth and PM2.5 over the United States: A geographical comparison by US Environmental Protection Agency Regions. *J. Air Waste Manag. Assoc.* **2009**, *59*, 1358–1369. [CrossRef] [PubMed]

28. Zhang, H.; Lyapustin, A.; Wang, Y.; Kondragunta, S.; Laszlo, I.; Ciren, P.; Hoff, R.M. A multi-angle aerosol optical depth retrieval algorithm for geostationary satellite data over the United States. *Atmos. Chem. Phys.* **2011**, *11*, 11977–11991. [CrossRef]

29. Weber, S.A.; Engel-Cox, J.A.; Hoff, R.M.; Prados, A.I.; Zhang, H. An improved method for estimating surface fine particle concentrations using seasonally adjusted satellite aerosol optical depth. *J. Air Waste Manag. Assoc.* **2010**, *60*, 574–585. [CrossRef] [PubMed]

30. Kumar, N.; Chu, A.D.; Foster, A.D.; Peters, T.; Willis, R. Satellite remote sensing for developing time and space resolved estimates of ambient particulate in Cleveland, OH. *Aerosol Sci. Technol.* **2011**, *45*, 1090–1108. [CrossRef] [PubMed]

31. Lee, H.J.; Liu, Y.; Coull, B.; Schwartz, J.; Koutrakis, P. PM2.5 prediction modeling using MODIS AOD and its implications for health effect studies. *Epidemiology* **2011**, *22*, S215. [CrossRef]

32. Lee, H.J.; Liu, Y.; Coull, B.A.; Schwartz, J.; Koutrakis, P. A novel calibration approach of MODIS AOD data to predict PM2.5 concentrations. *Atmos. Chem. Phys.* **2011**, *11*, 7991–8002. [CrossRef]

33. Choi, Y.S.; Ho, C.H.; Chen, D.; Noh, Y.H.; Song, C.K. Spectral analysis of weekly variation in PM10 mass concentration and meteorological conditions over China. *Atmos. Environ.* **2008**, *42*, 655–666. [CrossRef]

34. Liu, Y.; Koutrakis, P.; Kahn, R. Estimating fine particulate matter component concentrations and size distributions using satellite-retrieved fractional aerosol optical depth: Part 1—Method development. *J. Air Waste Manag. Assoc.* **2007**, *57*, 1360–1369. [CrossRef] [PubMed]

35. Liu, Y.; Koutrakis, P.; Kahn, R.; Turquety, S.; Yantosca, R.M. Estimating fine particulate matter component concentrations and size distributions using satellite-retrieved fractional aerosol optical depth: Part 2—A case study. *J. Air Waste Manag. Assoc.* **2007**, *57*, 1360–1369. [CrossRef] [PubMed]

36. Liu, Y.; Paciorek, C.J.; Koutrakis, P. Estimating regional spatial and temporal variability of PM2.5 concentrations using satellite data, meteorology, and land use information. *Environ. Health Perspect.* **2009**, *117*, 886–892. [CrossRef] [PubMed]

37. Liu, Y.; Chen, D.; Kahn, R.A.; He, K. Review of the applications of multiangle imaging spectroradiometer to air quality research. *Sci. China Ser. D-Earth Sci.* **2009**, *52*, 132–144. [CrossRef]

38. Liu, Y.; Kahn, R.A.; Chaloulakou, A.; Koutrakis, P. Analysis of the impact of the forest fires in August 2007 on air quality of Athens using multi-sensor aerosol remote sensing data, meteorology and surface observations. *Atmos. Environ.* **2009**, *43*, 3310–3318. [CrossRef]

39. Liu, Y.J.; Harrison, R.M. Properties of coarse particles in the atmosphere of the United Kingdom. *Atmos. Environ.* **2011**, *45*, 3267–3276. [CrossRef]

40. Liu, Y.; He, K.; Li, S.; Wang, Z.; Christiani, D.C.; Koutrakis, P. A statistical model to evaluate the effectiveness of PM2.5 emissions control during the Beijing 2008 Olympic Games. *Environ. Int.* **2012**, *44*, 100–105. [CrossRef] [PubMed]

41. Lyamani, H.; Olmo, F.J.; Alcantara, A.; Alados-Arboledas, L. Atmospheric aerosols during the 2003 Heat Wave in southeastern Spain I: Spectral optical depth. *Atmos. Environ.* **2006**, *40*, 6453–6464. [CrossRef]

42. Pelletier, B.; Santer, R.; Vidot, J. Retrieving of particulate matter from optical measurements: A semiparametric approach. *J. Geophys. Res.-Atmos.* **2007**, *112*. [CrossRef]

43. Wang, Q.; Shao, M.; Liu, Y.; William, K.; Paul, G.; Li, X.; Liu, Y.; Lu, S. Impact of biomass burning on urban air quality wstimated by organic tracers: Guangzhou and Beijing as cases. *Atmos. Environ.* **2007**, *41*, 8380–8390. [CrossRef]

44. Natunen, A.; Arola, A.; Mielonen, T.; Huttunen, J.; Komppula, M.; Lehtinen, K.E.J. A multi-year comparison of PM2.5 and AOD for the Helsinki region. *Boreal Environ. Res.* **2010**, *15*, 544–552.

45. Paciorek, C.J.; Liu, Y.; Moreno-Macias, H.; Kondragunta, S. Spatiotemporal associations between GOES aerosol optical depth retrievals and ground-level PM2.5. *Environ. Sci. Technol.* **2008**, *42*, 5800–5806. [CrossRef] [PubMed]

46. Paciorek, C.J.; Liu, Y. Limitations of remotely sensed aerosol as a spatial proxy for fine particulate Matter. *Environ. Health Perspect.* **2009**, *117*. [CrossRef]

47. Paciorek, C.J.; Liu, Y. *Assessment and Statistical Modeling of the Relationship between Remotely Sensed Aerosol Optical Depth and PM2.5 in the Eastern United States*; Research Report; Health Effects Institute: Boston, MA, USA, 2012.

48. Rajeev, K.; Parameswaran, K.; Nair, S.K.; Meenu, S. Observational evidence for the radiative impact of Indonesian smoke in modulating the sea surface temperature of the equatorial Indian Ocean. *J. Geophys. Res.-Atmos.* **2008**, *113*. [CrossRef]

49. Schaap, M.; Apituley, A.; Timmermans, R.M.A.; Koelemeijer, R.B.A.; de Leeuw, G. Exploring the relation between aerosol optical depth and PM2.5 at Cabauw, The Netherlands. *Atmos. Chem. Phys.* **2009**, *9*, 909–925. [CrossRef]

50. Tian, D.; Wang, Y.; Bergin, M.; Hu, Y.; Liu, Y.; Russell, A.G. Air quality impacts from prescribed forest fires under different management practices. *Environ. Sci. Technol.* **2008**, *42*, 2767–2772. [CrossRef] [PubMed]

51. Van de Kassteele, J.; Koelemeijer, R.B.A.; Dekkers, A.L.M.; Schaap, M.; Homan, C.D.; Stein, A. Statistical mapping of PM10 concentrations over western Europe using secondary information from dispersion modeling and MODIS satellite observations. *Stoch. Environ. Res. Risk Assess.* **2006**, *21*, 183–194. [CrossRef]

52. Zhang, J.; Reid, J.S. An analysis of clear sky and contextual biases using an operational over ocean MODIS aerosol product. *Geophys. Res. Lett.* **2009**, *36*. [CrossRef]

53. Lary, D.J.; Remer, L.A.; MacNeill, D.; Roscoe, B.; Paradise, S. Machine learning and bias correction of MODIS aerosol optical depth. *IEEE Geosci. Remote Sens. Lett.* **2009**, *6*, 694–698. [CrossRef]

54. Hyer, E.J.; Reid, J.S.; Zhang, J. An over-land aerosol optical depth data set for data assimilation by filtering, correction, and aggregation of MODIS Collection 5 optical depth retrievals. *Atmos. Meas. Tech.* **2011**, *4*, 379–408. [CrossRef]

55. Shi, Y.; Zhang, J.; Reid, J.S.; Hyer, E.J.; Hsu, N.C. Critical evaluation of the MODIS Deep Blue aerosol optical depth product for data assimilation over North Africa. *Atmos. Meas. Tech. Discuss.* **2012**, *5*, 7815–7865. [CrossRef]

56.  Reid, J.S.; Hyer, E.J.; Johnson, R.S.; Holben, B.N.; Yokelson, R.J.; Zhang, J.; Campbell, J.R.; Christopher, S.A.; Girolamo, L.D.; Giglio, L.; *et al.* Observing and understanding the Southeast Asian aerosol system by remote sensing: An initial review and analysis for the Seven Southeast Asian Studies (7SEAS) program. *Atmos. Res.* **2013**, *122*, 403–468. [CrossRef]

57.  Ho, T.K. The random subspace method for constructing decision forests. *IEEE Trans. Pattern Anal. Mach. Intell.* **1998**, *20*, 832–844. [CrossRef]

58.  Breiman, L. Random forests. *Mach. Learn.* **2001**, *45*, 5–32. [CrossRef]

International Journal of
isprs *Geo-Information*

MDPI

*Article*

# Dasymetric Mapping and Spatial Modeling of Mosquito Vector Exposure, Chesapeake, Virginia, USA

Haley Cleckner and Thomas R. Allen *

Department of Geography, Environment & Planning, East Carolina University, Greenville, NC 27858, USA;
hsbcleckner@gmail.com
* Author to whom correspondence should be addressed; allenth@ecu.edu;
  Tel.: +1-252-328-6624; Fax: +1-252-328-6054.

Received: 14 November 2013; in revised form: 21 June 2014; Accepted: 2 July 2014; Published: 14 July 2014

**Abstract:** Complex biophysical, social, and human behavioral factors influence population vulnerability to vector-borne diseases. Spatially and temporally dynamic environmental and anthropogenic patterns require sophisticated mapping and modeling techniques. While many studies use environmental variables to predict risk, human population vulnerability has been a challenge to incorporate into spatial risk models. This study demonstrates and applies dasymetric mapping techniques to map spatial patterns of vulnerable human populations and characterize potential exposure to mosquito vectors of West Nile Virus across Chesapeake, Virginia. Mosquito vector abundance is quantified and combined with a population vulnerability index to evaluate exposure of human populations to mosquitoes. Spatial modeling is shown to capture the intersection of environmental factors that produce spatial hotspots in mosquito vector abundance, which in turn poses differential risks over time to humans. Such approaches can help design overall mosquito pest management and identify high-risk areas in advance of extreme weather.

**Keywords:** risk mapping; mosquito-borne disease; dasymetric mapping

---

## 1. Introduction

The spread of vectors and growing number of vector-borne diseases pose a major threat to human health. In order to prevent the spread of disease, it is advantageous to predict the risk of disease transmission, both spatially and temporally. Geospatial technologies are commonly used to evaluate patterns of vector or human case distributions, as well as estimate the risk of disease transmission based on entomological, epidemiological, and environmental factors [1]. One limitation is that prediction systems are often static and only predict risk at one particular time and place. This study has addressed this shortcoming by predicting the risk of disease transmission from mosquitoes across Chesapeake, Virginia, for the summer of 2003, following a broad scale field surveillance campaign by this city's Mosquito Control Commission. In a situation where a disease or hazard is rare (and control-case studies or cohort studies are impractical), analyzing trends in threats and coexistence of risk factors with disease are fruitful approaches [2]. Using GIS techniques, spatial analyses were conducted in this study to track the trends in competent vector species abundance and identify coincident population most susceptible to mosquito-borne diseases and estimate relative risk of disease transmission to humans.

Predictions of human sensitivity and potential disease transmission have been recommended to incorporate vector abundance and the immune status of the host population [3]. Such epidemiological evidence is also translated into surveillance and control of vector species by local public health agencies. In addition, the risk of exposure to mosquito-borne disease vectors may be estimated using mosquito abundance values and the broad physiological factors of human vulnerability to disease infection. Mosquito abundance values for this project were derived from a previously published study measuring the abundance of competent mosquito vector species *Culiseta melanura* as well as the combined abundance of *Aedes vexans* and *Psorophora columbiae* [4,5]. *A. vexans* and *P. columbiae* share a habitat preference for ephemeral pools and therefore are referred to as the "ephemeral species" throughout this paper. *C. melanura* is an important species in this region because it is the primary enzootic vector of Eastern Equine Encephalitis (EEE) and is also a potential vector of West Nile Virus (WNV). *A. vexans* is also another important potential epizootic vector for WNV. While several studies use vector and environmental variables to estimate risk, many studies do not take into account human vulnerability to these diseases or the temporal and spatial variability of vector abundance or human demographics and activity spaces.

A primary objective of this study is to estimate human exposure to mosquito vector species capable of disease transmission and human infection across a landscape. The City of Chesapeake, Virginia, offers this setting with a populous urban to rural gradient and low topographic coastal location adjoining extensive mosquito habitats that are exceedingly productive in the humid, subtropical summer climate, including abundance of ephemeral pools, agricultural ditches, and container breeding sites in urbanized areas. The second objective is to map the vulnerable human population in a manner that depicts the most accurate distribution of the population and facilitates assessing potential vector exposure risk. Using monthly mosquito abundance values and human population vulnerability data, a monthly risk index is calculated which estimates the exposure to mosquito vector species. These objectives underscore the overall goal of this study, to improve prediction of the risk of exposure to mosquitoes across Chesapeake, Virginia, during the peak mosquito breeding summer months of June to August. In addition, an achievable and practical result of this analysis includes a spatial risk model that can be adapted to other regions to identify vector exposure for surveillance and control, demonstrating geographic-targeting and more sophisticated, micro-level control measures [6].

A growing number of studies have predicted disease risk using geospatial methods (remote sensing, spatial analysis, geovisualization, and other methods applying and integrating Geographic Information Systems (GIS)). Statistical approaches to predicting outbreaks of West Nile Virus (WNV) using land cover and environmental factors, for instance, have been developed for Indianapolis, Indiana [7]. A spatial-temporal model was created that predicted disease risk from 2002 through 2007. This study found that the highest clusters of WNV outbreaks were located in agriculture and grassland areas. Such cover types exhibited a relatively high moisture content and provided favorable habitats for mosquito breeding. Discriminant analysis was used to find the relationship between environmental variables and disease. The results indicated that certain variables had a high impact on WNV dissemination. Wetland size, agriculture area percentage, and stream length were all positively correlated with WNV outbreaks. WNV infection potential has also been studied using a wide sample of dead birds and human population data with cluster analysis to infer potential disease infection hotspots [8]. In another empirical study combining LandScan gridded population data and *Aedes aegypti* mosquito species distribution maps, correlation and monte carlo simulation analyses related human mobility, vector abundance, and disease transmission [9]. Indeed, GIS and remote sensing techniques have been used to assess mosquito-borne disease risk widely, even globally as in the case for malaria risk [10]. NASA scientists at the Goddard Space Flight Center's Healthy Planet program developed a malaria transmission model that incorporates parasites, hosts, vectors, human factors, and environmental factors. They also developed a risk model that predicts transmission intensity using meteorological data. Remotely sensed imagery was used to identify potential breeding sites of mosquito vectors in order to better focus mosquito control applications. Similar work pioneered by Beck *et al.* investigated satellite remote sensing and GIS techniques to identify the risk of malaria transmission in Chiapas, Mexico, for identifying villages with high vector-human contact risk [11].

Although GIS has been widely used for disease risk estimation, one limitation is the dearth of human health data incorporated into risk models. However, vulnerability is an important factor that can guide estimations of the risk of disease infection. Factors such as age, immune suppression, activity space and behavior patterns, and genetics influence a human's risk of infection. The spatial pattern of infectious and susceptible people to vector-borne diseases may be a critical determinant of exposure and disease risk [12]. Using demographic and GIS data, this study estimates the population that is most vulnerable to mosquito-borne exposure across the City of Chesapeake. Initial assessment of vulnerability utilizes mapped U.S. Census block group demographics. Aggregating population density to geographic units such as Census blocks, block groups, or tracts is a common method for creating choroplethic maps of population data. One constraint of these maps is that the reader is left to assume that all areas within a given geographic unit have equal population densities. However, this is usually not the case and presents an impediment to understanding finer scale patterns. If the spatial units are too large, a commonplace occurrence in suburban to rural settings, the data spatial variation tends to be reduced or overgeneralized [13]. Statistical techniques such as dasymetric mapping can be used to show a more accurate distribution of the population by disaggregating spatial data in polygonal areas to a finer unit of analysis using ancillary data to help refine locations of the population [14]. The technique has been succinctly described cartographically, "dasymetric mapping depicts quantitative areal data using boundaries that divide the mapped area into zones of relative homogeneity with the purpose of best portraying the underlying statistical surface" ([15], p. 125). This method of mapping transforms data from the arbitrary or choroplethic zones of data aggregation to a dasymetric map in order to enhance and depict the spatial pattern of human population. The transformation of data from the arbitrary zones of the source data to the meaningful zones incorporates the use of an ancillary data set that is separate from, but related to, the variation in the statistical surface [16].

The most common dasymetric technique is the grid binary method in which ancillary classes are regarded as either populated or unpopulated [15]. This particular study uses the "intelligent" dasymetric mapping (IDM) approach to map human vulnerability [17]. The IDM approach takes data mapped in polygonal "source" zones and a second, categorical grid data set, then redistributes the polygonal population data to a set of target zones formed by the intersection of the source and ancillary grid units. In this project, intelligent dasymetric mapping is used to display the vulnerable population according to land cover classes. The final output is a raster surface, which displays vulnerability per pixel unit. This representation of the population provides a more accurate spatial representation of the vulnerable population throughout Chesapeake as compared to coarser observational units in choroplethic maps.

Tradeoffs in data collection, cartographic output, and spatial data representation must be considered when selecting methods for spatial analysis and map symbolization. Most often, cadastral boundaries or census data polygons are selected by default. There are several reasons why these frequently chosen units are not always the most effective for mapping population data. For instance, this approach may not be appropriate for showing population data over time, as census boundaries may change over each censusing period in tandem with the population settlement pattern and density [18]. Census data also pose a problem because these data often have an uneven spatial distribution. Oftentimes, the cartographer may want to estimate the population within a smaller unit of a census block. Dasymetric maps, on the other hand, can show a population at a much finer scale than a tract, block group or block, such as a grid derived from a satellite image in raster cells. Another issue with census or vector data is that in some parts of the world, these data may not be current or readily available [14]. Satellite imagery is available for all parts of the world and can be used to create dasymetric maps of the population. Although dasymetric techniques are effective for creating thematic maps, the lack of standardization of production methods has reduced the use of dasymetric maps in GIS [15].

Because vector-borne disease incidence is intrinsically dependent on exposure to the disease pathogen and vectors, human vulnerability indices must necessarily derive from overlays with the mosquito abundance values in order to estimate risk. In such an approach, spatial overlay becomes the critical proxy for potential spatial interaction between vectors and humans. The factors affecting disease transmission are temporally dynamic, thus a mosquito vector exposure index is calculated on a monthly basis for June through August of 2003. Predictions were limited to the summer of 2003 due to the ample mosquito trap data available for these months. The selected year also provided an ideal opportunity to analyze a large volume of local mosquito data collected during a relatively wet, increasingly favorable breeding season. Further, the season preceded a landfalling hurricane (Isabel, 17–18 September 2003) whose aftermath elicited widespread aerial adulticide spraying (insecticide for adult mosquitoes.) Hence, the summer of 2003 in Chesapeake provides a rich spatial data set for surveillance of mosquitoes, before subsequently profound changes in mosquito surveillance reduced the volume and spatial extent of mosquito observations (light traps, larval dips, nuisance reports, and landing counts) in the study area. The summer months also represent the prime breeding period for these mosquitoes, as the high temperatures and abundant precipitation create an ideal habitat for mosquito populations to thrive. Six indices are calculated which represent the monthly potential exposure to mosquito vectors from both groups of mosquito species. The exposure risk indices are calculated on a raster grid on a pixel-by pixel basis.

## 2. Study Area

Chesapeake is an independent city, which comprises 88,000 hectares (340 square miles) of Southeastern Virginia and a population of approximately 220,000 (estimated 2008) during the study. The city is located in the coastal plain of Virginia and contains the northeastern portion of the Great Dismal Swamp (Figure 1). Although it serves as a large reservoir of bird and mosquito vectors, the Great Dismal Swamp was excluded from the study area because there are no permanent residents in the swamp. Nonetheless, mosquito observations are reported from immediate surroundings, and the extensive wetlands and non-tidal creeks, agricultural and stormwater ditches within Chesapeake are conducive to mosquito breeding and therefore an extensive suitable habitat. The proximity of these mosquito habitats to the urbanized areas of Chesapeake allows mosquitoes to easily encounter and potentially transmit diseases to humans. Mosquito abundance was predicted to be high across Chesapeake due to Hurricane Isabel, which made landfall in North Carolina on 18 September 2003 [19]. Chesapeake was also chosen as the study area for its noteworthy mosquito surveillance program. In 2003, the City of Chesapeake Mosquito Control Commission (CMCC) [20] collected mosquitoes using 56 $CO_2$ baited light traps distributed among 28 permanent sites. Bi-weekly, monthly, and seasonal totals of all species were collected and were made available to the research project. Mosquito captures were collected and counted weekly from April through November of 2003 by a set of staff trained in taxonomy. The mosquito counts were used to predict the mosquito abundance values that are incorporated in the risk model. Focusing on the summer peak breeding months prior to Hurricane Isabel, only the mosquito counts from June through August were used, as these months had sufficient capture data for statistical analysis. In addition, as concerns over potential WNV outbreak evolved, dead birds, cases of Eastern Equine Encephalitis (EEE has historic prevalence in this area), and public mosquito abatement service requests were georeferenced for possible analysis and comparison to mosquito vector trap data and human population exposure.

## 3. Data and Methodology

### 3.1. Predicting Population Vulnerability to Disease Risk

To assess the population most susceptible to disease, a choroplethic map of population vulnerability was created. Using 2000 U.S. decennial Census data and a set of augmented GIS point vulnerabilities, the vulnerable population was estimated across Chesapeake. A choroplethic map was created to depict the population mapped according to Census block groups. The vulnerable population data were then partitioned into land use zones using dasymetric mapping techniques.

Census block group statistics provided the initial baseline vulnerable population data. The elderly are at greatest risk of developing severe disease after infection by a mosquito [21]. In addition, children are also at high risk of disease infection owing to their underdeveloped immune systems [22]. To predict the age density, a 2000 decennial Census polygon shapefile was obtained and clipped to the extent of Chesapeake, Virginia. The shapefile included 2004 population estimates per block group. 2004 estimates were used due to the absence of Census data specific to 2003. To tally the elderly and childhood population in our estimates, the sum of population was calculated for persons less than 5 years of age and greater than 50 years of age within each Census block group (Figure 2). Although the typical age for elderly would classify at 65 years or above, there is research supporting the prevalence of persons over 50 to exhibit symptoms, receive hospitalization, and incur a higher case-fatality ratio [23]. The combined total of these age groups per block group is shown in the choroplethic map in Figure 3.

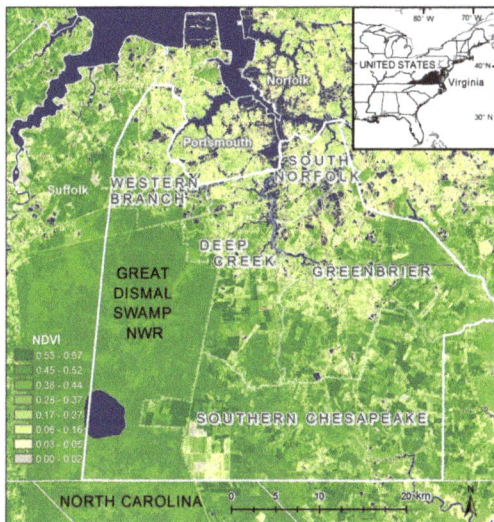

**Figure 1.** Study area location City of Chesapeake, Virginia, a low-lying coastal plain situated adjacent to the Great Dismal Swamp and extensive estuaries of the Chesapeake Bay. Map shows the Chesapeake Mosquito Control District boroughs superimposed on a Normalized Difference Vegetation Index (NDVI) image from Landsat Thematic Mapper, 29 July 2002. NDVI shows brighter green tones for healthy vegetation.

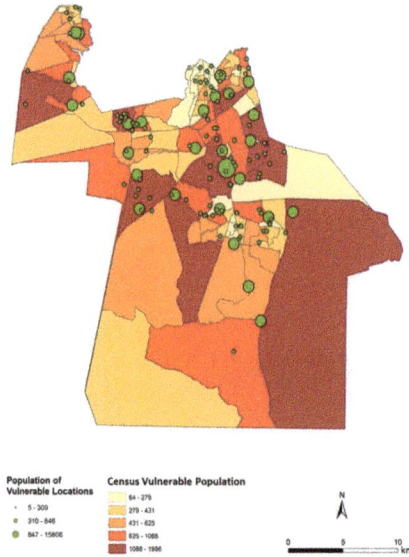

**Figure 2.** Census 2000 population age mapped as choropleths by block groups with point locations of vulnerable populations (hospitals, daycares, schools, *etc.*) displayed using proportional symbols for discrete population concentrations.

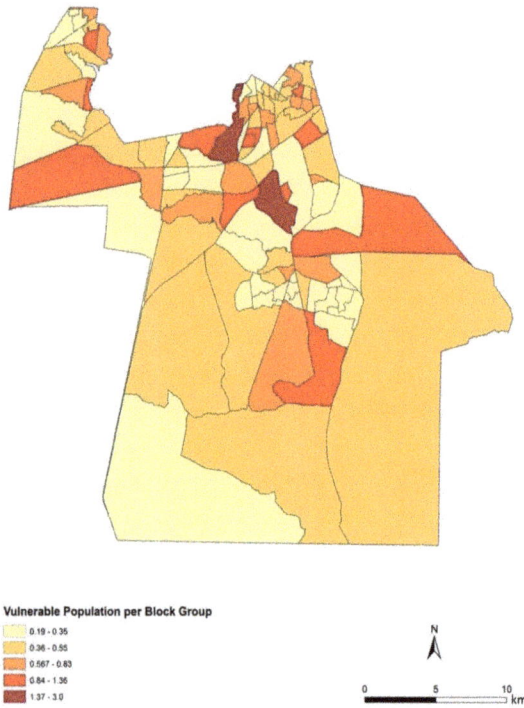

**Figure 3.** Vulnerable populations derived for Census block group in persons per hectare (estimated using Equation (1)).

Census data are also limiting in that they reflect a "nighttime" population distribution. This cannot account for daytime locations. Urban areas such as Chesapeake are known to experience diurnal surging of population density with workplace and suburban residential commuting. In addition, sites of frequent visitation that are non-domicile are not reflected in the Census, such as hospitals, child care facilities, schools, and outpatient clinics. To account for this discrepancy, points of higher possible vulnerable populations were sought for inclusion in the vulnerable population analysis. These locations are considered particularly important given the large proportion of children and elderly populations at these sites. Hospitals are also considered vulnerable due to the high number of people with compromised immune systems. To complete the second step of vulnerability prediction, vulnerable point facility location data was obtained from the City of Chesapeake Information Technology Department. The data was obtained in the form of a GIS points shapefile in which each feature represents a vulnerable point location (Figure 2). The population of each hospital and elderly care facility were included within the GIS data and counted as permanent residents. The population of Chesapeake General Hospital, for example, was calculated by summing the total number of inpatients and outpatients. School and daycare centers were also included, however, population counts for these facilities were so sparse that additional count inquiry was limited to primary, elementary, and intermediate schools. Determining the population of every daycare was cost-prohibitive across this extensive area having over 250 permitted facilities. Nonetheless, inclusion of site-specific vulnerable populations using these ancillary data and summing with Census data is believed to provide a robust first-order estimate, conservatively estimating the gross vulnerable population and reflecting a slightly higher population that Census nighttime and undercounts would tend to include.

To calculate the final vulnerable population within each block group, the population of each vulnerable point was added to the previously calculated block group vulnerable population based on point-in-polygon overlay. The final output was stored as a polygon shapefile (Equation (1)). In order to reflect the total population in each block group, the vulnerable population was normalized using the following equation and subsequently areally transformed from persons per block group to persons per hectare:

$$Population_n = \frac{Population_v}{Population_{Total}}$$

(1)

$Population_n$ = the normalized vulnerable population, per block group;
$Population_v$ = the calculated vulnerable population, per block group;
$Population_{Total}$ = total population of the block group.

*3.2. Dasymetric Mapping of Population Vulnerability*

In order to provide a more accurate depiction of how the demographic data is distributed within each block group, dasymetric mapping techniques were used to redistribute the population based on levels of urbanization. These methods were adapted from USGS work in the San Francisco Bay region according to land cover type [12]. A 30 m pixel land cover grid from NOAA's Coastal Change Analysis Program (C-CAP) for 2001 was used as the ancillary layer for the dasymetric map. Initially, the land cover data set consisted of 22 land cover classes. In order to map vulnerability according to residential land use, the land cover types were separated into four classes: highly intensity developed, low intensity developed, non-urban, and water (Figure 4). Using the ArcGIS dasymetric Mapping Tool developed by the U.S. Geological Survey [24], the population density (Figure 3) was mapped according to the reclassified and slightly aggregated C-CAP land cover types (Figure 4). The data were redistributed based on a combination of areal weighting and the relative densities of ancillary classes (*i.e.*, the different land cover categories) for populating the target zones (Equation (2)).

$$Y_t = Y_s \left[ \frac{A_t D_t}{\sum_{i=1}^{n}(A_i D_i) \dots (A_n D_n)} \right] \left\{ \begin{array}{c} where\ each \\ ancillary\ class\ from\ 1\ to\ n \\ (found\ in\ the\ 'source'region) \\ is\ considered \end{array} \right.$$

(2)

$Y_t$ = the estimated count for target zone $t$;

$Y_s$ = the count of a source zone, which overlaps the target zone;

$A_t$ = the area of the given target zone;

$D_t$ = the estimated density of ancillary class $c$ associated with the target zone.

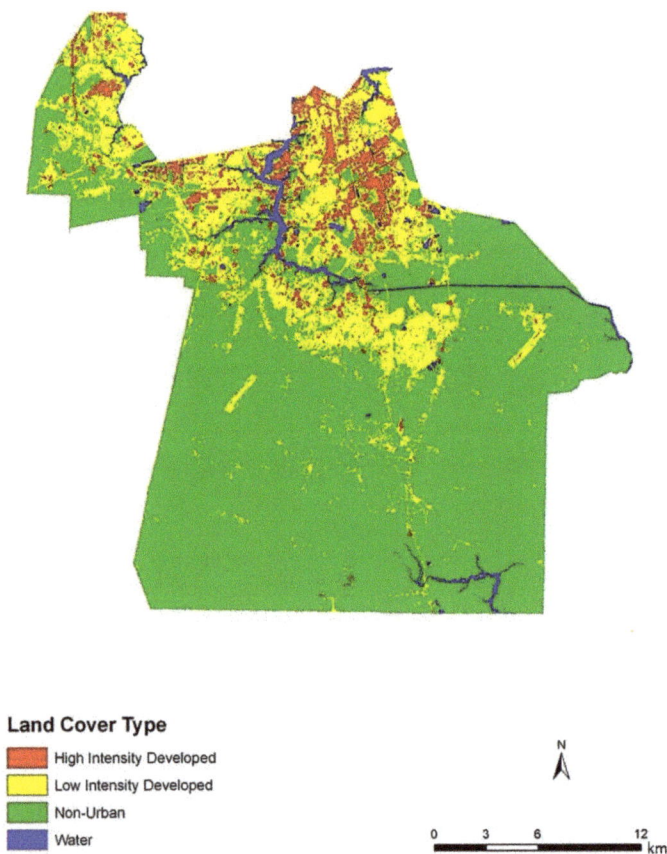

**Land Cover Type**

- High Intensity Developed
- Low Intensity Developed
- Non-Urban
- Water

N

0    3    6              12
                          km

**Figure 4.** Simplified Coastal Change Analysis Program (C-CAP) 2001 land cover types used as ancillary spatial units for dasymetric mapping.

In some cases, cartographers may use their own domain knowledge to specify the value of $D_c$. In this case, the value of $D_c$ is computed using the percent cover method. This option allows the analyst to select a threshold percentage and the zones whose percentage of coverage equals or exceeds that threshold. In this case, the percent cover used was 80%. Once the $n$ sample zones are selected, the estimated density of the ancillary class is derived (Equation (3)) in order to find an average density across representative zones.

$$D_c = \frac{\sum_{i=1}^{n} Y_s}{\sum_{i=1}^{n} A_s}$$

(3)

$D_c$ = the estimated density of ancillary class $c$;
$Y_s$ = the count of a source zone;
$A_s$ = the area of a source zone.

The population dasymetric map derived is computed on a 30 m pixel resolution grid (Figure 5). In order to clearly represent the population across Chesapeake, population density is converted from persons per pixel (in 30 m × 30 m pixels) to persons per hectare (Equation (4)).

$$Population_{ha} = \frac{Population_p \times 10000}{900 \text{ m}^2}$$

(4)

$Population_{ha}$ = the population per hectare;
$Population_p$ = the population per 30 m × 30 m pixel.

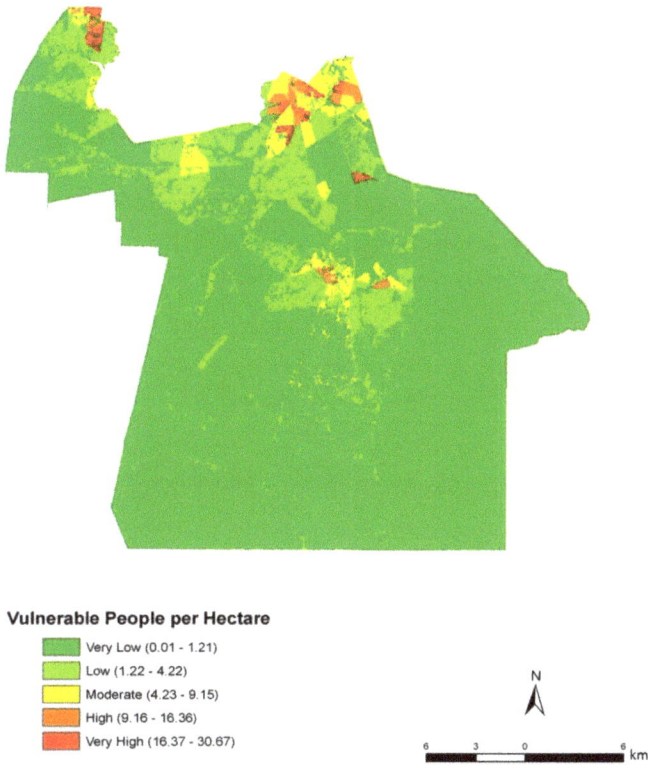

**Vulnerable People per Hectare**

- Very Low (0.01 - 1.21)
- Low (1.22 - 4.22)
- Moderate (4.23 - 9.15)
- High (9.16 - 16.36)
- Very High (16.37 - 30.67)

**Figure 5.** Dasymetric map of the composite population vulnerable to mosquito-borne diseases (natural breaks classification from very low to very high vulnerable population).

### 3.3. Mapping Exposure to Mosquito Vectors

Using mosquito abundance values and the dasymetric map of the vulnerable population, a monthly risk index could be calculated to quantify the exposure of human population to mosquito vector species. Vector abundance maps were available by statistical estimation from multiple regressions of trap data on environmental factors and habitat suitability and are published in prior works [3,4]. The vulnerability index was overlaid onto predicted mosquito abundance grids to calculate a monthly exposure potential index for both groups of species per pixel, with mosquito abundance and human vulnerability input indices weighted equally in the risk formulas. All raster values are represented in the form of co-registered 30 m pixel resolution grids. The mosquito abundance values used to predict risk are shown in Figure 6.

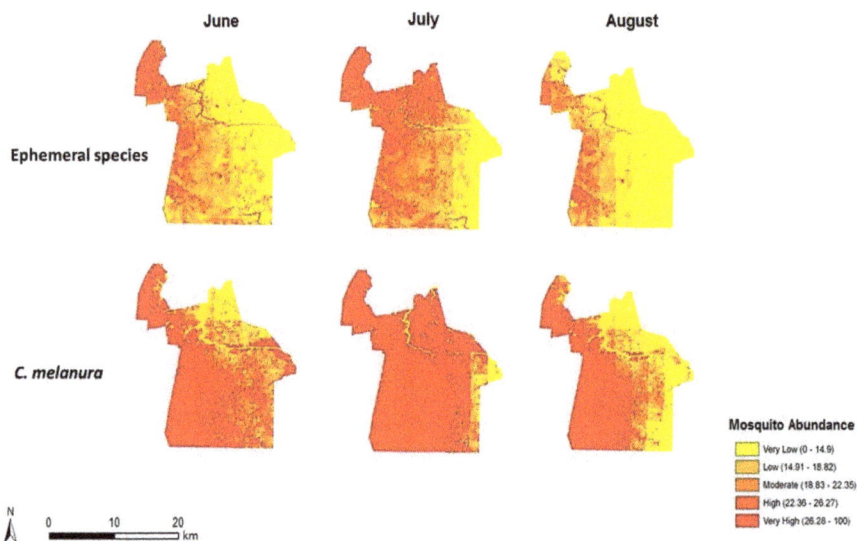

**Figure 6.** Predicted monthly mosquito abundance (classified in quantiles).

A model was created which multiplies each pixel within the vulnerable population grid to the corresponding mosquito abundance pixel. The model results in six raster grids representing the monthly risk of disease infection from both groups of mosquitoes. The final risk values for both groups of mosquitoes were calculated using Equations (5) and (6).

$$\text{Exposure}_{ep} = (\text{Population}_h \times \text{Abundance}_{ep}) \tag{5}$$

$$\text{Exposure}_{Cm} = (\text{Population}_{ha} \times \text{Abundance}_{Cm}) \tag{6}$$

$\text{Exposure}_{ep}$ = the risk of exposure to the ephemeral species for a particular month;
$\text{Exposure}_{Cm}$ = the risk of exposure to the *C. melanura* for a particular month;
$\text{Population}_{ha}$ = the vulnerable population per hectare;
$\text{Abundance}_{ep}$ = the rescaled abundance of the ephemeral species for the corresponding month;
$\text{Abundance}_{Cm}$ = the rescaled abundance of the *C. melanura* for the corresponding month.

## 4. Results and Discussion

### 4.1. Human Vulnerability and Mosquito Vector Abundance

To assess the sensitivity and accuracy of the dasymetric mapping techniques, the raster surface of vulnerability can be compared to the data mapped by block groups. The dasymetric map was expected to provide a more spatially precise representation of population vulnerability as compared to the Census block group choropleths. Indeed, the block groups show few distinct patterns of vulnerability across Chesapeake (Figure 4). In general, the block groups indicate vulnerability mapped at a coarse scale. Regions of high vulnerability are dispersed and highly concentrated at various locations throughout Chesapeake in block groups, primarily highlighting the urbanized northern and suburban central Chesapeake (Figure 2). Land cover data corroborate this gradient between northern and central, suburban Chesapeake (Figure 3). However, compared to the block group map, the dasymetric map shows a more detailed and fine-scale spatial representation of vulnerability (Figure 7). Again, the highly vulnerable regions are concentrated in the northern portion of the city, but a greater fine-scale pattern and interspersion representing variations are found along arterial transportation routes, major suburban developments, and extensive creek and low floodplain areas. This pattern is mainly attributed to the clustering of vulnerable locations such as schools and daycare centers and surrounding neighborhoods in northern Chesapeake. Northern, urban Chesapeake is more developed than other regions of the city and consequently has a greater population density.

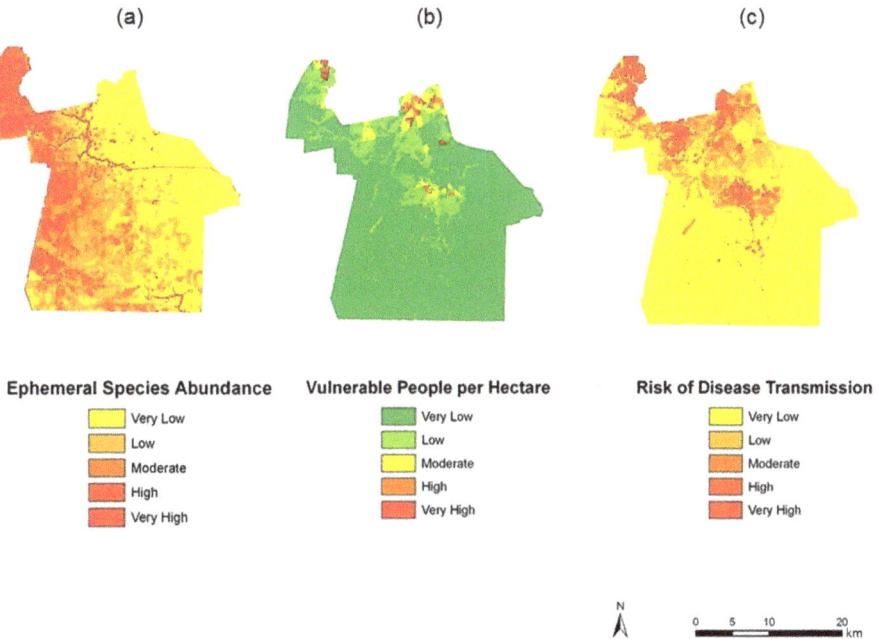

**(a)**　　　　　　　　**(b)**　　　　　　　　**(c)**

| Ephemeral Species Abundance | Vulnerable People per Hectare | Risk of Disease Transmission |
|---|---|---|
| Very Low | Very Low | Very Low |
| Low | Low | Low |
| Moderate | Moderate | Moderate |
| High | High | High |
| Very High | Very High | Very High |

**Figure 7.** Spatial overlay used to predict potential exposure to ephemeral species for June (**a–c**). The exposure in June (c) is the product of (a) ephemeral species abundance for that month; and (b) the dasymetric surface of vulnerable population in quantiles.

Two finer scale patterns of vulnerability are evident across the city. First, regions covered in water or rural areas are populated with a smaller number of vulnerable people. This was expected since these regions are less developed and contain lower disease susceptible populations. Mosquito control personnel interpret the fine-scale patterns of mosquito abundance, vulnerable populations, and composite exposure to assist operational surveillance and abatement activities. In some cases, the proximity of high abundance and high human populations may be amenable to spraying from trucks. In other cases where risk is low to moderate, less frequent spraying or larvicide applications are typically pursued. Second, the high risk indicated in the Western Branch area of Figure 7c has typically spurred routine seasonal spraying, with particular attention to neighborhoods adjoining tidal creeks and the zone between this part of the city and the Great Dismal Swamp. High to very high risk is also noted in the north-central area of Chesapeake, the most urbanized area of the city adjoining the City of Norfolk. Although ephemeral species abundances in the month of June are low, the total population and areal density of human population is extremely high (much higher than the density in the western and southern suburbs.)

Public health and mosquito control specialists treat highly urbanized and dense population centers slightly differently from suburban populations. In this instance, the urbanized areas have permanent surveillance operations in light trapping and inspection of urban ditches, culverts, storm drains, and creeks prevalent in this area. In contrast, suburban surveillance typically comprises fewer and wider spread light traps and the use of periodic roving trapping efforts, as well as a greater dependence on nuisance reports and abatement requests to cover the much greater suburban area. The indices developed here are thus trending toward a persistently higher risk in the more populous urban centers, with more changeable risk in the suburban and rural zones. However, this index could indicate bias that could be construed as a false positive in low mosquito abundance periods or a lack of sensitivity during a rare high abundance bloom event. This concern is somewhat muted in the instance of northern Chesapeake, which also has not only the densest human population, but also a concentration of elderly. Such population densities can have profound effects on potential disease exposure [3].

Other studies have used dasymetric techniques to map population density and have obtained similar results. Many studies have found that dasymetric mapping provides a more accurate representation of population distribution compared to conventional mapping techniques. One study used dasymetric mapping to map population density in five counties within southeast Pennsylvania [16]. Using areal weighting, block group demographic data was mapped according to three classes of urban land cover. The dasymetric map raster was compared to vector population data within block groups. In a similar finding [16], it was observed that within the urban core areas, the choroplethic and dasymetric maps did not differ significantly. However, in areas with parks and cemeteries, the dasymetric map was significantly more detailed. Similar methods were also used to map the San Francisco Bay population by land cover type [13], but rather than using a three-tier classification system, that study used four classes of land cover. Correlation coefficients indicated that the daysmetric mapping method of representing block-group population density was more accurate than the choropleth mapping method. Recent developments have also used address weighting (AW) and parcel distribution (PD) to map rural populations within three North Carolina counties with a focus on using local sources of ancillary data [25]. These mapping techniques are modifications of the existing dasymetric mapping techniques, street weighting (SW) and limiting variable (LV) algorithms. Statistical results indicated that AW and PD were appropriate for mapping rural populations, while SW and LV methods were more effective for mapping the population within urban areas. Yet another approach uses binary dasymetric mapping methods to interpolate populations across regions, such as demonstrated in the county of Leicestershire, England [26]. ArcGIS spatial analyst tools were used to reclassify raster pixel maps as populated or unpopulated. Weighting methods to map Census data have also focused on streets to adjust population across areal units in Los Angeles County, California [27]. Compared to simple aerial interpolation techniques, the dasymetric surface showed a 20% increase in population accuracy. An even finer scale and 3D cadastral-based data have been applied to map the population of New York City [14]. This method disaggregates population data at a high spatial resolution using cadastral data. For this study, residential units (RU) and residential area (RA) were aggregated to the tax lot, block group, and census tract level. The census tract values were then disaggregated to the tax lot level and then re-aggregated to predict the values within each block group. Cadastral-based Expert Dasymetric System (CEDS) derived population values are thus more finely allocated population than census values within block groups. Our results using the dasymetric mapping method corroborate these previously published techniques, highlighting the technique's potential to illustrate spatial variation and pattern among populations as compared to choroplethic mapping.

### 4.2. Risk Maps of Mosquito Vector Exposure

The risk of exposure to mosquito vectors across Chesapeake from both groups of mosquitoes is shown in Figure 8. The risk indices were scaled to show the level of risk for each month. To provide an effective visualization of the risk indices, risk values were classified into quantiles while maintaining each monthly index on the same relative scale. The actual units of risk are arbitrary and represent a relative index ranging from low to high risk. It is apparent on examining the maps that the threat of disease does not change significantly from June through August of 2003. This is due to the relatively static climate (warm and humid) conditions that persisted throughout the summer of 2003. The threat of disease transmission was also very similar between the two species groups. A longer time series or a period with more variable precipitation or weather events inducing mosquito population cycles would have elicited greater temporal variability, as documented in preceding habitat suitability and environmental modeling [4].

In particular, northern and central Chesapeake have a high risk of exposure to mosquitoes across all summer months. These high risk areas are reflective of the dasymetric map of vulnerability (Figure 5). The high risk regions coincide with regions predicted to have a relatively high vulnerability to disease infection. Despite the elevated risk in these regions, absolute mosquito abundance was predicted to be low in northern Chesapeake (Figure 6). The low number of mosquitoes may be due to the high level of urbanization and less prevalent suitable mosquito habitats or human adaptations to mitigate their abundance (especially reducing container breeding sites.) Although mosquito abundance was estimated to be low in northern Chesapeake, disease transmission levels may nonetheless be relatively high due the greater level of contact with humans in these developed areas. Indeed, cities may generally exacerbate disease transmission by bringing a large number of people into intimate contact with mosquito vectors [28]. Even with a supply of clean water, adequate shelter, and access to health care, a high population density would greatly facilitate the spread of transmissible disease. Despite the low number of swamps and floodplains in these developed regions, mosquitoes may still come into close proximity of humans by other sources of standing water. *Ae. Vexans* and *P. columbiae* are typically found in flood plains where rivers overflow their banks, but significant numbers can be produced from virtually any area where water accumulates on an intermittent basis [29]. These mosquitoes may breed in temporary water sources such as drainage ditches and tire ruts, which are extensive in low-lying coastal plains such as Chesapeake.

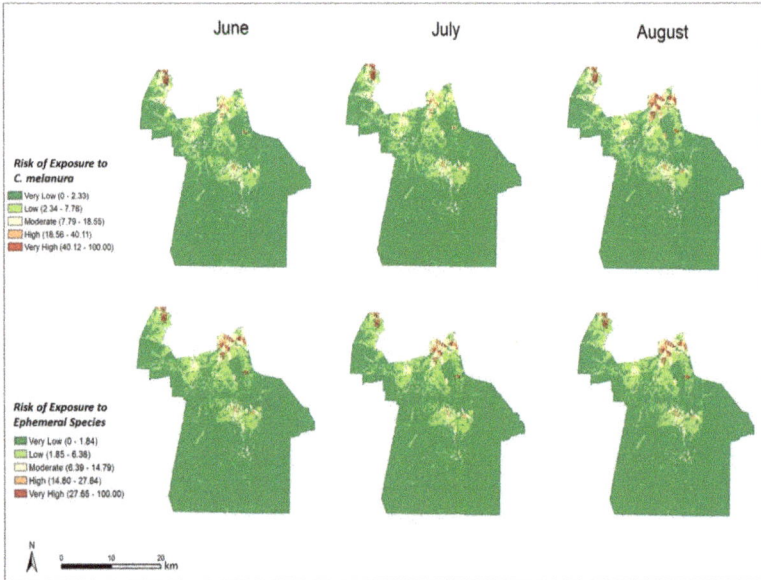

**Figure 8.** Monthly indices representing the risk of exposure to mosquito vectors for *C. melanura* and ephemeral species (values classified using natural breaks and fixed class breaks through time).

Prior to estimating the vulnerable population, it was expected that regions of high mosquito abundance would coincide with regions of high risk. However, many regions observed to have a high number of mosquitoes such as southern Chesapeake were predicted to have a lower relative risk of exposure. The area surrounding the Great Dismal Swamp, for instance, was observed, as expected, to exhibit high mosquito abundance. Nonetheless, this area exhibits a relatively low disease risk owing to low resident populations. Recreational activities in such swamplands should be considered a significant caveat, as recreational activity spaces (boaters, hikers, bird-watchers, and paddlers) are not characterized in this modeling. Such mosquito-abundant areas are mostly undeveloped and therefore capable of supporting mosquito populations. The climatic variables such as temperature and rainfall were calculated to have a greater effect on the mosquito trap counts in western Chesapeake compared to the rest of the city [4]. These western areas are covered in either wetlands or croplands. The immature stages of mosquitoes require water and therefore are often found in wetlands [30]. The runoff from croplands can also support mosquito presence [31]. Despite these favorable conditions, many of these rural areas have a low human population density and exhibit less total population at risk of disease exposure. Hence, these results highlight that although mosquitoes can transmit disease to humans, without exposure to the pathogen, the likelihood of disease transmission is decreased.

In order to compare the variation in risk from June through August, the differences in the monthly risk indices were calculated. Each month's risk values were subtracted from the following month's values to calculate the difference over each month. The results of this calculation are shown in Figure 9 It is clear that the most significant changes in risk occurred in northern Chesapeake. The remaining portion of the city did not show a significant change in risk over the three-month period. From June through July, the risk of exposure from *C. melanura* increased drastically in northern Chesapeake. Conversely, the risk values for the ephemeral species decreased over this time period. Modest and opposite trends are evident from July through August. For *C. melanura*, risk decreased in northern Chesapeake, while the risk values for the ephemeral species increased. Overall, there was little variation in risk between June and August. *C. melanura* showed minimal change, while the ephemeral species showed almost no change in risk values.

Validation of trends in abundance and population exposure is a challenging proposition in the instance of rare disease and a complex mosaic landscape such as Chesapeake. No human cases of West Nile Virus or EEE were reported in Chesapeake during the season studied. However, The Virginia Department of Public Health later reported (2004) [32] that 20 pools of mosquitoes were found to be positive for *C. melanura* infected with EEE, while 10 pools were positive for West Nile. Additional surveillance and nuisance abatement request data were provided by the City of Chesapeake and offer limited affirmation of the risk mapped. While not validated, some studies have found that animal cases can be accurate indicators of human disease prevalence [33]. Figure 10 depicts a composite of available information, including total vector species trap counts over the season, dead bird reports (a majority of crows), and cases of Eastern Equine Encephalitis (EEE) among horse farms. Although an exhaustive map of horse farms was not available, EEE incidents are coincident with the highest vector abundance data. Further, the abatement service requests and pattern of dead birds collected (initially many of these were tested for WNV), also spatially conforms to the dasymetric population map and human exposure risk. Figure 9 also illustrates the dilemma of solely relying upon discovery of surveillance tools such as dead birds or resident reported abatement requests, as these patterns reflect the abundance of humans and not necessarily the abundance of mosquito vector species.

It is difficult to validate such risk models due to the low number of disease cases and underreporting. Disease surveillance data do not record any actual human cases of WNV or EEE occurring in Chesapeake in 2003. However, in 2003, 20 pools of mosquitoes were found to test positive for *C. melanura* infected with EEE, while 10 pools were infected with *C. melanura* positive for WNV [32]. These corroborating data, combined with the usual under-reported and underdiagnosed nature of WNV, suggest that *C. melanura* mosquitoes were an emerging health threat to Chesapeake in 2003. In addition, bird, equine, and sentinel flock cases of WNV and EEE were reported for this year. These cases may not provide a validation of the model since the model estimates risk to humans, rather than animal hosts, but they do lend some credence and affirmation to the risk mapping approach. A plausible spatial relationship can be seen between EEE cases and *C. melanura* abundance. In 2003, the majority of EEE cases occurred in western Chesapeake, surrounding the Dismal Swamp, and our abundance mapping of the EEE vector *C. melanura* was also predicted to be high in western Chesapeake from June through August. The high predicted abundance of mosquito vectors, observations in traps, and resident abatement requests also lend credence to inference of higher risk exposure.

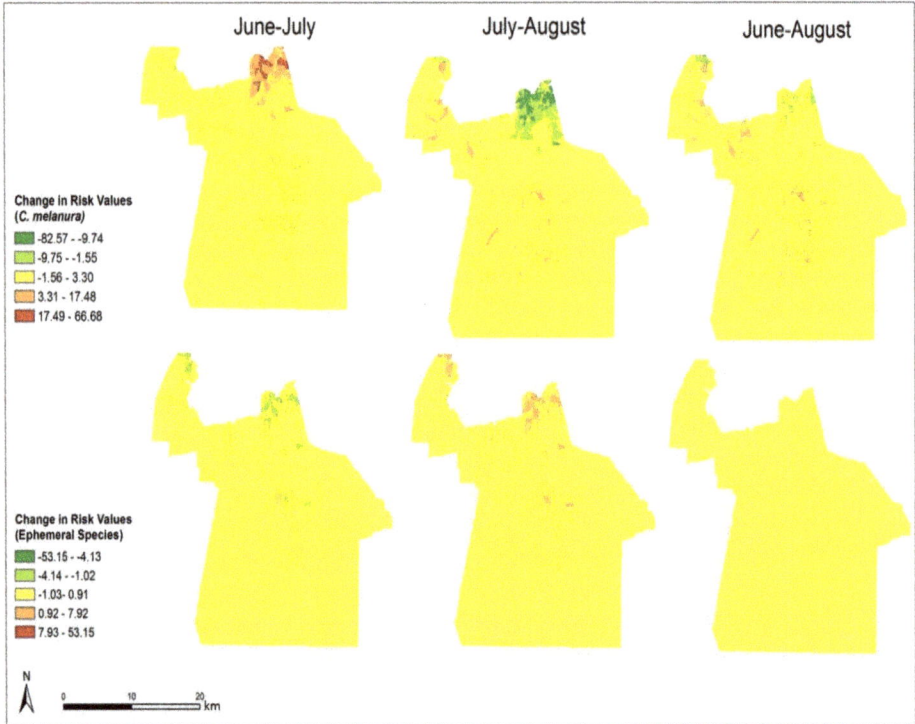

**Figure 9.** Change in monthly exposure risk values through the summer derived by calculating the difference in the risk indices shown in Figure 8.

Dasymetric mapping is shown in this investigation to be effective at representing population data as compared to mapping by areal choropleths such as Census block groups for population vulnerability and exposure assessment. The choroplethic map of vulnerability in Census units (Figure 9) gives the impression that the population is homogenously distributed within each block group, yet proportions of each block group are extensively uninhabited. The dasymetric map on the other hand (Figure 8), shows vulnerability at a finer scale and with many localized patterns and concentrations among urban, suburban, and rural settlements. By using pixels as the areal units rather than block groups, the raster shows a continuous and variable surface of vulnerability. By integrating the "nighttime" population values from the Census data with the point-based nodes of vulnerability to produce the vulnerability index, the resulting population risk map is likely to have a higher accuracy than results generated from considering the Census data alone.

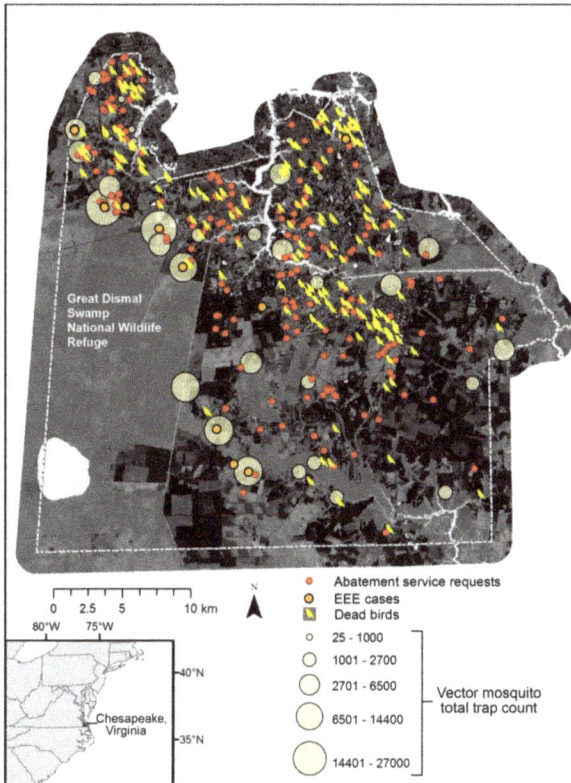

**Figure 10.** Seasonal mosquito vector trap counts, enzootic disease reports (dead birds and veterinary surveillance of positive EEE horses), and public abatement service requests within Chesapeake, over Landsat TM tasseled cap wetness index image for 29 July 2002.

This study affirms the role of demographic data and spatial analysis for mapping and predicting the risk of mosquito vector exposure. Due to the many factors affecting disease transmission, vector abundance and exposure may not be positively correlated with general areal risk of exposure. Although exposure to mosquitoes may be high in some regions across Chesapeake, disease transmission may not necessarily be high due to the low population density in these regions. This study supports the importance of incorporating human data in addition to climate data when predicting risk and may be used as a guide to strategically allocate surveillance and control activities over space. Ancillary enzootic data from dead birds and EEE cases as well as nuisance abatement requests support the inference of geographic gradients of vector abundance and possible exposure risk. Future studies may consider incorporating other human factors such as behavior, disease resistance, and socioeconomic data. According to Daily and Ehrlich [28], "human demographic factors are key variables in epidemiology, influencing the rate at which a population is invaded by new parasites, their chances of becoming established, the rate of their spread, the evolution of their virulence, and the capacity of human social structures (and other cultural traits) to coevolve in defense". The lack of fine-scale human population data was considered a limiting factor in the initial development of dasymetric maps, yet future studies should more carefully assess the role of scale in both vector abundance as well as susceptible populations.

Further considerations of disease transmission factors could improve similar research. For instance, it is possible that scale issues of human settlements and activity species and the vectors could be individually better understood and that fine-scale data (of either factor) may not add additional value to surveillance or control. Other factors, which influence vector exposure and disease transmission, may also be incorporated into future studies. Deforestation, for instance, can lead to environmental changes conducive to mosquito survival [30]. Vector competence, for example, could also be used to help predict the rate of transmission, and studies have begun to explore the relationship between West Nile virus, climate change [34], and specific vector competence of *Culiseta incidens* and *Culex thriambus* [35] and the variation in vector competence of *Culex pipens* across California [36]. Further, climate change, particularly extreme heat, rainfall, and rising sea levels, is poised to alter the habitats, suitability, and potential human settlement patterns in low-lying areas such as Chesapeake.

The geospatial methods used in this study may be beneficial to other cities afflicted by vector-borne diseases and extreme weather events. In rich and poor nations alike, cities are afflicted by a lack of control over disease reservoirs and vectors [28]. Predicting where mosquito-borne diseases pose the greatest risk to human health could improve spatial precision of mosquito control efforts and thereby reduce costs and impacts of broader insecticide use. GIS has been demonstrated as highly useful in emergency operational assessment of control methods, such as estimating areal needs for aerial or ground-based spraying, including distance and drive-times [37].

## 5. Conclusions

As vector-borne diseases continue to persist, many researchers and healthcare officials are concerned with estimating and mapping disease risk. Many spatial risk assessments rely solely on vector presence to predict disease risk. Using only vector presence to estimate risk can be constraining because mosquitoes are influenced by many interacting factors, particularly climatic variables [33]. By incorporating human vulnerability into a spatial model of disease risk, geographic variations in potential disease transmission can provide health surveillance and mosquito control officials greater information compared to monitoring vectors and environmental factors alone. By identifying high-risk areas in advance, health officials may also reduce disease transmission rates. Emergency managers can target where to implement early-warning systems and risk management procedures in high mosquito abundance habitats in proximity to vulnerable populations following hurricane landfall and flooding (e.g., aerial spraying, larviciding, and emergency shelter site selection). Knowing human exposure and where infectious diseases are likely to emerge could also aid health practitioners in diagnosing and treating patients promptly. With improved data and spatial analytic tools public health managers may apply such spatial risk maps and models to their jurisdictions to reduce the chances of disease.

**Acknowledgments:** The authors gratefully acknowledge the cooperation of the City of Chesapeake Mosquito Control Commission and research funding from the Virginia Space Grant Consortium initiative led by George Mason University. C. John Neely inspired early concepts of spatial epidemiology applicable to the project. A. Scott Bellows assimilated field data and GIS base maps in support of the work. Lab space and software were provided to the Laboratory for Remote Sensing and Environmental Analysis (LaRSEA) by Old Dominion University and the Center for Geographic Information Science at East Carolina University. Three anonymous reviewers' comments greatly improved the presentation and inferences in the submitted manuscript.

**Author Contributions:** Haley M. Cleckner acquired and analyzed the data herein as part of a Master's Thesis in Geography at East Carolina University. Thomas R. Allen supervised this thesis and led subsequent manuscript editing, compilation, and the additional maps and analyses during the manuscript review.

**Conflicts of Interest:** The authors declare no conflict of interest.

## References

1.  Kitron, U. Risk maps: Transmission and burden of vector-borne diseases. *Parasitol. Today* **2000**, *16*, 324–325. [CrossRef]
2.  Mather, F.J.; White, L.E.; Langlois, E.C.; Shorter, C.F.; Swalm, C.M.; Shaffer, J.G.; Hartley, W.R. Statistical methods for linking health, exposure, and hazards. *Environ. Health Perspect.* **2004**, *112*, 1440–1445. [CrossRef]

3. Sutherest, R.W. Global change and human vulnerability to vector-borne diseases. *Clin. Microbiol. Rev.* **2004**, *17*, 136–173. [CrossRef]

4. Cleckner, H.L.; Allen, T.A.; Bellows, A.S. Remote sensing and modeling of mosquito bundance and habitats in coastal Virginia, USA. *Remote Sens.* **2011**, *3*, 2663–2681. [CrossRef]

5. Bellows, A.S. Modeling Habitat and Environmental Factors Affecting Mosquito Abundance in Chesapeake, Virginia. Ph.D. Dissertation, Old Dominion University, Norfolk, VA, USA, 2007.

6. Ozdenerol, E.; Taff, G.N.; Akkus, C. Exploring the spatio-temporal dynamics of reservoir hosts, vectors, and human hosts of West Nile Virus: A review of the recent literature. *Int. J. Environ. Res. Public Health* **2013**, *10*, 5399–5432. [CrossRef]

7. Liu, H.; Weng, Q.; Gaines, D. Spatio-temporal analysis of the relationship between WNV dissemination and environmental variables in Indianapolis, USA. *Int. J. Health Geogr.* **2008**, *7*, 66. [CrossRef]

8. McKnight, K.P.; Messina, J.P.; Shortridge, A.M.; Burns, M.D.; Pigozzi, B.W. Using volunteered geographic information to assess the spatial distribution of West Nile Virus in Detroit, Michigan. *Int. J. Appl. Geospatial Res.* **2011**, *2*, 72–85. [CrossRef]

9. Behrens, J.J.; Moore, C.G. Using geographic information systems to analyze the distribution and abundance of *Aedes aegypti* in Africa: The potential role of human travel in determining the intensity of mosquito infestation. *Int. J. Appl. Geospatial Res.* **2013**, *4*, 9–38. [CrossRef]

10. Ceccato, P.; Connor, S.J.; Jeanne, I.; Thomson, M.C. Application of geographical information systems and remote sensing technologies for assessing and monitoring malaria risk. *Parassitologia* **2005**, *47*, 81–96.

11. Beck, L.R.; Rodriguez, M.H.; Dister, S.W.; Rodriguez, A.D.; Washino, R.K.; Roberts, D.R.; Spanner, M.A. Assessment of a remote sensing-based model for predicting malaria transmission risk in villages of Chiapas, Mexico. *Am. J. Trop. Med. Hyg.* **1994**, *56*, 96–106.

12. Wilson, M.L. Emerging and vector-borne diseases: Role of high spatial resolution and hyperspectral images in analyses and forecasts. *J. Geogr. Syst.* **2002**, *4*, 31–42. [CrossRef]

13. Sleeter, R. Dasymetric Mapping Techniques for the San Francisco Bay region, California. In *Proceedings of the Annual Conference on Urban and Regional Information Systems Association*; Reno, NV, USA, 7–10 November 2004. Available online: http://geography.wr.usgs.gov/science/dasymetric/data/URISA_Journal.pdf (accessed on 13 November 2013).

14. Maantay, J.A.; Maroko, A.R.; Herrmann, C. Mapping population distribution in the urban environment: The cadastral-based expert dasymetric system (CEDS). *Cartogr. Geogr. Inf. Sci.* **2007**, *34*, 77–102.

15. Eicher, C.L.; Brewer, C.A. Dasymetric mapping and areal interpolation: Implementation and evaluation. *Cartogr. Geogr. Inf. Sci.* **2001**, *28*, 125–138. [CrossRef]

16. Mennis, J. Generating surface models of population using dasymetric mapping. *Prof. Geogr.* **2003**, *55*, 31–42.

17. Mennis, J.; Hultgren, T. Intelligent dasymetric mapping and its application to areal interpolation. *Cartogr. Geogr. Inf. Sci.* **2006**, *33*, 179–194. [CrossRef]

18. Gregory, I.N. An Evaluation of the Accuracy of the Areal Interpolation of Data for the Analysis of Long-Term Change in England and Wales. In Proceedings of the 5th International Conference on GeoComputation, University of Greenwich, London, UK, 23–25 August 2000.

19. Smith, C.M.; Graffeo, C.S. Regional impact of Hurricane Isabel on emergency departments in coastal southeastern Virginia. *Vet. Res. Commun.* **2005**, *12*, 1201–1205.

20. Chesapeake Mosquito Control Commission. 2010. Available online: http://www.chesapeake.va.us/services/depart/mosquito/index.shtml (accessed on 27 March 2012).

21. Zielinksi-Gutierrez, E. West Nile Virus & Other Mosquito-Borne Infections. In *The Health Care of Homeless Persons, Part 1, Mosquito-Borne Infections*; O'Connell, J.J., Ed.; Boston Health Care for Homeless Program: Boston, MA, USA, 2004; pp. 181–186.

22. Shea, K.M. Global climate change and children's health. *Pediatrics* **2007**, *120*, e1359–e1367. [CrossRef]

23. Lindsey, N.P.; Staples, J.E.; Lehman, J.A.; Fischer, M. Surveillance for human West Nile virus disease—United States, 1999–2008. *Morb. Mortal. Wkly. Rep.* **2010**, *59*, 1–17.

24. US Geological Survey. Dasymetric Mapping Tool—ArcGIS 10. 2012. Available online: http://geography.wr.usgs.gov/science/dasymetric/data.htm (accessed on 29 March 2013).

25. Tapp, A.F. Areal interpolation and dasymetric mapping methods using local ancillary data sources. *Cartogr. Geogr. Inf. Sci.* **2010**, *37*, 215–228. [CrossRef]

26. Langford, M. Rapid facilitation of dasymetric-based population interpolation by means of raster pixel maps. *Comput. Environ. Urban Syst.* **2007**, *31*, 19–32. [CrossRef]

27. Reibel, M.; Bufalino, M.E. Street weighted interpolation techniques for demographic count estimation in incompatible zone systems. *Environ. Plan.* **2005**, *37*, 127–131. [CrossRef]

28. Daily, G.C.; Ehrlich, P.R. Global change and human susceptibility to disease. *Annu. Rev. Energy Environ.* **1996**, *21*, 125–144. [CrossRef]

29. Crans, W.G. A classification system for mosquito life cycle: Life cycle types for mosquitoes of the northeastern United States. *J. Vector Ecol.* **2004**, *29*, 1–10.

30. Dale, P.E.R.; Knight, J.M. Wetlands and mosquitoes: A review. *Wetl. Ecol. Manag.* **2008**, *16*, 255–276. [CrossRef]

31. Ward, M.P.; Wittich, C.A.; Fosgate, G.; Srinivasan, R. Environmental risk factors for equine West Nile Virus disease cases in Texas. *Vet. Res. Commun.* **2009**, *33*, 461–471. [CrossRef]

32. Virginia Department of Health. Arbovirus Data 2003. Available online: http://www.vdh.virginia.gov/epidemiology/DEE/Vectorborne/arboviral/documents/testresults/2003.html (accessed on 12 March 2004).

33. Eidson, M.; Komar, N.; Sorhage, F.; Nelson, R.; Talbot, T.; Mostashari, F.; Mclean, R. West Nile Virus Avian Surveillance Group. Crow deaths as a sentinel surveillance system for West Nile Virus in the northeastern United States. *Emerg. Infect. Dis.* **1999**, *7*, 615–620.

34. Gage, K.L.; Burkot, T.R.; Eisen, R.R.; Hayes, E.B. Climate and vectorborne diseases. *Am. J. Prev. Med.* **2008**, *35*, 436–450. [CrossRef]

35. Reisen, W.K.; Fang, Y.; Martinez, V.M. Vector competence of *Culiseta incidens* and *Culex thriambus* for West Nile virus. *J. Am. Mosq. Control Assoc.* **2006**, *22*, 662–665. [CrossRef]

36. Vaidyanathan, R.; Scott, T.W. Geographic variation in vector competence for West Nile Virus in the *Culex pipiens* (Diptera: Culicidae) Complex in California. *Vector Borne Zoonotic Dis.* **2007**, *7*, 193–198.

37. Mak, S.; Buller, M.; Furnell, A.; MacDougall, L.; Henry, B. Use of geographic information systems to assess the feasibility ground- and aerial-based adulticiding for West Nile Virus control in British Columbia, Canada. *J. Am. Mosq. Control Assoc.* **2007**, *23*, 396–404. [CrossRef]

International Journal of
**_Geo-Information_**

MDPI

_Article_

# Modeling Properties of Influenza-Like Illness Peak Events with Crossing Theory

**Ying Wang \*, Peter R. Waylen and Liang Mao**

Department of Geography, University of Florida, Gainesville, FL 32611, USA; prwaylen@ufl.edu (P.R.W.); liangmao@ufl.edu (L.M.)

\* Author to whom correspondence should be addressed; happywy17@gmail.com; Tel.: +1-352-328-9314; Fax: +1-352-392-8855.

Received: 6 February 2014; in revised form: 17 April 2014; Accepted: 15 May 2014; Published: 26 May 2014

**Abstract:** The concept of "peak event" has been used extensively to characterize influenza epidemics. Current definitions, however, could not maximize the amount of pertinent information about the probabilities of peak events that could be extracted from the generally limited available records. This study proposes a new method of defining peak events and statistically characterizing their properties, including: annual event density, their timing, the magnitude over prescribed thresholds and duration. These properties of peak events are analyzed in five counties of Florida using records from the Influenza-Like Illness Surveillance Network (ILINet). Further, the identified properties of peak events are compared between counties to reveal the geographic variability of influenza peak activity. The results of this study illustrate the proposed methodology's capacity to aid public health professionals in supporting influenza surveillance and implementing timely effective intervention strategies.

**Keywords:** influenza-like illness (ILI); peak event; properties of peak events; crossing theory; generalized Pareto distribution (GPD)

## 1. Introduction

Influenza, widely known as the flu, is a highly contagious and acute respiratory disease. For a typical season, influenza activity often peaks in one or more weeks when the observed number of cases is noticeably higher than other weeks. These peak weeks incorporate a high proportion of influenza cases during the entire epidemic and are referred to variously as "peak events" or "peak weeks". The properties of influenza peak events, such as timing, magnitude and duration, offer critical implications in disease surveillance, dynamics and control policies [1,2]. For example, the potential magnitude of peak events provides crucial information about the scale of an outbreak and suggests the amount of health resources in response to the disease [3]. The frequency, timing and duration of peak events offer a statistical basis for health insurance companies and long-term public health planning [4], e.g., the risk of more than one such event in a year and the length of time that each may persist. Because of their significance in epidemiology and planning, the study of such events has received increasing attention in recent years [5–7].

Although the concept of peak events is widely used in influenza-related studies, their workable definition remains under-studied. The traditional definition of "peak event" is an annual maximum. Smith [8] defined the "peak event" as the week with the greatest number of weekly influenza cases during an influenza season. This widely used definition is straightforward; however, it has the potential to exclude other events with epidemiological importance that may have occurred in a year and to include the annual maximum in the sample, which really does not constitute an event of epidemiological importance. Valuable information concerning epidemics, such as spatial variations, dynamics and periodicity, cannot be derived from the generally short historic records by this definition. Sakai _et al._ [9] slightly modified this approach by identifying the annual maximum of the smoothed data. They incorporated information from weeks before and after the peak event by smoothing data with a five-week unweighted moving average of weekly reported cases. The risk of this approach is that it obscures the important characteristics of the greatest number of influenza-like illness (ILI) cases in a week and may induce apparent periodic behavior in what could, in reality, be a random process. More importantly, these existing definitions of peak events offer little consideration to spatial heterogeneity, for instance, differences in demographics, and thus, the peak events alone are not comparable between geographic areas.

The limitations of existing definitions call for a more sophisticated approach employing spatially differentiated data that characterize weekly influenza activity and maximizing the pertinent information that may be extracted from the limited available records. As the first step in an on-going study that seeks to establish associations between ILI peak events and potential factors, this study aims to define ILI peak events and statistically characterize their properties: annual event density, their timing, magnitude over prescribed thresholds and duration.

## 2. Materials and Methods

### 2.1. Study Area and Data

Florida experienced an average of 2900 estimated deaths per year from influenza over the past decade [10]. In 2004, for example, influenza and pneumonia together were the eighth leading cause of death reported by the Florida Department of Health (DOH) [11]. The Florida DOH estimates that an influenza pandemic could infect up to 10 million [12]. Several factors encourage the rapid transmission of influenza in the state: its developed tourism industry, high inter- and intra-national immigration and high proportion of aged population and their living styles. Despite its sub-tropical location and peninsular nature, much of Florida experiences periods of relatively low temperatures and low humidity in winter. Nearly one third of the population, including a large proportion of immigrants, resides in urban or suburban areas of three southeastern counties. Several interstates and 13 international airports, including Orlando and Miami, bring in tens of thousands of tourists each year (38 million used air travel in 2000 alone).

Data employed in this study are obtained from the Influenza-Like Illness Surveillance Network (ILINet), which conducts surveillance of weekly ILI outpatient cases [13]. The ILI case is defined as any combination of fever ($\geq$38 °C) and cough or sore throat, which may embed influenza along with other conditions, such as colds and pneumonia. ILI activity collected through outpatient illness surveillance provides important epidemiologic information for monitoring influenza activity and supports influenza surveillance [14,15]. Weekly reports from ILINet are available dating back to 2001 in some counties; however, most counties did not have the necessary continuity of reporting at the earliest stages. As representatives of environmental, demographic and social conditions in Florida, five counties are selected for extensive study (Figure 1): the lengthy (2001–2012) historic data from dominantly urban Broward (Fort Lauderdale), Duval (Jacksonville) and Miami-Dade (Miami) counties, as well as shorter (2006–2012) records for Orange (Orlando) and Hillborough (Tampa) counties.

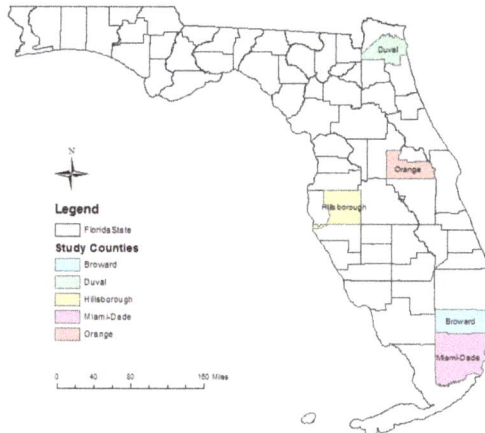

**Figure 1.** The five selected counties in Florida.

### *2.2. Methodology*

Crossing theory states that the number of crossings of a threshold by a Gaussian process become Poisson distributed the further the threshold lies from the mean of the process [16,17]. Results have also been extended to non-Gaussian processes [18] and can be applied to estimate the characteristics of ILI events. The magnitudes of events over the threshold and their durations can be approximated by an exponential-like distribution, such as the generalized Pareto distribution (GPD) [19,20], which can represent such data exhibiting both greater and lesser skew than the exponential itself. In combination with the Poisson assumption, it implies that the annual maximum or "peak week" [8] follows a generalized extreme value (GEV) distribution [19], the properties of which can be estimated from this approach, if desired. In general, the criteria for adopting a specific distribution are; the goodness-of-fit, a strong theoretical basis and the relative ease of computation and interpretation. Although few influenza-related studies have focused on the statistical properties of peak events, the proposed approach has been extensively used in studies modeling extremes in various fields, including floods, stock market returns and daily maximum temperatures [19–21].

The statistical properties of influenza events may all be defined by the prescription of a specific threshold. To facilitate spatial comparisons, this study defines the threshold in terms of common percentiles of historic weekly ILI cases (*i.e.*, defined in the frequency domain); although for epidemiological or planning purposes, the threshold could be defined in the magnitude domain, in terms of the total number of ILI cases of particular interest. Results extracted above the 80th percentile level (0.20 probability of occurring in any week) are extrapolated to the more rarely experienced levels equivalent to the 90th (0.10 probability) and 95th percentiles (0.05 probability) and compared to the small available sample of historic events that exceed these higher levels. In this way, a larger proportion of the limited available historic records can be utilized to characterize properties of ILI events above levels commonly witnessed.

Definitions of events and the flu year are established first. Then, the variables of interest are identified: (1) annual event density (events per year); (2) the timing ($t$) of each event; (3) the magnitude of the peak event ($\tau = x - q_0$) during an event when the observed number of weekly cases ($x$) surpasses the thresholds ($q_0$); and (4) the duration of events.

#### 2.2.1. Definitions of Events

Although thresholds ($q_0$) are considered in terms of percentiles of historic weekly ILI cases throughout the study, two definitions of the magnitudes of events are investigated. The first includes all weeks with an absolute weekly ILI count greater than the prespecified threshold. In Figure 2a, for example, all the eight ILI observations greater than the defined threshold would be considered (one or more observations of the magnitude per event). The second definition considers only the week with the highest ILI count above the threshold within the period between successive up- and down-crossings of the threshold level (one observation of the magnitude per event)—a local maximum. In Figure 2b, only the three observations of local maxima would be considered. The properties of ILI peak events are examined above a commonly witnessed 80th percentile level, although this approach is applicable for any other reasonably high thresholds.

2.2.2. Definition of Flu Year

Since all ILI cases are likely to be recorded during the winter season, the use of a calendar year definition would arbitrarily bisect a flu season, producing a misleading aggregation of events from two halves of consecutive and distinct seasons. To determine when flu is least likely to occur in the historic record (an appropriate point to start and finish a "flu year"), the occurrence of ILI events throughout Florida is analyzed using the mean ($\mu$) and defined fractions of the standard deviations ($0.25\sigma$, $0.32\sigma$ and $0.5\sigma$) of all weekly ILI cases. In terms of the mean weekly occurrence in all counties in Florida (Figure 3), Week 29 (starting 15 July) is the week in which ILI cases are least likely by this measure and is thus defined as the beginning of the "flu year", noted hereafter as "Week 1".

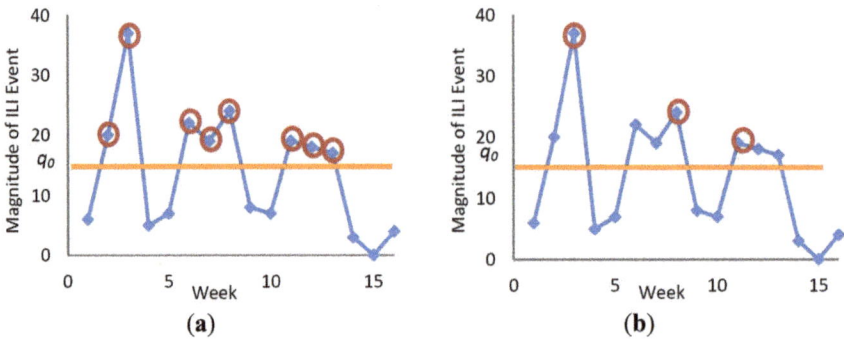

**Figure 2.** Definitions of magnitudes above the threshold. (a) Definition 1. (b) Definition 2. The blue curve represents weekly influenza-like illness (ILI) cases; the red circle represents the selected peak; and the orange line represents the threshold ($q_0$).

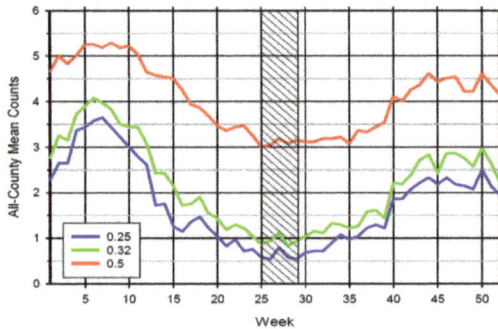

**Figure 3.** The mean weekly occurrence of ILI events in all counties based on the defined standard deviations ($0.25\sigma$, $0.32\sigma$ and $0.5\sigma$).

### 2.2.3. Annual Event Density

Annual event density is defined as the number of events per flu year. The probability mass function of the Poisson distribution is:

$$P(M) = e^{-\Lambda} \times \Lambda^{M} / M! \tag{1}$$

where $M$ is the number of events in a flu year and $\Lambda$ is estimated using the method of moments as the mean number of events per flu year:

$$\Lambda = K / N \tag{2}$$

$K$ is the total number of events in $N$ flu years with complete yearly data in the historic records.

### 2.2.4. Timing of Events within a Flu Year

Due to the strong seasonal nature of ILI cases, the Poisson distribution is modified to exhibit a time-dependent rate of event occurrence, $\lambda(t)$:

$$P(m(t) = n) = e^{-\lambda(t)} \times \lambda(t)^{n} / n! \tag{3}$$

where $P(m(t) = n)$ is the probability of having experienced $n$ events up to and including week $t$ and $\lambda(t)$ is the mean number of events expected up to that time. As influenza outbreaks generally occur in a particular season with some interannual variability, the timings of events are modeled by a Gaussian distribution and $\lambda(t)$, estimated as,

$$\lambda(t) = G(t: \mu, \sigma) \times \Lambda \tag{4}$$

where $G(t: \mu, \sigma)$ is a Gaussian distribution fitted to the observed timing of ILI events, with $\mu$ being the mean week of occurrences and $\sigma$ their standard deviation.

### 2.2.5. Event Magnitude

The distribution of event magnitudes is fit with a GPD [19,20]:

$$F(x \geq X) = 1 - \left(1 - k\frac{X}{\alpha}\right)^{1/k} \qquad k \neq 0 \tag{5}$$

where $X$ is the magnitude of the event over the predetermined threshold of interest. Completely characterized by a scale parameter, $\alpha$, and a shape parameter, $k$, the GPD is a generalization of both the exponential ($k = 0$) and Pareto distributions ($k < 0$), which provides greater flexibility in matching the heavier ($k < 0$) and thinner ($k > 0$) upper tails of the distribution. The parameters are estimated via the method of moments from the sample mean, $\hat{\mu}$, and variance $\hat{\sigma}^2$ as Equations (6) and (7):

$$\hat{\alpha} = \frac{1}{2}\hat{\mu}\left(\frac{\hat{\mu}^2}{\hat{\sigma}^2} + 1\right) \tag{6}$$

$$\hat{k} = \frac{1}{2}\left(\frac{\hat{\mu}^2}{\hat{\sigma}^2} - 1\right) \tag{7}$$

### 2.2.6. Event Duration

The duration of events is also represented by the GPD [20]:

$$F(d \geq D) = 1 - \left(1 - k\frac{D}{\alpha'}\right)^{1/k'} \qquad k' \neq 0$$

(8)

where $D$ is the duration of the event, representing total weeks related to a peak event. Similarly, $\alpha'$ and $k'$ are scale and shape parameters, which are estimated via the method of moments by Equations (6) and (7), using appropriate means and variances.

### 2.2.7. Independence of Events

A period of two consecutive weeks in which the weekly ILI cases fall below the threshold level is employed as the criterion to separate independent peak events. Occasions when weekly cases dropped marginally below the threshold level only to exceed it again in the next week are probably the result of the same event. Parallel considerations in the definition of flood and heat wave events can be found in Rosbjerg *et al.* [19] and Keellings and Waylen [20]. "Events" failing to meet this independence criterion are combined and included in subsequent analysis as if they constitute a single event.

### 2.2.8. Extrapolation of Properties to Higher Thresholds

An ability to derive the stochastic properties of ILI events above higher, less commonly experienced levels, from the larger sample sizes available at the lower, less epidemiologically-important thresholds, would be useful. Any portion of a GPD is itself GPD-distributed; thus, the process of raising the threshold is effectively "cutting off" the lower end of GPD and leaving only that portion that rises above the new level. Estimation of the mean ($\mu_1$) and variance ($\sigma_1$) of the remaining portion of the distribution yields revised estimates of $\alpha_1$ and $k_1$ (Equations (6) and (7)). The proportion of events expected to exceed the higher threshold represented by the area under the original distribution of magnitudes, which lies beyond the new level, yields the parameter, $\Lambda_1$, of the Poisson distribution. The probability of the annual number of crossings above the corresponding threshold, or events, can then be estimated. Assuming that the timing and magnitudes of ILI events are independent, the distribution of the timing of censored events should remain unchanged.

## 3. Results

### 3.1. Annual Numbers of Events, Their Timings and Durations

The one-sample Kolmogorov–Smirnov test is applied to examine the goodness-of-fit of all models. All results show no significant differences between fitted and observed distributions at the 0.05 level of significance in any of the five study counties. The assumption of normality of the timing of peak events is reasonable graphically and statistically. Data from the longer-term record of Duval County are examined as an example.

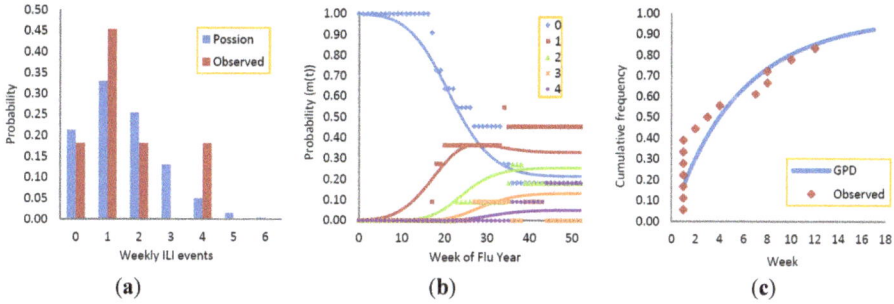

**Figure 4.** Annual event density, timing and duration at the 80th percentile level in Duval County. (a) Annual event density. (b) Timing plots. The probabilities of zero, one, two, three and four events having occurred, in Duval County, up to any week during the flu year. (c) The duration of peak events. GPD, generalized Pareto distribution.

Historic ILI events are most likely to occur during the late fall and early spring (Weeks 20 to 32 of the flu year), coincident with conducive meteorological conditions and the early weeks of the spring semester of school. The Poisson probability function is fitted to the numbers of events exceeding the 80th percentile level annually (Figure 4a), and the non-homogeneous Poisson function is applied in order to estimate probabilities of experiencing zero, one, two, three and four events up to any week of the flu year (Figure 4b). This reproduces well the observed patterns of occurrence during late fall and early spring. Taking Week 26 in the flu year (the second week of January) as an example, the probability of having experienced no peak events up to that time ($m(t) = 0$) is 0.43; the probability of exactly one peak event is 0.36, *etc.* The probability of an ILI event occurring in a particular week, $t$, can be computed as $\{[P(m(t-1)=0)] - [P(m(t)=0)]\}$. The generalized Pareto distribution provides a reasonable approximation to the distribution of the likely durations of events at the 80th percentile level (Figure 4c).

### 3.2. Magnitudes of Events and Comparisons of Definitions

Figure 5 illustrates the GPD's ability to model the observed cumulative distribution function (CDF) based on either of the two definitions of magnitudes. The location parameter, $\alpha$, conveys information about the relative magnitude of the cases above the threshold in each county and could be standardized to some base, such as estimated total county population, while the values of $k$ can be compared directly between counties. As expected, the sample sizes derived using Definition 1 (Figure 5b) are much larger than that using Definition 2 (Figure 5c). Negative values of the shape parameter, $k$, imply that, at this relatively low threshold, the upper tail is particularly "heavy" (larger outliers in the right-hand tail of the distribution) in comparison to the bulk of observations.

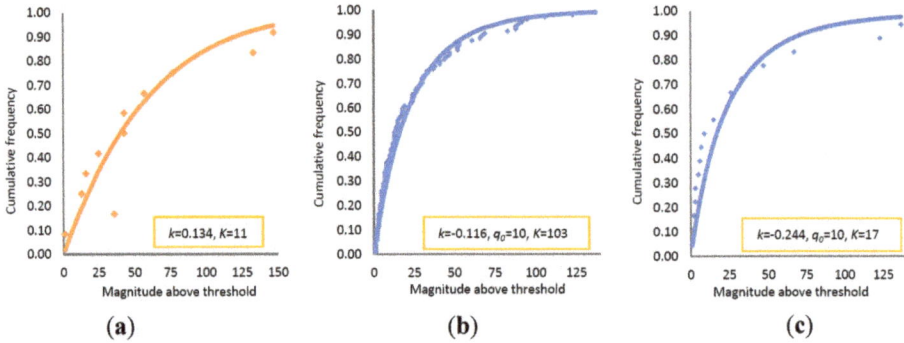

**Figure 5.** Cumulative distribution functions (CDFs) of observed magnitudes over the 80th percentile level based on three definitions in Duval County. (**a**) Traditional definition. (**b**) Definition 1. (**c**) Definition 2. Note: $q_0$ represents the threshold ILI cases for each level, and $K$ represents the total number of peak events in all flu years.

### 3.3. Extrapolation of Weekly ILI Cases to Higher Levels

The parameters of the above distributions are simply estimated by the application of moment estimators to data extracted at the 80th percentile level. The proposed methodology has the capacity to yield distributions of events exceeding higher, more rarely experienced levels (for example, here, the 90th and 95th percentile levels) from the larger sample sizes of observations gathered at the lower truncation level (the 80th percentile level) (Figure 6).

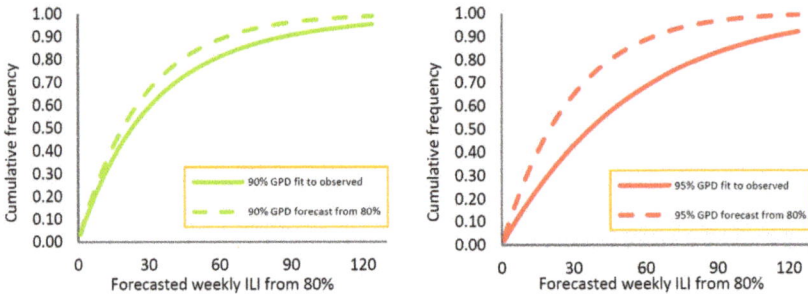

**Figure 6.** Forecasted magnitudes over the 90th and 95th percentile levels from the 80th percentile level in Duval County.

When the critical threshold for Duval county is raised from 10 cases (the 80th percentile level) to 23 (the 90th percentile level), the observed mean annual number of events drops from 1.55 per flu year to 0.91 (Table 1). The GPD fitted to the local maxima of events indicates that 58.5% of the original 17 events should exceed the increased threshold, yielding an anticipated mean annual number of events of 0.90. If the critical threshold is raised to 41 cases (the 95th percentile level), and the observed mean annual number of events drops to 0.45, while 30.8% of the original 17 events (0.48 events per year) are anticipated to exceed this increased threshold. The use of higher thresholds levels leads to the exclusion of the bulk of the lower magnitude events, reducing the "heaviness" of the tail of the surviving events and increasing the values of $k$.

## 4. Discussions

Grounded in theory, this approach has the ability to describe important statistical properties of such events and provides the necessary degree of flexibility in the definition of ILI events, while permitting spatial comparisons and the handling of various planning scenarios. Once the suitable probability distributions are identified, the probability of the occurrence of ILI events and their properties can be obtained for further specified purposes.

### 4.1. Comparisons of Definitions

The traditional definition of peak events only captures an annual maximum (magnitude) in each flu year, but discards other important properties of annual event density and duration and runs the risk of including "peak" events of no epidemiological significance. The proposed approach possesses the benefit of only including ILI events that meet the level of practical interest, while incorporating a potentially larger sample size from the short records currently available. The traditional "peak week" definition applied to Duval County yields 11 observations, while the application of the 80th percentile threshold increases the available sample size upon which risk can be estimated to 103 using Definition 1 and 17 for Definition 2 (Figure 5). Once estimated, the parameters of the GPD provide the basis for the estimation of properties above successively higher, more rarely observed, levels of ILI incidence. The threshold of interest can be expressed either in terms of acceptable risk (frequency domain) for spatial comparisons, or case numbers (magnitude domain) for planning purposes.

**Table 1.** Summary of observed and expected parameters of ILI peak events upon raising the thresholds to the 90th and 95th percentile levels (Definition 2).

| Name | Years of Record | 80th Percentile | | | | 90th Percentile | | | | | 95th Percentile | | | | |
|---|---|---|---|---|---|---|---|---|---|---|---|---|---|---|---|
| | | Timings | | Mean Duration $\bar{d}$ | Mean Number of Events $A$ | Timings | | Mean Duration $\bar{d}$ | Mean Number of Events $A$ (O/E) | | Timings | | Mean Duration $\bar{d}$ | Mean Number of Events $A$ (O/E) | |
| | | Mean $\mu$ | Standard Deviation $\sigma$ | | | Mean $\mu$ | Standard Deviation $\sigma$ | | | | Mean $\mu$ | Standard Deviation $\sigma$ | | | |
| Broward | 11 | 24.91 | 10.76 | 5.36 | 2.00 | 24.33 | 10.40 | 3.39 | 1.64 | 1.45 (72.3%) | 21.75 | 12.24 | 2.33 | 1.09 | 0.83 (41.5%) |
| Duval | 11 | 25.06 | 8.33 | 6.35 | 1.55 | 28.50 | 6.88 | 5.30 | 0.91 | 0.90 (58.5%) | 25.60 | 7.83 | 6.20 | 0.45 | 0.48 (30.8%) |
| Miami-Dade | 11 | 26.21 | 12.34 | 6.05 | 1.73 | 29.82 | 10.07 | 5.27 | 1.00 | 0.88 (50.8%) | 26.33 | 9.46 | 4.83 | 0.55 | 0.56 (32.2%) |
| Hillsborough | 6 | 22.63 | 10.57 | 7.63 | 1.33 | 23.56 | 12.65 | 3.22 | 1.50 | 1.02 (76.7%) | 28.50 | 8.81 | 5.25 | 0.67 | 0.59 (49.8%) |
| Orange | 6 | 29.90 | 11.79 | 6.50 | 1.67 | 31.40 | 14.79 | 6.00 | 0.83 | 0.85 (50.9%) | 23.00 | 10.44 | 4.67 | 0.50 | 0.45 (26.7%) |

**Table 2.** Observed and expected parameters of magnitudes upon raising the thresholds to the 90th and 95th percentile levels (Definition 1). Note: $q_0$, the threshold; $\alpha$, the scale parameter of GPD; $k$, the shape parameter of GPD; $K$, the total number of peak events in all flu years.

| Name | Years of Record | 80th Percentile | | | | 90th Percentile | | | | | | | 95th Percentile | | | | | | |
|---|---|---|---|---|---|---|---|---|---|---|---|---|---|---|---|---|---|---|---|
| | | $q_0$ | $\alpha$ | $k$ | $K$ | $q_0$ | $\alpha$ (O/E) | | $k$ (O/E) | | $K$ (O/E) | | $q_0$ | $\alpha$ (O/E) | | $k$ (O/E) | | $K$ (O/E) | |
| Broward | 11 | 2 | 4.27 | −0.30 | 93 | 4 | 5.83 | 5.36 | −0.29 | −0.20 | 54 | 60 | 8 | 6.85 | 7.19 | −0.29 | −0.19 | 25 | 41 |
| Duval | 11 | 10 | 22.52 | −0.12 | 103 | 23 | 25.84 | 34.33 | 0.06 | 0.02 | 51 | 59 | 41 | 44.32 | 27.50 | 0.27 | 0.07 | 27 | 29 |
| Miami-Dade | 11 | 18 | 13.76 | −0.16 | 102 | 27 | 16.75 | 15.55 | −0.15 | −0.05 | 53 | 55 | 34 | 26.55 | 17.94 | 0.06 | −0.03 | 27 | 35 |
| Hillsborough | 6 | 24 | 20.09 | −0.10 | 56 | 35 | 22.93 | 30.67 | 0.08 | 0.00 | 28 | 33 | 53 | 37.85 | 24.78 | 0.22 | 0.04 | 13 | 15 |
| Orange | 6 | 69 | 73.93 | 0.34 | 59 | 114 | 51.86 | 73.36 | 0.64 | 0.42 | 29 | 30 | 155 | 49.45 | 31.25 | 0.68 | 0.40 | 14 | 14 |

**Table 3.** Observed and expected parameters of magnitudes upon raising the thresholds to the 90th and 95th percentile levels (Definition 2). Note: $q_0$, the threshold; $\alpha$, the scale parameter of GPD; $k$, the shape parameter of GPD; $K$, the total number of peak events in all flu years.

| Name | Years of Record | 80th Percentile | | | | 90th Percentile | | | | | | | 95th Percentile | | | | | | |
|---|---|---|---|---|---|---|---|---|---|---|---|---|---|---|---|---|---|---|---|
| | | $q_0$ | $\alpha$ | $k$ | $K$ | $q_0$ | $\alpha$ (O/E) | | $k$ (O/E) | | $K$ (O/E) | | $q_0$ | $\alpha$ (O/E) | | $k$ (O/E) | | $K$ (O/E) | |
| Broward | 11 | 2 | 5.82 | −0.35 | 22 | 4 | 5.75 | 7.85 | −0.37 | −0.23 | 18 | 16 | 8 | 5.67 | 9.50 | −0.40 | −0.21 | 12 | 9 |
| Duval | 11 | 10 | 22.69 | −0.24 | 17 | 23 | 31.58 | 27.59 | −0.16 | 0.00 | 10 | 10 | 41 | 55.13 | 29.82 | 0.09 | 0.07 | 5 | 5 |
| Miami-Dade | 11 | 18 | 12.16 | −0.26 | 19 | 27 | 14.76 | 16.42 | −0.23 | −0.08 | 11 | 10 | 34 | 21.99 | 18.25 | −0.13 | −0.05 | 6 | 6 |
| Hillsborough | 6 | 24 | 41.54 | −0.00 | 8 | 35 | 22.88 | 41.43 | 0.05 | 0.16 | 9 | 6 | 53 | 46.93 | 39.60 | 0.10 | 0.19 | 4 | 4 |
| Orange | 6 | 69 | 68.15 | 0.07 | 10 | 114 | 195.14 | 40.64 | 1.74 | 0.25 | 5 | 5 | 155 | 1921.63 | 19.54 | 34.37 | 0.09 | 3 | 3 |

Although Definition 1 of magnitudes yields a larger sample size, their obvious serial auto-correlation results in less reliable estimates of the proportion of the observations surviving censoring to higher threshold levels; a task which is performed much better using magnitudes derived from Definition 2. Tables 2 and 3 display the values of observed and predicted parameters describing the distribution of magnitudes under both definitions above increased truncation levels.

*4.2. Geographic Variability and Potential Impacts*

This study provides flexible models that render probabilistic estimates of the variables associated with ILI events that can be adapted to various conditions. The robust statistical methodology may be implemented at any location, no matter the base (e.g., population) and critical thresholds established. Geographic variability in the parameters indicates differences in the potential influences on the occurrence of ILI peak events. For example, the observed spatial pattern of the shape parameter, $k$, at the 80th, 90th and 95th percentile levels in Figure 7 suggests that at higher thresholds, more counties exhibit positive values, no matter the definition. Since $k$ is, in the theory, independent of threshold and stable with respect to shifts in the threshold [22], it can be compared directly across spatially differentiated locations. Negative values of $k$ indicate large outliers in the right-hand tail of the distribution relative to the bulk of the observations. The values of $k$ become less negative at higher thresholds, because the comparative magnitudes of the outliers decrease as thresholds increase by progressive censoring. This is particularly noticeable when using Definition 1. It is likely to be over-interpreting due to the limited data, especially the small number of observations at a higher threshold. With increasing data availability in the future, a longer time period and a larger sample size can help to better represent these properties of ILI events.

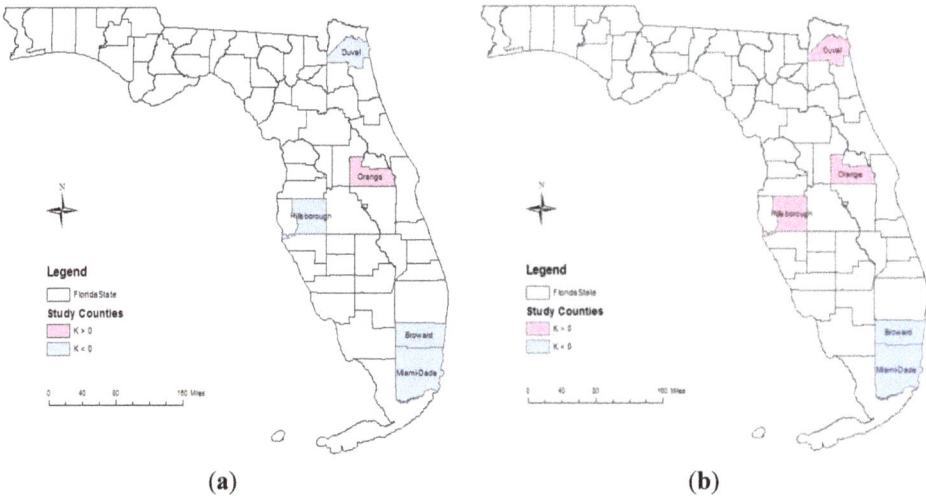

**(a)**                    **(b)**

**Figure 7.** *Cont.*

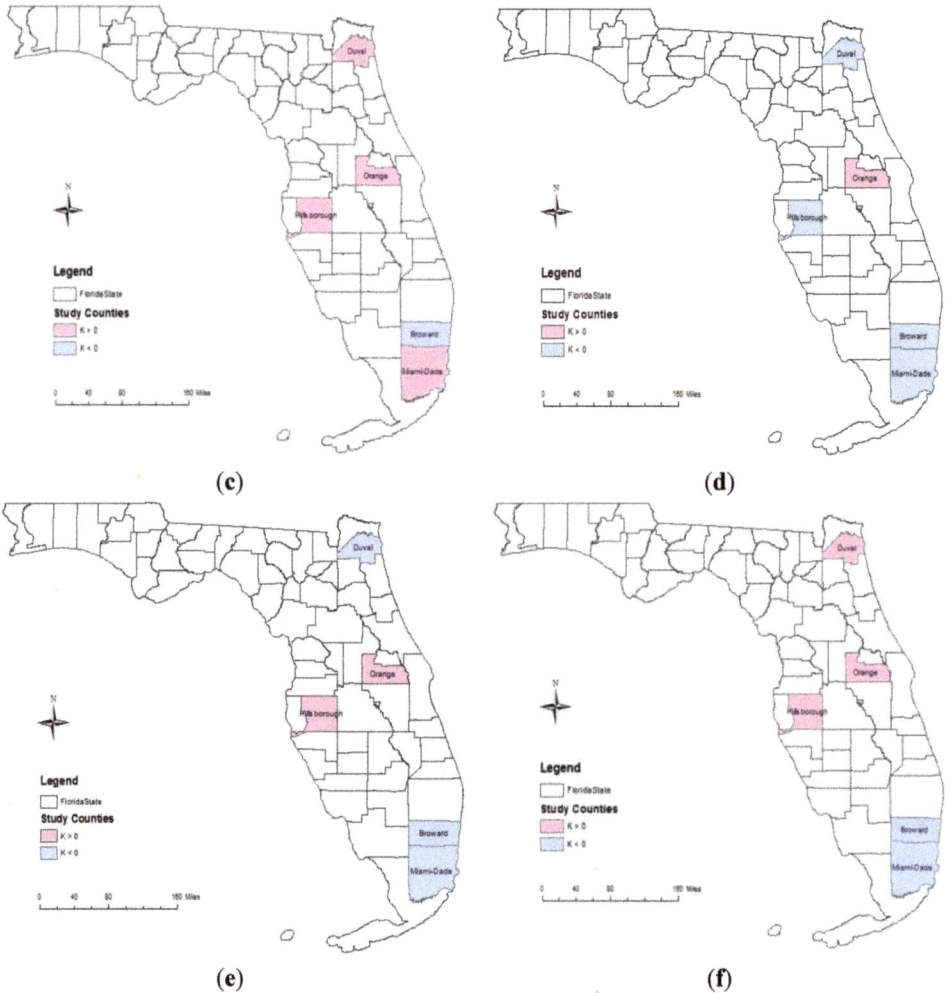

**Figure 7.** The observed spatial distributions of the parameter, $k$, values at three thresholds based on two definitions. (**a**) Definition 1, threshold 80%-$k$. (**b**) Definition 1, threshold 90%-$k$. (**c**) Definition 1, threshold 95%-$k$. (**d**) Definition 2, threshold 80%-$k$. (**e**) Definition 2, threshold 90%-$k$. (**f**) Definition 2, threshold 95%-$k$.

As a highly contagious and acute respiratory disease, the occurrence and properties of ILI peak events may be influenced by environmental (weather, *etc.*), demographic and social (urban, rural, transportation, *etc.*) factors. This approach identifies the average week of occurrence as late fall to early spring (Weeks 22 to 30, starting 10 December to 6 February with fairly large standard deviations of eight to 12 weeks). These observations are supported physically by studies that influenza outbreaks are sensitive to the weekly or bi-weekly average temperatures and humidity [5,23], particularly low temperature (optimum: 8 °C) and relative humidity [23–25]. Mean temperatures during the coldest month in Florida across the counties examined range from 10 °C to 16 °C, suggesting that weather conditions may have impacts on spatial patterns of peak events. However, no clear spatial pattern related to latitude and winter temperatures emerges. As the critical level of ILI cases of interest rises, the mean week of timing for events increases (later in the year) in almost all counties, except Broward, implying that peak events with greater weekly ILI cases tend to occur during late winter and early spring. In addition to weather conditions, each county possesses features that encourage influenza transmission: high population density, high proportions of their populations in sensitive age groups, international airports and ready access to major interstates, all of which have potentially profound effects on the occurrence of ILI events [26–29]. Similarly comparative use of public transport and immunization behavior may cause differences in the course of epidemics and modify their space-time spread [30,31]. Orange County, for example, exhibits relatively high values of the threshold ($q_0$) and total numbers of peak events ($K$) compared to the other four counties. This might be explained by its special characteristics of being located in the center of the state with comparatively low temperature and relative humidity with respect to the two southern counties (Miami-Dade and Broward). The major city in Orange County, Orlando, receives tens of thousands of tourists annually, especially during holidays in the late fall and winter, increasing the possibility of influenza transmission, for example compared to the northern county (Duval). To better understand the spatial patterns of the derived properties of events, the impacts of the above factors deserve to be further examined.

### 4.3. Application

The spatio-temporal visualizations of these statistical properties have the potential to deliver information in an efficient manner and assist decision-making within public health, such as the early warning of influenza peak activity, determining where and when to intervene, increasing the accessibility of health facilities, *etc.* For example, Figure 8 visualizes the spatial and temporal patterns of the timing of peak events, which represents the probability of having two ILI peak events up to Week 25, Week 26 and Week 27. These weeks are the first three weeks in January. The increasing probability of having two peak events in Week 26 and Week 27 may be due to the possible impacts of cold weather in January and the new semester of school on ILI activity. These visualizations can be expanded to the entirety of Florida in the future.

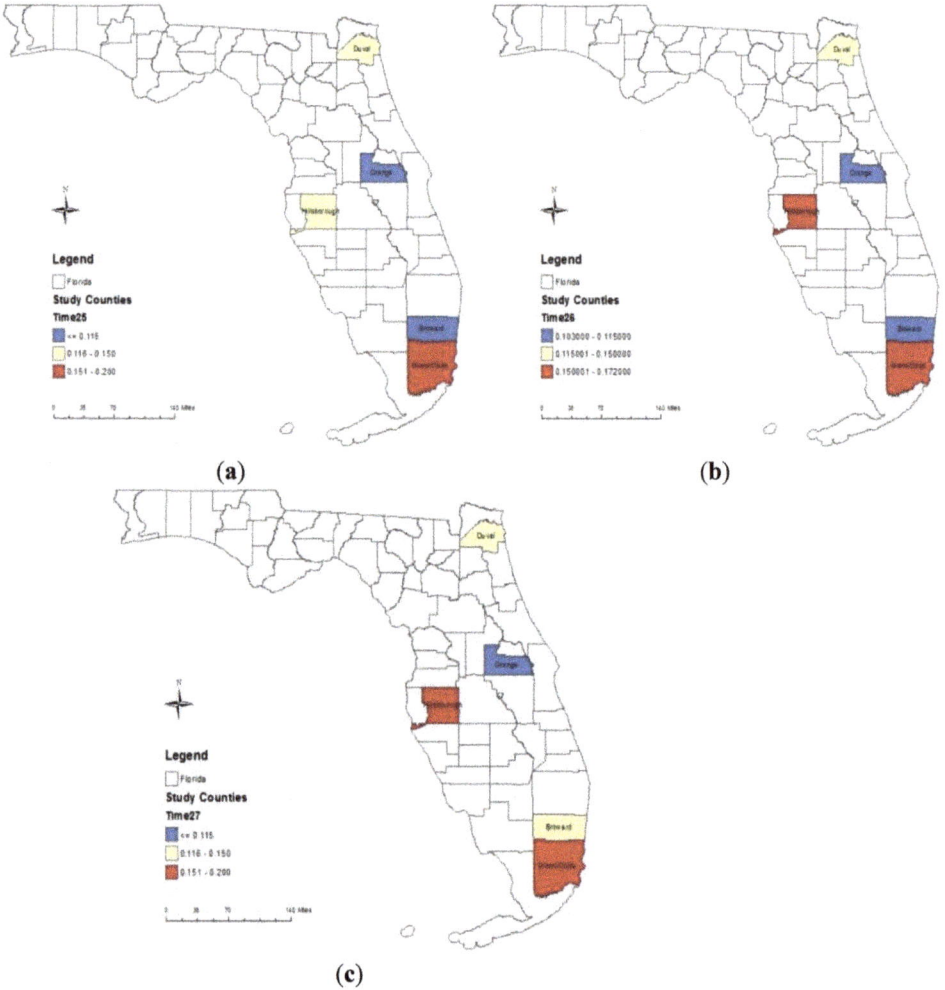

**Figure 8.** Application of timing in 18 selected counties in Florida. (**a**) Probability of experiencing two events up to Week 25. (**b**) Probability of experiencing two events up to Week 26. (**c**) Probability of experiencing two events up to Week 27.

## 5. Conclusions

This study innovatively applies an established method in hydrology and climatology to the field of epidemiology to describe the statistical properties of periods during which weekly ILI cases exceed critical thresholds. The new definition of events of interest beyond "peak events" considers only, and all, outbreaks of epidemiological interest and permits the estimation of the parameters of the distributions. The strong theoretical basis in crossing theory allows for the calculation of the properties of ILI events above various thresholds of interests. Another advantage of this approach is that it can be applied to spatially differentiated data to determine and compare risks associated with peak events, not defined by a common number of cases, but by a common frequency of outbreak regardless of the base population of the area (e.g., a weekly count that is only experienced in 20%, or 5%, of all the weeks of historic records in a county).

*ISPRS Int. J. Geo-Inf.* **2014**, *3*, 764–780

The methodology has the added flexibility of permitting the extrapolation of ILI event properties, especially the number of events and the magnitude, to other critical thresholds that vary in space and that are influenced by environmental, demographic and social factors. In the meantime, the potentially limited information contained in the standard ILI "peak event" (annual maximum) definition hinders public health professionals in efficiently implementing timing intervention strategies, such as vaccination and quarantine, thus leading to unnecessary socio-economic costs. This study can aid public health officials in supporting influenza surveillance and intervention by including the properties of the variables, annual event density, timing, magnitude and duration. The development and testing of these flexible models is the first step in an on-going study that seeks to establish associations between the statistical properties of ILI events and potential environmental factors. These associations can then be combined with vaccination and human mobility to give predictions of influenza transmission and to determine optimal periods to implement influenza vaccination programs among priority regions. Importantly, the models in this study could be easily extended to other infectious diseases in a further modification.

**Author Contributions:** Ying Wang designed the study, performed data collection and analyses, and drafted the manuscript. Peter Waylen contributed to the study design and revisions. Liang Mao revised the manuscript. All authors read and approved the final manuscript.

**Conflicts of Interest:** The authors declare no conflict of interest.

## References

1. Fleming, D.M.; Zambon, M.; Bartelds, A.I.M.; de Jong, J.C. The duration and magnitude of influenza epidemics: A study of surveillance data from sentinel general practices in England, Wales and the Netherlands. *Eur. J. Epidemiol.* **1999**, *15*, 467–473. [CrossRef]
2. Bock, D.; Andersson, E.; Frisén, M. Statistical surveillance of epidemics: Peak detection of influenza in Sweden. *Biom. J.* **2008**, *50*, 71–85. [CrossRef]
3. Cooper, D.L.; Verlander, N.Q.; Elliot, A.J.; Joseph, C.A.; Smith, G.E. Can syndromic thresholds provide early warning of national influenza outbreaks? *J. Public Health* **2009**, *31*, 17–25.
4. Cowling, B.J.; Wong, I.O.; Ho, L.M.; Riley, S.; Leung, G.M. Methods for monitoring influenza surveillance data. *Int. J. Epidemiol.* **2006**, *35*, 1314–1321.
5. Charland, K.M.L.; Buckeridge, D.L.; Sturtevant, J.L.; Melton, F.; Brownstein, J.S. Does climate predict the timing of peak influenza activity in the United States? *Adv. Dis. Surveill.* **2008**, *5*, 169.
6. Greene, S.K.; Ionides, E.L.; Wilson, M.L. Patterns of influenza-associated mortality among US elderly by geographic region and virus subtype, 1968–1998. *Am. J. Epidemiol.* **2006**, *163*, 313–326.
7. Paget, J.; Marquet, R.; Meijer, A.; van Der Velden, K. Influenza activity in Europe during eight seasons (1999–2007): An evaluation of the indicators used to measure activity and an assessment of the timing, length and course of peak activity (spread) across Europe. *BMC Infect. Dis.* **2007**, *7*. [CrossRef]
8. Smith, L.P. Numerical forecasting of epidemics of influenza in Great Britain and Northern Ireland. *Rev. Epidemiol. Sante Publique* **1982**, *30*, 413–422.
9. Sakai, T.; Suzuki, H.; Sasaki, A.; Saito, R.; Tanabe, N.; Taniguchi, K. Geographic and temporal trends in influenza like illness, Japan, 1992–1999. *Emerg. Infect. Dis.* **2004**, *10*, 1822–1826. [CrossRef]
10. CDC WONDER Online Database. Underlying Cause of Death 1999–2009. Available online: http://wonder.cdc.gov/ucd-icd10.html (accessed on 11 November 2012).
11. Florida Department of Health. Available online: http://www.doh.state.fl.us/disease_ctrl/epi/htopics/flu/FSPISN/influenza_sentinels.html (accessed on 18 November 2012).
12. Florida Department of Health. Available online: http://www.doh.state.fl.us/disease_ctrl/epi/htopics/flu/panflu.htm (accessed on 15 November 2012).
13. Florida Department of Health. Available online: http://www.doh.state.fl.us/floridaflu/FSPISN/influenza_sentinels.html (accessed on 19 November 2012).
14. Centers for Disease Control and Prevention. Available online: http://www.cdc.gov/flu/weekly/overview.htm (accessed on 10 January 2013).

15. Cooley, P.; Ganapathi, L.; Ghneim, G.; Holmberg, S.; Wheaton, W.; Hollingsworth, C.R. Using influenza-like illness data to reconstruct an influenza outbreak. *Math. Comput. Model.* **2008**, *48*, 929–939. [CrossRef]

16. Cramer, H.; Leadbetter, M.R. *Stationary and Related Stochastic Processes*; Wiley: New York, NY, USA, 1967.

17. Rice, S.O. Mathematical analysis of random noise. *Bell Syst. Tech. J.* **1945**, *24*, 46–156. [CrossRef]

18. Desmond, A.F.; Guy, B.T. Crossing theory for Non-Gaussian processes with an application to hydrology. *Water Resour. Res.* **1991**, *279*, 2791–2797. [CrossRef]

19. Rosbjerg, D.; Madsen, H.; Rasmussen, P.F. Prediction in partial duration series with generalized pareto-distributed exceedances. *Wat. Resour. Res.* **1992**, *28*, 3001–3010. [CrossRef]

20. Keellings, D.; Waylen, P.R. The stochastic properties of high daily maximum temperatures applying crossing theory to modeling high-temperature event variables. *Theor. Appl. Climatol.* **2012**, *108*, 579–590. [CrossRef]

21. Straetmans, S.T.M.; Verschoor, W.F.C.; Wolff, C.C.P. Extreme US stock market fluctuations in the wake of 9/11. *J. Appl. Econom.* **2008**, *23*, 17–42.

22. Beisel, C.J.; Rokyta, D.R.; Wichman, H.A.; Joyce, P. Testing the extreme value domain of attraction for distributions of beneficial fitness effects. *Genetics* **2007**, *176*, 2441–2449. [CrossRef]

23. Lowen, A.C.; Mubareka, S.; Steel, J.; Palese, P. Influenza virus transmission is dependent on relative humidity and temperature. *PLoS Pathog.* **2007**, *3*, e151. [CrossRef]

24. Shaman, J.; Kohn, M. Absolute humidity modulates influenza survival, transmission, and seasonality. *Proc. Natl. Acad. Sci. USA* **2009**, *106*, 3243–3248. [CrossRef]

25. Tsuchihashi, Y.; Yorifuji, T.; Takao, S.; Suzuki, E.; Mori, S.; Doi, H.; Tsuda, T. Environmental factors and seasonal influenza onset in Okayama city, Japan: Case crossover study. *Acta Med. Okuyama* **2011**, *65*, 97–103.

26. Ertek, M.; Durmaz, R.; Guldemir, D.; Altas, A.B.; Albayrak, N.; Korukluoglu, G. Epidemiological, demographic, and molecular characteristics of laboratory-confirmed pandemic influenza A (H1N1) virus infection in Turkey. *Jpn. J. Infect. Dis.* **2010**, *63*, 239–245.

27. Olson, D.R.; Heffernan, R.T.; Paladini, M.; Konty, K.; Weiss, D.; Mostashari, F. Monitoring the impact of influenza by age: Emergency department fever and respiratory complaint surveillance in New York city. *PLoS Med.* **2007**, *4*, e247. [CrossRef]

28. Viboud, C.; Boëlle, P.Y.; Cauchemez, S.; Lavenu, A.; Valleron, A.J.; Flahault, A.; Carrat, F. Risk factors of influenza transmission in households. *Br. J. Gen. Pract.* **2004**, *54*, 684–689.

29. Rivas, A.L.; Chowell, G.; Schwager, S.J.; Fasina, F.O.; Hoogesteijn, A.L.; Smith, S.D.; Bisschop, S.P.; Anderson, K.L.; Hyman, J.M. Lessons from Nigeria: The role of roads in the geo-temporal progression of avian influenza (H5N1) virus. *Epidemiol. Infect.* **2009**, *138*, 192–198.

30. Lim, W.-Y.; Chen, C.-H.; Ma, Y.; Chen, M.-I.; Lee, V.-J.; Cook, A.-R.; Tan, L.W.; Flores Tabo, N., Jr.; Barr, I.; Cui, L.; *et al.* Risk factors for pandemic (H1N1) 2009 seroconversion among adults, Singapore, 2009. *Emerg. Infect. Dis.* **2011**, *17*, 1455–1462.

31. Yang, Y.; Sugimoto, J.D.; Halloran, M.E.; Basta, N.E.; Chao, D.L.; Matrajt, L.; Potter, G.; Kenah, E.; Longini, I.M., Jr. The transmissibility and control of pandemic influenza A (H1N1) virus. *Science* **2009**, *326*, 729–733. [CrossRef]

International Journal of
*Geo-Information*

isprs

MDPI

Article

# Correlating Remote Sensing Data with the Abundance of Pupae of the Dengue Virus Mosquito Vector, *Aedes aegypti*, in Central Mexico

Max J. Moreno-Madriñán [1,*], William L. Crosson [2], Lars Eisen [3], Sue M. Estes [4], Maurice G. Estes Jr. [4], Mary Hayden [5], Sarah N. Hemmings [2,6], Dan E. Irwin [7], Saul Lozano-Fuentes [3], Andrew J. Monaghan [5], Dale Quattrochi [7], Carlos M. Welsh-Rodriguez [8] and Emily Zielinski-Gutierrez [9]

[1] Department of Environmental Health, Fairbanks School of Public Health, Indiana University, IUPUI, Indianapolis, IN 46202, USA
[2] Science and Technology Institute, Universities Space Research Association (USRA), Huntsville, AL 35805, USA; bill.crosson@nasa.gov (W.L.C.); sarah.n.hemmings@nasa.gov (S.N.H.)
[3] Department of Microbiology, Immunology and Pathology, Colorado State University, Fort Collins, CO 80523, USA; lars.eisen@colostate.edu (L.E.); saul.lozano-fuentes@colostate.edu (S.L.-F.)
[4] Earth and System Science Center, University of Alabama in Huntsville, Huntsville, AL 35805, USA; sue.m.estes@nasa.gov (S.M.E.); maury.estes@nsstc.uah.edu (M.G.E.)
[5] Research Applications Laboratory, National Center for Atmospheric Research, Boulder, CO 80307, USA; monaghan@ucar.edu (M.H.); mhayden@ucar.edu (A.J.M.)
[6] Earth Science Division, Applied Sciences Program, NASA Headquarters, Washington, DC 20024-3210, USA
[7] Earth Science, NASA Marshall Space Flight Center, Huntsville, AL 35811, USA; daniel.irwin@nasa.gov (D.E.I.); dale.quattrochi@nasa.gov (D.Q.)
[8] Earth Sciences Center, Veracruz University, 91090 Xalapa, Mexico; cwelsh@uv.mx
[9] Division of Vector-Borne Diseases, National Center for Emerging and Zoonotic Infectious Diseases, Centers for Disease Control and Prevention, Fort Collins, CO 80521, USA; Ebz0@cdc.gov
* Author to whom correspondence should be addressed; mmorenom@iu.edu; Tel.: +1-317-274-3170; Fax: +1-317-274-3443.

Received: 24 March 2014; in revised form: 23 April 2014; Accepted: 29 April 2014; Published: 20 May 2014

**Abstract:** Using a geographic transect in Central Mexico, with an elevation/climate gradient, but uniformity in socio-economic conditions among study sites, this study evaluates the applicability of three widely-used remote sensing (RS) products to link weather conditions with the local abundance of the dengue virus mosquito vector, *Aedes aegypti* (*Ae. aegypti*). Field-derived entomological measures included estimates for the percentage of premises with the presence of *Ae. aegypti* pupae and the abundance of *Ae. aegypti* pupae per premises. Data on mosquito abundance from field surveys were matched with RS data and analyzed for correlation. Daily daytime and nighttime land surface temperature (LST) values were obtained from Moderate Resolution Imaging Spectroradiometer (MODIS)/Aqua cloud-free images within the four weeks preceding the field survey. Tropical Rainfall Measuring Mission (TRMM)-estimated rainfall accumulation was calculated for the four weeks preceding the field survey. Elevation was estimated through a digital elevation model (DEM). Strong correlations were found between mosquito abundance and RS-derived night LST, elevation and rainfall along the elevation/climate gradient. These findings show that RS data can be used to predict *Ae. aegypti* abundance, but further studies are needed to define the climatic and socio-economic conditions under which the correlations observed herein can be assumed to apply.

**Keywords:** MODIS; TRMM; DEM; Aqua; remote sensing; elevation; mosquito; rainfall; temperature

## 1. Introduction

Environmental changes potentially impacting the geographical ranges or local abundance of arthropod vectors transmitting infectious disease agents are among the important concerns linked to climate [1–6]. Associations reported in the literature show that climate-related variables can be used to predict local abundance and the potential for the expansion of arthropod vectors, such as mosquitoes or ticks [7–10]. Since field surveys are both costly and time consuming, remote sensing (RS) technology is increasingly used to estimate habitat suitability for a variety of vector species [11–14]. Temperature and rainfall are the weather parameters of special interest, because they impact both the distribution of suitable vector habitat and the potential for local vector proliferation. Although terrain elevation is strongly associated with temperature, urban heat islands might cause slight differences in the associations between vector abundance and climate parameters in studies conducted within urban environments. Thus, elevation is included among the variables of interest for this study. *Aedes aegypti* (*Ae. aegypti*), the primary mosquito vector of dengue and yellow fever viruses and an important vector of chikungunya virus to humans in urban settings, is most abundant in urban environments [15].

Dengue is one of the most important mosquito-borne viral diseases in the subtropics and tropics, with one estimate of the global infection burden reaching approximately 390 million virus infections and nearly 100 million cases with disease manifestations per year, over three times that estimated by the World Health Organization [16]. Although the presence and abundance of the mosquito vector is strongly influenced by the human peridomestic environment (e.g., access to water-holding containers serving as larval development sites and the potential for intrusion into homes to engage in indoor biting), these are also affected by meteorological variables, such as temperature, rainfall, humidity and solar radiation. Several studies have addressed the relationship between weather or climate variability and the incidence of dengue disease cases [17–28]. Such relationships, however, may be influenced by additional factors, such as the exposure of humans to mosquitos and the intensity of virus transmission [29].

The strength of the association between RS-based climate parameters and vector abundance may be limited by the spatial resolution of the satellite products that are freely available and more commonly used [7]. Environmental and socio-economic conditions can change drastically over distances of 10–100 m of meters; therefore, spatial models designed to estimate vector presence and abundance developed at the regional or state level cannot reliably be down-scaled for locally-relevant risk predictions. To develop models to predict vector presence and abundance at the local (community or neighborhood) scale using RS-based environmental inputs requires consistent monitoring of recent local environmental conditions with RS imagery that can distinguish the differences between adjacent communities or neighborhoods. The present study tests if three widely used RS-based environmental products are able to distinguish those differences at the local level, despite having spatial resolutions equal or larger than 90 m. Our motivation is to evaluate the potential for using RS-based environmental products that are freely-available to decision-makers in developing countries, to monitor the presence and abundance of *Ae. aegypti* at the local scale. For a geographic transect of approximately 330 km by road, corresponding to an area of approximately 245 km (west-east) by 98 km (north-south) in central Mexico, we describe the associations between the presence and abundance of the pupal life stage of *Ae. aegypti* and environmental conditions estimated from RS products, including land surface temperature (LST), rainfall, land surface properties and elevation.

## 2. Methodology

### 2.1. Study Site

This study used sites from a previously published study [30] on the occurrence of *Ae. aegypti* along the elevation gradient between Veracruz at sea level and Puebla at more than 2000 m in Central Mexico (Figure 1). Sites were composed of groups of homes with low to middle income and small to medium-sized yards, distributed among 12 communities along the elevation gradient described.

## 2.2. Field Survey Mosquito Data

Data on *Ae. aegypti* pupal abundance were generated from field surveys conducted in the cities of Córdoba, Orizaba, Rio Blanco, Ciudad Mendoza, Acultzingo, Maltrata, Puebla City and Atlixco from 11 July to 20 August 2011 and from the cities of Coatepec, Xalapa and Perote between 23 August and 1 September of the same year. Approximately 50 study premises were examined for each one of the 12 communities; these premises were contained within 3–4 spatially distinct clusters within each community. The methodologies for the selection of premises to examine, the collection of immature mosquitoes (larvae and pupae) from indoor and outdoor water-holding containers on the premises, the subsequent rearing to adults and species identification and, finally, the estimation of pupal abundance for *Ae. aegypti* in the study sites were described previously by Lozano-Fuentes [30]. About 73% of collected pupae were successfully reared to adults and identified to species, compared to just 16% for collected larvae. Consequently, we focus only on the more robust estimates for pupal abundance.

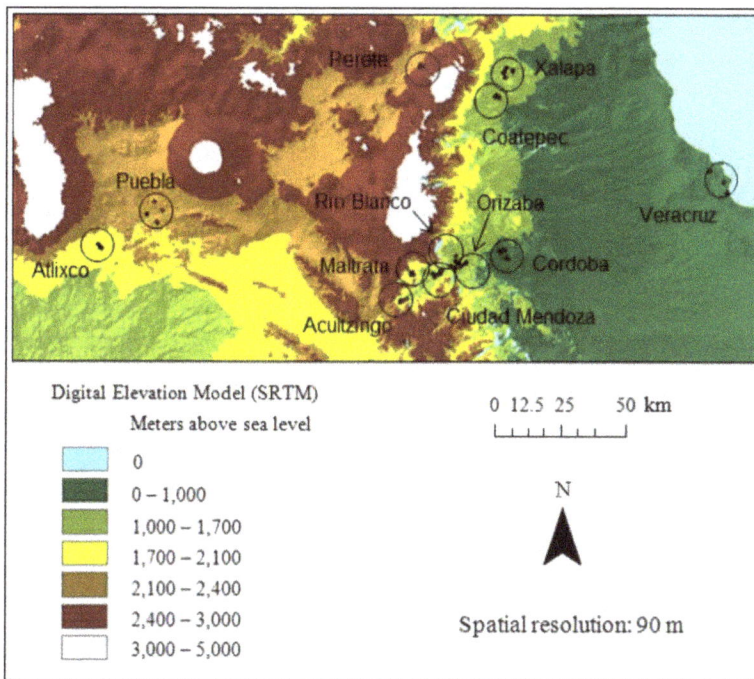

**Figure 1.** The study area with the locations of communities in Central Mexico in relation to elevation, as estimated by the digital elevation model (DEM) of the Shuttle Radar Topographic Mission (SRTM).

## 2.3. Remotely Sensed Data

### 2.3.1. Visible Infrared Scanner (VIRS)

Data on precipitation were estimated with product 3B42 V7 derived from the Visible Infrared Scanner (VIRS) sensor onboard the TRMM satellite and retrieved from the TRMM Online Visualization and Analysis System (TOVAS) [31]. This system is maintained by the NASA Goddard Earth Science Data and Information Services Center. The 3B42 V7 data cover the tropical and subtropical regions between 50°N and 50°S with a daily temporal resolution adjusted from a 3-hourly temporal resolution and a spatial resolution of 0.25° by 0.25°, roughly equivalent to 27 km in the study area (Figure 2). TRMM is a joint mission between the U.S. National Aeronautics and Space Administration (NASA) and the Japan Aerospace Exploration Agency (JAXA), launched on 27 November of 1997 and designed to measure rainfall for weather and climate research. Data were processed using ArcMap 10.2 software. TRMM has been shown to reasonably reproduce rainfall variability at monthly timescales that are analogous to the four-week timescales used in this study [32].

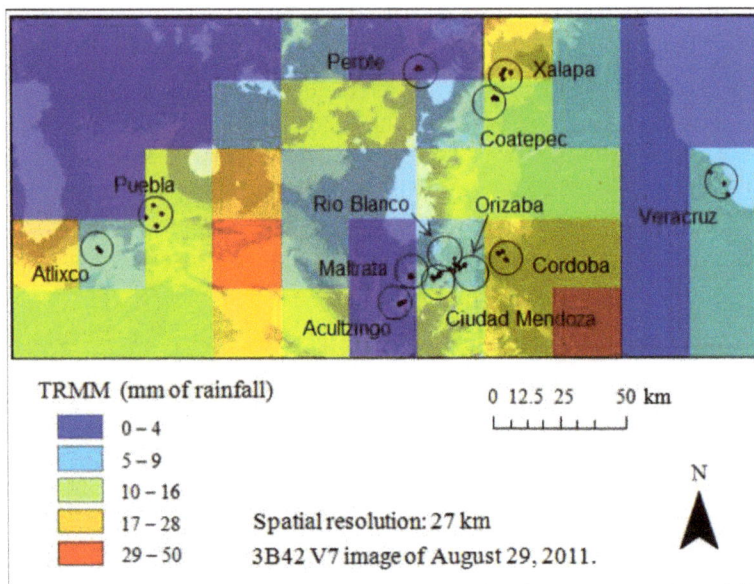

**Figure 2.** Tropical Rainfall Measuring Mission (TRMM), 3B42 V7 image of 29 August 2011.

### 2.3.2. Moderate Resolution Imaging Spectroradiometer (MODIS)

MODIS data were downloaded from the Reverb/Echo NASA EOS Data and Information System (EOSDIS) website [33] and were processed using the MODIS Reprojection Tool (MRT) and ArcMap 10.2 software. Onboard the Terra and Aqua satellites, MODIS has been one of the most used instruments for the Earth Observing System (EOS), a NASA international program, which, in turn, is a key component of NASA's Earth Science Enterprise [34]. Launched on 18 December 1999 (Terra), and 4 May 2002 (Aqua), Terra and Aqua are designed to monitor many conditions of the atmosphere, land, oceans, biosphere and cryosphere, although their foci are on land and ocean observations, respectively [35]. Both satellites have sun-synchronous orbits crossing the equator at an approximate local time of 10:30 AM and 10:30 PM in the case of Terra and at 1:30 PM and 1:30 AM for Aqua, in a northward and southward track, respectively.

LST estimates are from the MODIS Land Surface Temperature and Emissivity product (MYD11A1) from the Aqua satellite. This product provides temperature and emissivity values per-pixel. MYD11A1 measurements along with all data generated from sensors carried by Aqua can be obtained for the entire globe within two days [35]. The daily MYD11A1 product has a spatial resolution of 1 km (Figure 3). Previous studies have shown that MODIS LST products have generally been accurate within ±1 K compared to *in situ* temperature measurements, for a variety of sites and conditions [36,37].

### 2.3.3. Shuttle Radar Topography Mission (SRTM)

Expected to be strongly associated with ambient temperature, elevation was examined as a potential proxy for temperature. Elevation was estimated through the Digital Elevation Model (DEM) from the SRTM, a collaboration between NASA and the National Imagery and Mapping Agency

(NIMA) of the U.S. Department of Defense [38]. With a resolution of 90 m at the equator (Figure 1), the SRTM was designed to produce a DEM of the Earth's land surface approximately between latitudes 60°N and 56°S. In an 11-day flight around the world onboard the Space Shuttle Endeavour, the mission was completed on 22 February 2000. The data for this study were downloaded from the Global Data Explorer website [39], which is maintained by the United States Geological Survey (USGS) and NASA's Land Processes Distributed Active Archive Center (LPDAAC).

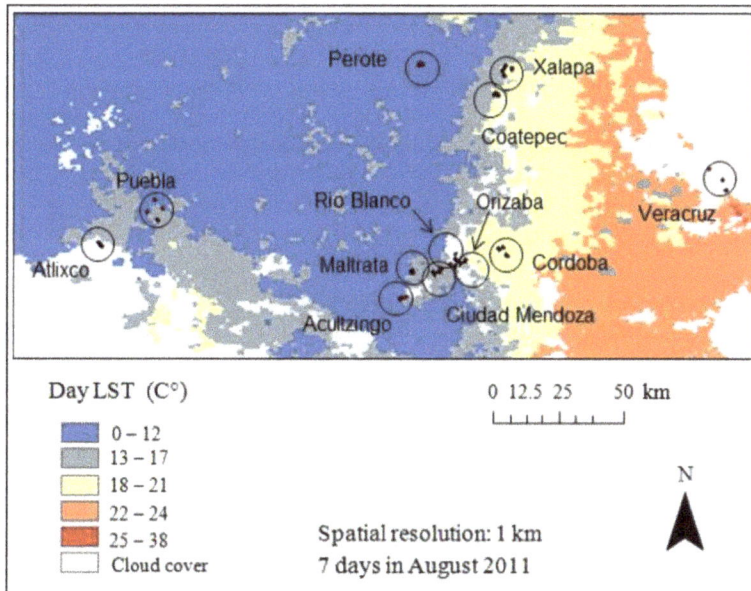

**Figure 3.** Composite of 7 days of day LST images (MYD11A1) from August 2011.

*2.4. Match-Up Data Procedure*

Field survey-based estimates for *Ae. aegypti* pupal abundance from the 607 examined premises were aggregated at the community and cluster levels through two approaches by estimating: (1) the percentages of premises with the presence of *Ae. Aegypti* pupae; and (2) the mean abundances of *Ae. aegypti* pupae per premises. Both types of estimates for pupal abundance were matched up with RS data derived from the location of the corresponding premises and, similarly, aggregated at the community and cluster level. RS data provided daily data for LST (nighttime and daytime), rainfall and elevation. The matched values of RS data were calculated as follows: Nighttime LST was calculated from the average nighttime cloud-free LST data for the 29 nights preceding the survey. The same process was repeated for daytime LST data, except that the average was calculated over the 28 days preceding the day of the survey plus the day of the survey. After removing cloudy days, the average number of days with daily data used to obtain the average LST values was 7.9 for nighttime LST and 6.8 for daytime LST. RS estimates of rainfall were calculated using the accumulated amount over the period comprising the day of the survey plus the 28 preceding days. No special calculation was required for elevation data, since these are a snapshot of the DEM data of the 2000 flight. Each set of matched pairs between pupae abundance and LST (day and night), rainfall and elevation were analyzed for correlation at both levels of aggregation: community and cluster, using the SAS 9.3 proc corr method.

We also tested using a two-week period instead of a four-week period preceding the field mosquito surveys, and in all cases, the correlations between climatic data and mosquito presence/abundance were lower (data not shown). A period longer than four weeks was not analyzed, since no improvement was noticed by Lozano-Fuentes *et al.* [30] when analyzing for similar correlations, but using *in situ* temperature data from HOBO meteorological stations (Onset Computer Corporation, Bourne, MA, USA) and rainfall from the Climate Prediction Center Morphing Technique (CMORPH) dataset.

## 3. Results

Table 1 summarizes the aggregated values per community for: (1) the percentage of premises with the presence of *Ae. aegypti* pupae; and (2) mean abundances of pupae per premises, along with the corresponding aggregated RS values for the climate variables LST (nighttime and daytime), rainfall and elevation. The table is ordered by decreasing estimated mean abundance of *Ae. Aegypti* pupae per premises. In total, a mean number of 3.8 *Ae. Aegypti* pupae were found per examined premises. The overall percentage of premises with the presence of *Ae. aegypti* pupae was 19.44%. Perote, the community at the highest elevation (2400 m) out of the 12 study communities, was the only one where no *Ae. aegypti* were found. As reported by Lozano [30] using the same data set, this was also the case when including larvae. In general, a lower percentage of presence and mean abundance values for *Ae. aegypti* occurs at the cooler, drier sites at higher elevations.

**Table 1.** Estimates for the abundance of *Aedes aegypti*, by study community, in relation to climate variables and elevation.

| Community | Percentage | Mean | Night LST (C°) | Day LST (C°) | Rainfall (mm) | Elevation (m) |
|---|---|---|---|---|---|---|
| Orizaba | 35 | 14.29 | 16.0 | 33.2 | 354 | 1236 |
| Rio Blanco | 41 | 10.43 | 15.0 | 30.9 | 334 | 1258 |
| Cordoba | 24 | 6.04 | 16.7 | 33.1 | 507 | 860 |
| Veracruz | 28 | 4.91 | 21.7 | 40.7 | 372 | 18 |
| Coatepec | 25 | 3.26 | 16.1 | 32.2 | 208 | 1203 |
| Ciud. Mendoza | 21 | 2.63 | 14.6 | 31.9 | 348 | 1338 |
| Acultzingo | 14 | 1.68 | 12.8 | 28.9 | 217 | 1695 |
| Xalapa | 24 | 0.98 | 15.4 | 31.8 | 209 | 1419 |
| Atlixco | 12 | 0.28 | 15.0 | 36.6 | 124 | 1831 |
| Maltrata | 4 | 0.06 | 10.9 | 29.9 | 210 | 1714 |
| Puebla | 4 | 0.04 | 11.4 | 35.2 | 112 | 2141 |
| Perote | 0 | 0.00 | 8.1 | 28.5 | 203 | 2417 |

Percentage: the percentage of premises with the presence of *Ae. aegypti* pupae; mean: the mean abundance of *Ae. aegypti* pupae per premise; night LST: the MODIS estimated LST (MYD11A1, night); day LST: the MODIS estimated LST (MYD11A1, day); rainfall: the TRMM estimated precipitation (3B42 V7); elevation: the SRTM's DEM estimated elevation.

### 3.1. Correlations among RS Estimated Climate Variables

A summary of the correlation analysis of RS estimated variables with each other is presented in Table 2. An obviously expected significant correlation between the estimated values of elevation and LST was detected when using nighttime LST data, both at the community and cluster levels. However, this relationship was not significant at either level of aggregation when using daytime LST data. Elevation was significantly and inversely correlated with estimated precipitation at both levels of aggregation, community and cluster. Furthermore, as expected, there was a significant positive correlation between the LST estimates for night and day at both scale levels. Finally, for either level of aggregation, significant associations between estimated precipitation and LST were only detected with nighttime LST, but not with daytime LST.

**Table 2.** Summary of the Spearman correlations among the RS estimated climate variables.

| Climate Variables | Community | Cluster |
|---|---|---|
| | N = 12 | N = 43 |
| Elevation and night LST | −0.91 ** | −0.87 ** |
| Elevation and day LST | −0.39 | −0.29 |
| Elevation and rainfall | −0.80 ** | −0.80 ** |
| Night LST and day LST | 0.59 * | 0.55 ** |
| Rainfall and night LST | 0.60 * | 0.60 * |
| Rainfall and day LST | 0.11 | −0.008 |

$** p < 0.01, * p < 0.05$.

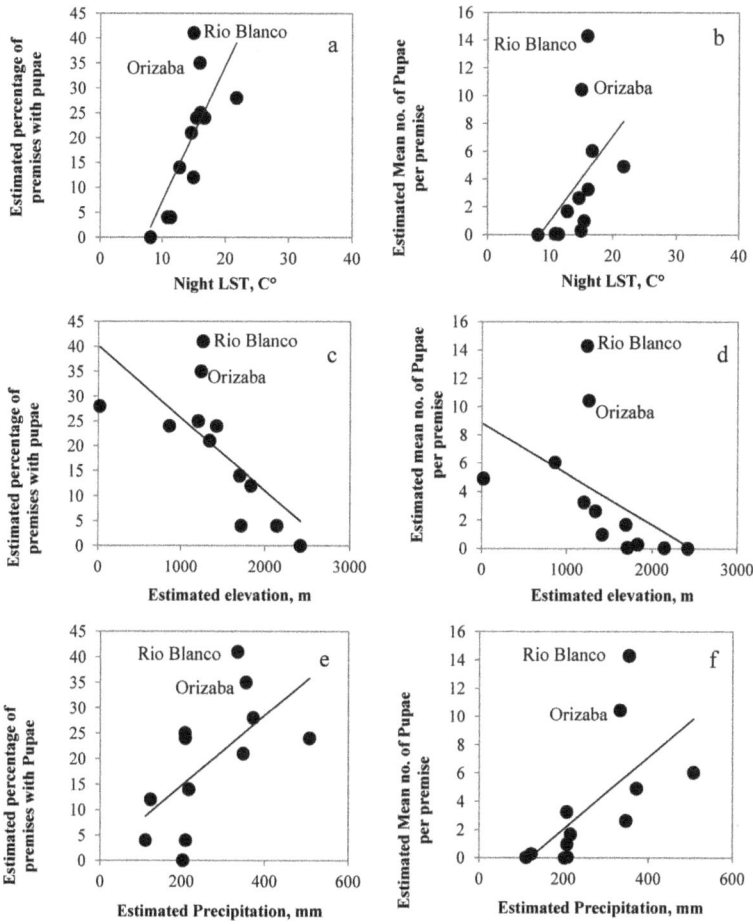

**Figure 4.** Plots at the community level. Relationships between: (**a**) night LST and the estimated percentage of premises ("sites") with *Aedes aegypti* pupae present; (**b**) night LST and the estimated mean number of pupae per site; (**c**) elevation and the estimated percentage of premises with pupae present; (**d**) elevation and the estimated mean number of pupae per site; (**e**) the estimated precipitation and estimated percentage of premises with pupae present; (**f**) the estimated precipitation and estimated mean number of pupae per premises.

*3.2. Correlations among RS-Estimated Climate Variables and Mosquito Presence and Abundance*

At the community and cluster levels, Table 3 summarizes the correlation results between RS estimated climate variables and both measures of *Ae. aegypti* populations: the percentage of premises with the presence of *Ae. aegypti* pupae and mean abundances of *Ae. aegypti* pupae per premises. MODIS-estimated nighttime LST was positively and significantly correlated with the percentage of homes with *Ae. aegypti* pupae or the mean abundance of *Ae. aegypti* pupae per premises at the community (Figure 4a,b) and cluster level (plot not shown). Similarly, elevation estimated through SRTM showed a significant, although inverse, correlation with the percentage of premises with the presence of *Ae. aegypti* pupae or a mean abundance of *Ae. aegypti* pupae per premises at both levels of aggregation, community (Figure 4c,d) and cluster (plot not shown). Positive and significant correlations were detected between TRMM-estimated precipitation and the percentage of premises with the presence of *Ae. aegypti* pupae or mean abundances of *Ae. aegypti* pupae per premises at the community (Figure 4e,f) and cluster level (plot not shown). No correlation was detected between MODIS-estimated daytime LST and the percentage of premises with the presence of *Ae. aegypti* pupae or mean abundances of *Ae. aegypti* pupae per premises at any level of aggregation (plots not shown).

**Table 3.** Summary of Spearman correlations between RS estimated climate variables and the abundance of *Aedes aegypti* pupae.

| Climate Variables | Community | Cluster |
|---|---|---|
| | N = 12 | N = 43 |
| Night LST and percentage | 0.82 ** | 0.56 ** |
| Night LST and mean | 0.78 ** | 0.64 ** |
| Elevation and percentage | −0.84 ** | −0.67 ** |
| Elevation and mean | −0.87 ** | −0.75 ** |
| Rainfall and percentage | 0.61 * | 0.50 ** |
| Rainfall and mean | 0.79 ** | 0.55 ** |
| Day LST and percentage | 0.33 | 0.12 |
| Day LST and mean | 0.29 | 0.21 |

$** p < 0.01, * p < 0.05.$

## 4. Discussion

Associations between extreme weather events and mosquito outbreaks [40,41], as well as the weather-mediated seasonal dynamics of mosquito abundance [42] have been reported in the literature for *Ae. aegypti*. Although seasonal weather fluctuations can in large part explain intra-annual fluctuations in the abundance of *Ae. aegypti*, studies conducted at the local scale and, therefore, under very similar weather conditions, have revealed differences in vector abundance among nearby urban locations [42–44]. This likely reflects the effect of anthropogenic modifications in the urban environments that this mosquito prefers to inhabit [42,44]. Using a study design that takes advantage of a geographic transect with similar socio-economic conditions in the specific field survey areas, a previous cross-sectional analysis using this dataset for the abundance of *Ae. aegypti* and *in situ* weather data also reported significant associations between mosquito abundance and weather variables (temperature or rainfall) and elevation [30]. The uniformity in socio-economic conditions in the specific study areas allows the opportunity to minimize the influence of the anthropogenic sources of variability, while at the same time studying elevation and weather variables at approximately the same point in time. The present study further capitalizes on such advantages by specifically focusing on testing the applicability of widely-used RS products to explore weather linkages with *Ae. aegypti* abundance at the local scale.

Despite the relatively coarse spatial resolution of 1 km for nighttime LST and approximately 26 km for rainfall, strong correlations were found between data from the MYD11A1 product or 3B42 V7, respectively, and the local abundance of *Ae. aegypti* pupae. Conversely, the results were not encouraging for daytime LST. Honório *et al.* [45] showed a positive linear relationship between mosquito abundance and ambient temperature within a range of 18 °C to 24 °C. The authors did not detect any variation in the abundance of *Ae. aegypti* above this temperature threshold. Further, Eisen *et al.* [46] reported a positive linear relationship between water temperature and the development rate for immature *Ae. aegypti* between 15 °C to 30 °C. In that study, lower developmental zero temperatures were estimated to be in the 10 °C to 14 °C range for eggs and immature *Ae. aegypti* and the upper developmental zero temperatures in the 38 °C to 42 °C range. It is interesting to notice that in our study, the mean cluster level daytime LST ranged from 27.5 °C to 40.7 °C while the nighttime LST ranged from 7.8 °C to 24.2 °C, which included the range of positive correlation reported by Honório *et al.* [45] and partially overlapped with that reported by Eisen *et al.* [46]. Additional work is needed to assess the correlations between the abundance of *Ae. aegypti* and RS data for a broader range of temperatures than examined in the present study, especially for the higher end temperatures. Similar to our results with RS data, positive linear relationships with nighttime LST, but not with daytime LST, were detected in malaria prediction models [47].

As expected from the strong positive correlation between the abundance of *Ae. aegypti* and night LST, there was also a strong, but negative, correlation between mosquito abundance and elevation. In general, a negative relationship is expected between elevation and temperature [48]; in our case, however, this was detected only when considering night LST, but not when considering day LST. Indeed, daytime LST was only associated with nighttime LST and not associated with any other variable, including elevation. This may be related to more cloud cover during the daytime than at nighttime, as suggested by the lower mean number of cloud-free images available in the daytime LST data (6.8 images, Figure 3) compared to 7.9 images available for nighttime LST data. The effect of minimum temperatures may be another contributing factor; Lozano-Fuentes *et al.* [30] identified the mean minimum daily temperature and the mean minimum daily winter temperature among the weather parameters with potential relevance for the biology of *Ae. aegypti*. Although our study did not consider these specific parameters, it is possible that minimum temperatures (which occur in our night LST data) in general may play an important role in the biology of *Ae. aegypti*.

Finally, a strong correlation has been reported between median family income and surface temperature during the daytime, with a much weaker correlation at nighttime [49]. The specific areas (*i.e.*, neighborhoods within communities) used in this study were selected under the criteria of being urban low- to middle-income homes with small- to medium-sized yards. The study premises often harbored considerable vegetation, and it is known that leaf temperatures and other vegetative features are a major aspect related to urban temperature [50]. Although, the possibility exists that income factors play a role in the differences detected between daytime and nighttime LST, such a possibility is not evident in our study, since many of our communities were small and not likely to influence the urban heat island effect as detected in a 1-km × 1-km grid box.

While it is possible to explain the positive correlation between RS estimated rainfall and the abundance of *Ae. aegypti* in terms of more containers being filled with rain water (or some containers having a greater volume of water) and, thus, more potential larval development sites when it rains, this could be a casual association, due to the fact that the areas with higher temperature and lower elevation also typically have a higher rainfall. Indeed, rainfall and nighttime LST were weakly, but significantly and positively correlated. It also should be noted that the impact of rainfall on the presence of larval development sites for *Ae. aegypti* is complicated by human water storage practices, with the importance of containers filled by rain *versus* human action varying both among geographic areas and within the year in single locations. However, the results found in this study are consistent with those of Lozano-Fuentes *et al.* [30] using meteorological data from weather stations along the same transect. Additionally, our results are consistent with process-based life-cycle models of *Ae. aegypti*; mosquito populations become largest when temperatures are warm and rainfall is abundant [51,52]. They are also consistent with studies indicating that dengue risk in humans is positively correlated with temperature and rainfall [53], which is likely in part due to the impacts of temperature and rainfall on vector populations [54,55].

The apparent outlier of Rio Blanco and Orizaba depicted in Figure 4 is also clear in table of Lozano-Fuentes *et al.* [30], when using data on both pupae and larvae. We found that in general, the relationship between *Ae. aegypti* pupal abundance and temperature/rainfall is robust in the higher elevation regions (above ~1300 m ASL) and is weaker at lower elevations (below ~1300 m ASL). We believe the weaker relationship between temperature/rainfall and pupal abundance below ~1300 m in our study region is an indicator that once temperature and rainfall increase to a certain point, pupae are abundant no matter how much warmer or wetter it becomes; *i.e.*, the climate is ideal for immature *Ae. aegypti* development anywhere in the warm, wet regions below ~1300 m. From a mathematical standpoint, this observation indicates that the relationship between pupal abundance and climate variability likely becomes more asymptotic below ~1300 m, so the linear fits shown in Figure 4 may oversimplify what is in reality a non-linear relationship. However, for the purposes of our paper, in which we are simply trying to show that LST, rainfall and elevation are correlated with *Ae. aegypti* pupal abundance, these linear fits are adequate. It may also be possible that the comparatively greater urban density of Orizaba and Rio Blanco (Rio Blanco is a dormitory city of Orizaba) may be more favorable for higher numbers of *Ae. aegypti*, perhaps because of the sheer numbers of container habitats. In support of this conjecture, Coatepec has similar characteristics of elevation and temperature (although slightly lower rainfall) compared to Orizaba and Rio Blanco, but lower urban density; and it has also a lower presence and abundance of *Ae. aegypti* pupae (Table 1).

The findings of this study suggest the promise for future RS-based predictive models of *Ae. aegypti* population fluctuations and other applications, such as dengue outbreak prediction [56]. These results are even more promising when considering future remote sensing products with enhanced capabilities that may be available soon via new or planned NASA and/or partner missions. Of special note are the Visible Infrared Imaging Radiometer Suite (NPP-VIIRS), the Global Precipitation Measurement (GPM) mission and the Soil Moisture Active Passive (SMAP) mission. Carried onboard the Suomi National Polar-orbiting Partnership (NPP) satellite, the VIIRS sensor will use visible and infrared wavelengths to study land, atmosphere, ice and ocean. Besides LST, VIIRS will make observations on active fires, vegetation, ocean color, sea surface temperature and other surface features available to the scientific community [57]. These observations will facilitate the study of climate change, clouds and aerosols; phytoplankton and sediment in the seas; forest cover and productivity; and changes in polar sea ice. The similarities between VIIRS and MODIS will provide continuity for monitoring programs conducted with MODIS. The Global Precipitation Measurement (GPM) mission will provide continuity to TRMM data to monitor precipitation and for other hydrological applications [58]. Measuring soil moisture present at the Earth's land surface, the Soil Moisture Active Passive (SMAP) mission will be critical to flood assessment and drought monitoring, as well as to the study of the global carbon balance [59]. SMAP is expected to be another undertaking that will play an important role in the monitoring of environmental variables associated not just with vector transmitted diseases in general, but also, and especially, soil transmitted helminthes infections. Soil moisture is a critical determinant for the survival of helminth eggs in the soil [60–62].

The application of the remotely-sensed products in this study was useful for identifying correlations between environmental variables and the presence and abundance of *Ae. aegypti*. This finding is encouraging for using these products to identify areas most at risk of high pupal abundance at local scales on the order of 1-km. It is tempting to speculate that if these products work at the local scale for cities along our topographically complex transect in Mexico, they would work even better in areas where the topographic variability is comparatively smaller (and therefore, environmental variables do not vary as much in space), for example, in the southern part of the continental United States. However, since the abundance of *Ae. aegypti* is also closely linked to the characteristics of the human environment, it is imperative to also consider the potential confounding effects of socioeconomic and human ecology (e.g., housing style, water storage, *etc.*) factors associated with mosquito establishment and proliferation [3,63]. In addition to affecting mosquito abundance, socio-economic factors can play a role in the vulnerability of human inhabitants to dengue virus infection; for example, Hagenlocher *et al.* [63] constructed a composite index of socioeconomic vulnerability to dengue that included both indicators of susceptibility, as well as a lack of resilience.

An overarching goal, whether focusing on mosquito vector abundance, vector infection rates and the risk of virus transmission to humans, socioeconomic vulnerability or the risk for dengue epidemics, is to give public health professionals and the general public time to prepare for and attempt to prevent or mitigate disease outbreaks. For example, researchers have determined that the optimal lead time for dengue early warning for officials in Singapore would be three months in order to suppress an epidemic [64], and global risk maps have been developed to estimate the risk of dengue in Europe [65]. Better defining the linkages between environmental and climatic conditions and the incidence and geographic spread of mosquito vectors and dengue, together with the ability to use remotely-sensed observations to detect conditions signaling increased risk, would be of great value for future outbreak prediction and disease suppression.

## 5. Conclusions

Strong correlations were found between the abundance of the dengue virus mosquito vector, *Ae. aegypti*, and RS-derived nighttime LST, elevation or rainfall along a geographic climate/elevation gradient in Central Mexico. The results were consistent with what was found by Lozano-Fuentes *et al.* [30] using data from the same mosquito field surveys, but meteorological data from weather stations. These findings show that data from the three analyzed RS products can be used to predict *Ae. Aegypti* abundance, but further studies are needed to define the climatic and socio-economic conditions under which the correlations observed herein can be assumed to apply.

**Acknowledgments:** This study was funded by grants from the National Aeronautic and Space Administration (NASA) and the National Science Foundation (NSF). The NASA funds were granted through ROSES-2010: Earth Science Applications Feasibility Studies: Public Health (A.31); and to the Applied Earth Sciences group at the Marshall Space Flight Center (MSFC) (10-PHFEAS10-0010). The NSF funds were granted to the University Corporation for Atmospheric Research (GEO-1010204). The National Center for Atmospheric Research is partially funded by NSF. A partial contribution was provided by an appointment to the NASA Postdoctoral Program at the Marshall Space Flight Center/National Space Science and Technology Center/NASA Global Hydrology and Climate Center in Huntsville, AL, USA; administered by Oak Ridge Associated Universities through a contract with NASA. We express our appreciation to Douglas Rickman and Mohammad Z. Al-Hamdan for their guidance in the processing of remote sensing applications, as well as to SERVIR for providing logistics and support. We thank Gina Wade for helping with the proposal. We also thank Carolina Ochoa-Martinez and Berenice Tapia-Santos for running the field teams and to Kevin Kobylinski and Chris Uejio for helping out in the field. Lastly but not the least we thank the students from Veracruz University who helped with the field work completing the survey and collecting the *in situ* data.

**Author Contributions:** Max J. Moreno-Madriñán is the main author who conceived and designed the manuscript, processed most of the remote sensing applications, performed the analyses, wrote most of the manuscript and participated in a portion of the field *in situ* data collection. Lars Eisen, and Andrew J. Monaghan contributed in the writing of the manuscript and analysis of results. William L. Crosson revised the manuscript and along with the first author and Andrew J. Monaghan, Saul Lozano-Fuentes, Sue M. Estes, Maurice G. Estes, Mary Hayden and Emily Zielinski-Gutierrez participated in the field collecting *in situ* data. Sarah N. Hemmings contributed in the writing of the manuscript. Mary Hayden, Lars Eisen and Saul Lozano-Fuentes conceived the initial study while Sue Estes, William L. Crosson, Maurice G. Estes and Dale Quattrochi conceived the extension to remote sensing applications. Carlos M. Welsh-Rodriguez coordinated the survey for *in situ* data collection and facilitated the logistics needed. Saul Lozano-Fuentes wrote the field protocol, trained the field teams, and leaded most of the field work. William L. Crosson, Maurice G. Estes, Dan E. Irwin and Dale Quattrochi contributed in the processing of remote sensing applications. Dan E. Irwin facilitated the logistics required for the remote sensing work, the connection with end-users and the means for this study to happen. All authors read and approved the final manuscript.

**Conflicts of Interest:** The authors declare no conflict of interest.

## References

1. Epstein, P.R. Climate change and emerging infectious disease. *Microbes Infect.* **2001**, *3*, 747–754. [CrossRef]

2.  Githeko, A.K.; Lindsay, S.W.; Confalonieri, U.E.; Patz, J.A. Climate change and vector-borne diseases: A regional analysis. *Bull. World Health Organ.* **2000**, *78*, 1136–1147.

3.  Gubler, D.J.; Reiter, P.; Ebi, K.L.; Yap, W.; Nasci, R.; Patz, J.A. Climate variability and change in the United States: Potential impacts on vector and rodent-borne diseases. *Environ. Health Perspect.* **2001**, *109* (Suppl. S2), 223–233. [CrossRef]

4.  Haines, A.; Kovats, R.S.; Campbell-Lendrum, D.; Corvalan, C. Climate change and human health: Impacts, vulnerability and public health. *Public Health* **2006**, *120*, 585–596. [CrossRef]

5.  Lafferty, K.D. The ecology of climate change and infectious diseases. *Ecol. Soc. Am.* **2009**, *90*, 888–900.

6.  McMichael, A.J.; Woodruff, R.E.; Hales, S. Climate change and human health: Present and future risk. *Lancet* **2006**, *367*, 859–869. [CrossRef]

7.  Ceccato, P.; Connor, S.J.; Jeanne, I.; Thomson, M.C. Application of geographical information systems and remote sensing technologies for assessing and monitoring malaria risk. *Parassitologia* **2005**, *47*, 81–96.

8.  Mills, J.N.; Gage, K.L.; Khan, A.S. Potential influence of climate change on vector-borne and zoonotic diseases: A review and proposed research plan. *Environ. Health Perspect.* **2010**, *118*, 1507–1514. [CrossRef]

9.  Parkinson, A.J.; Butler, J.C. Potential impacts of climate change on infectious diseases in the Arctic. *Int. J. Circumpolar Health* **2005**, *64*, 478–486.

10. Semenza, J.C.; Menne, B. Climate change and infectious diseases in Europe. *Lancet Infect. Dis.* **2009**, *9*, 365–375. [CrossRef]

11. Allen, T.R.; Wong, D.W. Exploring GIS, spatial statistics and remote sensing for risk assessment of vector-borne diseases: A West Nile virus example. *Int. J. Risk Asmt. Mgmt.* **2006**, *6*, 253–275. [CrossRef]

12. Hay, S.I.; Packer, M.J.; Rogers, D.J. The impact of remote sensing on the study and control of invertebrate intermediate hosts and vectors for disease. *Int. J. Remote Sens.* **1997**, *18*, 2899–2930. [CrossRef]

13. Hay, S.I.; Snow, R.W.; Rogers, D.J. From predicting mosquito habitat to malaria seasons using remotely sensed data: Practice, problems and perspectives. *Parasitol. Today* **1998**, *14*, 306–313. [CrossRef]

14. Kalluri, S.; Gilruth, P.; Rogers, D.; Szczur, M. Surveillance of arthropod vector-borne infectious disease using remote sensing techniques: A review. *PLoS Pathog.* **2007**, *3*, 1361–1371.

15. Gubler, D.J.; Clark, G.G. Dengue/Dengue hemorrhagic fever: The emergence of a global health problem. *Emerg. Infect. Dis.* **1995**, *1*, 55–57. [CrossRef]

16. Bhatt, S.; Gething, P.W.; Brady, O.J.; Messina, J.P.; Farlow, A.W.; Moyes, C.L.; Drake, J.M.; Brownstein, J.S.; Hoen, A.G.; Sankoh, O.; *et al.* The global distribution and burden of dengue. *Nature* **2013**, *496*, 504–507. [CrossRef]

17. Brunkard, J.M.; Robles Lopez, J.L.; Ramirez, J.; Cifuentes, E.; Rothenberg, S.J.; Hunsperger, E.A.; Moore, C.G.; Brussolo, R.M.; Villarreal, N.A; Haddad, B.M. Dengue fever seroprevalence and risk factors, Texas-Mexico border, 2004. *Emerg. Infect. Dis.* **2007**, *13*, 1477–1483. [CrossRef]

18. Chowell, G.; Sanchez, F. Climate-based descriptive models of dengue fever: The 2002 epidemic in Colima, Mexico. *J. Environ. Health* **2006**, *68*, 40–44.

19. Chowell, G.; Cazelles, B.; Broutin, H.; Munayco, C.V. The influence of geographic and climate factors on the timing of dengue epidemics in Peru, 1994–2008. *BMC Infect. Dis.* **2011**, *11*. [CrossRef]

20. Colon-Gonzalez, F.J.; Lake, I.R.; Bentham, G. Climate variability and dengue fever in warm and humid Mexico. *Am. J. Trop. Med. Hyg.* **2011**, *84*, 757–763. [CrossRef]

21. Degallier, N.; Favier, C.; Menkes, C.; Lengaigne, M.; Ramalho, W.M.; Souza, R.; Servain, J.; Boulanger, J.P. Toward an early warning system for dengue prevention: Modeling climate impact on dengue transmission. *Clim. Chang.* **2010**, *98*, 581–592. [CrossRef]

22. Fuller, D.O.; Troyo, A.; Beier, J.C. El Nino Southern Oscillation and vegetation dynamics as predictors of dengue fever cases in Costa Rica. *Environ. Res. Lett.* **2009**, *4*, 140111–140118.

23. Herrera-Martinez, A.D.; Rodriguez-Morales, A.J. Potential influence of climate variability on dengue incidence registered in a western pediatric hospital of Venezuela. *Trop. Biomed.* **2010**, *27*, 280–286.

24. Hurtado-Diaz, M.; Riojas-Rodriguez, H.; Rothenberg, S.J.; Gomez-Dantes, H.; Cifuentes, E. Short communication: Impact of climate variability on the incidence of dengue in Mexico. *Trop. Med. Int. Health* **2007**, *12*, 1327–1337. [CrossRef]

25. Johansson, M.A.; Cummings, D.A.T.; Glass, G.E. Multiyear climate variability and dengue-El Nino Southern Oscillation, weather, and dengue incidence in Puerto Rico, Mexico, and Thailand: A longitudinal data analysis. *PLoS Med.* **2009**, *6*. [CrossRef]

26. Johansson, M.A.; Dominici, F.; Glass, G.E. Local and global effects of climate on dengue transmission in Puerto Rico. *PLoS Negl. Trop. Dis.* **2009**, *3*. [CrossRef]

27. Lowe, R.; Bailey, T.C.; Stephenson, D.B.; Graham, R.J.; Coelho, C.A.S.; Carvalho, M.S.; Barcellos, C. Spatiotemporal modelling of climate-sensitive disease risk: Towards an early warning system for dengue in Brazil. *Comput. Geosci.* **2011**, *37*, 371–381. [CrossRef]

28. Machado-Machado, E.A. Empirical mapping of suitability to dengue fever in Mexico using species distribution modeling. *Appl. Geogr.* **2012**, *33*, 82–93. [CrossRef]

29. Eisen, L.; Moore, C.G. *Aedes (Stegomyia) aegypti* in the Continental United States: A vector at the cool margin of its geographic range. *J. Med. Entomol.* **2013**, *50*, 467–478. [CrossRef]

30. Lozano-Fuentes, S.; Hayden, M.H.; Welsh-Rodriguez, C.; Ochoa-Martinez, C.; Tapia-Santos, B.; Kobylinski, K.C.; Uejio, C.K.; Zielinski-Gutierrez, E.; Monache, L.D.; Monaghan, A.J.; *et al.* The dengue virus mosquito vector *Aedes aegypti* at high elevation in México. *Am. J. Trop. Med. Hyg.* **2012**, *87*, 902–909. [CrossRef]

31. Online Visualization and Analysis System (TOVAS). Available online: http://disc.sci.gsfc.nasa.gov/precipitation/tovas (accessed on 6 September 2013).

32. Huffman, G.J.; Coauthors. The TRMM multisatellite Precipitation Analysis (TMPA): Quasi-global, multiyear, combined-sensor precipitation estimates at fine scales. *J. Hydrometeorol.* **2007**, *8*, 38–55. [CrossRef]

33. Reverb/Echo NASA EOS Data and Information System (EOSDIS) Website. Available online: http://reverb.echo.nasa.gov/reverb/ (accessed on 20 December 2013).

34. Xiong, X.; Barnes, W. An overview of MODIS radiometric calibration and characterization. *Adv. Atmos. Sci.* **2006**, *23*, 69–79. [CrossRef]

35. Parkinson, C.L. Aqua: An earth-observing satellite mission to examine water and other climate variables. *IEEE Trans. Geosci. Remote Sens.* **2003**, *41*, 173–183. [CrossRef]

36. Coll, C.; Caselles, V.; Galve, J.M.; Valor, E.; Niclòs, R.; Sánchez, J.M.; Rivas, R. Ground measurements for the validation of land surface temperatures derived from AATSR and MODIS data. *Remote Sens. Environ.* **2005**, *97*, 288–300. [CrossRef]

37. Wan, Z.; Zhang, Y.; Zhang, Y.Q.; Li, Z-L. Validation of the land-surface temperature products retrieved from Moderate Resolution Imaging Spectroradiometer data. *Remote Sens. Environ.* **2002**, *83*, 163–180. [CrossRef]

38. Farr, T.G.; Rosen, P.A.; Caro, E.; Crippen, R.; Duren, R.; Hensley, S.; Kobrick, M.; Paller, M.; Rodriguez, E.; Roth, L.; *et al.* The shuttle radar topography mission. *Rev. Geophys.* **2007**, *45*. [CrossRef]

39. Global Data Explorer. Available online: http://gdex.cr.usgs.gov/gdex/ (accessed on 7 October 2013).

40. Chaves, L.F.; Morrison, A.C.; Kitron, U.D.; Scott, T.W. Nonlinear impacts of climatic variability on the density-dependent regulation of an insect vector of disease. *Glob. Chang. Biol.* **2012**, *18*, 457–468.

41. Chaves, L.F.; Scott, T.W.; Morrison, A.C.; Takada, T. Hot temperatures can force delayed mosquito outbreaks viasequential changes in *Aedes aegypti* demographic parameters inautocorrelated environments. *Acta Trop.* **2014**, *129*, 15–24. [CrossRef]

42. Lana, R.M.; Carneiro, T.G.S.; Honório, N.A.; Codeco, C.T. Seasonal and nonseasonal dynamics of *Aedes aegypti* in Rio de Janeiro, Brazil: Fitting mathematical models to trap data. *Acta Trop.* **2014**, *129*, 25–32. [CrossRef]

43. Marteis, L.; Makowski, L.; Conceicao, K.; La Corte, R. Identification and spatial distribution of key premises for *Aedes aegypti* in the Porto Dantas neighborhood, Aracaju, Sergipe State, Brazil. *Cad. Sause Publica* **2013**, *29*, 368–378. [CrossRef]

44. Rubio, A.; Cardo, M.V.; Carbajo, A.E.; Vezzani, D. Imperviousness as a predictor for infestation levels of container-breeding mosquitoes in a focus of dengue and Saint Louis encephalitis in Argentina. *Acta Trop.* **2013**, *128*, 680–685. [CrossRef]

45. Honório, N.A.; Codeco, C.T.; Alves, F.C.; Magalhães, M.A.F.M.; Lourenco-de-Oliveira, R. Temporal distribution of *Aedes aegypti* in different districts of Rio de Janeiro, Brazil, measured by two types of traps. *J. Med. Entomol.* **2009**, *46*, 1001–1014. [CrossRef]

46. Eisen, L.; Monaghan, A.J.; Lozano-Fuentes, S.; Steinhoff, D.F.; Hayden, M.H.; Bieringer, P.E. The impact of temperature on the bionomics of *Aedes (Stegomyia) aegypti*, with special reference to the cool geographic range margins. *J. Med. Entomol.* **2014**, in press.

47. Riedel, N.; Vounatsou, P.; Miller, J.M.; Gosoniu, L.; Chizema-Kawesha, E.; Mukonka, V.; Steketee, R.W. Geographical patterns and predictors of malaria risk in Zambia: Bayesian geostatistical modelling of the 2006 Zambia national malaria indicator survey (ZMIS). *Malar. J.* **2010**, *9*, 1–13. [CrossRef]

48. Gebremariam, Y.B. Formulating the Relationship between Temperature and Elevation and Evaluating the Influence of Major Land Covers on Surface Temperature (A Case Study in Part of Northern Ethiopia). Available online: http://www.scribd.com/doc/21553665/ (accessed on 20 March 2014).

49. Buyantuyev, A.; Wu, J. Urban heat islands and landscape heterogeneity: Linking spatiotemporal variations in surface temperatures to land-cover and socioeconomic patterns. *Landsc. Ecol.* **2010**, *25*, 17–33. [CrossRef]

50. O'Rourke, P.A.; Terjung, W.H. Relative influence of city structure on canopy photosynthesis. *Int. J. Biomet.* **1981**, *25*, 1–19. [CrossRef]

51. Focks, D.A.; Haile, D.G.; Daniels, E.; Mount, G.A. Dynamic life table model for *Aedes aegypti* (Diptera: Culicidae): Analysis of the literature and model development. *J. Med. Entomol.* **1993**, *30*, 1003–1017.

52. Focks, D.A.; Haile, D.G.; Daniels, E.; Mount, G.A. Dynamic life table model for *Aedes aegypti* (Diptera: Culicidae): Simulation results and validation. *J. Med. Entomol.* **1993**, *30*, 1018–1028.

53. Gomes, A.F.; Nobre, A.A.; Cruz, O.G. Temporal analysis of the relationship between dengue and meteorological variables in the city of Rio de Janeiro, Brazil, 2001–2009. *Cad. Saúde Pública* **2012**, *28*, 2189–2197. [CrossRef]

54. Alto, B.W.; Bettinardi, D. Temperature and dengue virus infection in mosquitoes: Independent effects on the immature and adult stages. *Am. J. Trop. Med. Hyg.* **2013**, *88*, 497–505. [CrossRef]

55. Lambrechts, L.; Paaijmans, K.P.; Fansiri, T.; Carrington, L.B.; Kramer, L.D.; Thomas, M.B.; Scott, T.W. Impact of daily temperature fluctuations on dengue virus transmission by *Aedes aegypti*. *Proc. Natl. Acad. Sci. USA* **2011**, *108*, 7460–7465. [CrossRef]

56. Buczak, A.L.; Koshute, P.T.; Babin, S.M.; Feighner, B.H.; Lewis, S.H. A data-driven epidemiological prediction method for dengue outbreaks using local and remote sensing data. *BMC Med. Inform. Decis. Mak.* **2012**, *12*, 1–20.

57. NPOESS Preparatory Project (NPP): Building a Bridge to a New Era of Earth Observations. Available online: http://www.nasa.gov/pdf/596329main_NPP_Brochure_ForWeb.pdf (accessed on 7 March 2014).

58. Kidd, C.; Hou, A. The Global Precipitation Measurement (GPM) Mission: Hydrological Applications. In Proceedings of the 2012 EUMETSAT Meteorological Satellite Conference, Sopot, Poland, 3–7 September 2012.

59. Entekhabi, B.D.; Njoku, E.G.; O'Neill, P.E.; Kellogg, K.H.; Crow, W.T.; Edelstein, W.N.; Entin, J.K.; Goodman, S.D.; Jackson, T.J.; Johnson, J.; *et al.* The soil moisture active passive (SMAP) mission. *Proc. IEEE* **2010**. [CrossRef]

60. Bethony, J.; Brooker, S.; Albonico, M; Geiger, S.M.; Loukas, A.; Diemert, D.; Hotez, P.J. Soil-transmitted helminth infections: Ascariasis, richuriasis, and hookworm. *Lancet* **2006**, *367*, 1521–1532. [CrossRef]

61. Brooker, S.; Michael, E. The potential of geographical information systems and remote sensing in the epidemiology and control of human helminth infections. *Adv. Parasitol.* **2000**, *47*, 245–270. [CrossRef]

62. Moreno-Madriñán, M.J.; Parajón, D.G.; Al-Hamdan, M.Z.; Martinez, R.; Rickman, D.L.; Parajon, L.; Estes, S. Use of Remote Sensing/Geographical Information Systems (RS/GIS) to Identify Environmental Limits of Soil Transmitted Helminthes (STHs) Infection in Boaco, Nicaragua. Proceedings of American Public Health Association (APHA) 139th Annual Meeting and Exposition, Washington DC, USA, 29 October–2 November 2011; Available online: http://ntrs.nasa.gov/archive/nasa/casi.ntrs.nasa.gov/20120001772.pdf (accessed on 9 May 2014).

63. Hagenlocher, M.; Delmelle, E.; Casas, I.; Kienberger, S. Assessing socioeconomic vulnerability to dengue fever in Cali, Colombia: Statistical *vs.* expert-based modeling. *Int. J. Health Geogr.* **2013**, *12*. [CrossRef]

64. Hii, Y.L.; Rocklöv, J.; Wall, S.; Ng, L.C.; Tang, C.S.; Ng, N. Optimal lead time for dengue forecast. *PLoS Negl. Trop. Dis.* **2012**, *6*. [CrossRef]

65. Rogers, D.J.; Suk, J.E.; Semenza, J.C. Using global maps to predict the risk of dengue in Europe. *Acta Trop.* **2014**, *129*, 1–14. [CrossRef]

isprs International Journal of
Geo-Information

MDPI

*Article*

# Canadian Forest Fires and the Effects of Long-Range Transboundary Air Pollution on Hospitalizations among the Elderly

George E. Le [1], Patrick N. Breysse [2], Aidan McDermott [3], Sorina E. Eftim [2,4], Alison Geyh [2,†], Jesse D. Berman [5] and Frank C. Curriero [2,3,6,*]

[1]   Department of Health Policy and Management, Johns Hopkins Bloomberg School of Public Health, 615 N. Wolfe Street, Baltimore, MD 21205, USA; gle@jhsph.edu
[2]   Department of Environmental Health Sciences, Johns Hopkins Bloomberg School of Public Health, 615 N. Wolfe Street, Baltimore, MD 21205, USA; pbreysse@jhsph.edu (P.N.B.); sorina.eftim@gmail.com (S.E.E.)
[3]   Department of Biostatistics, Johns Hopkins Bloomberg School of Public Health, 615 N. Wolfe Street, Baltimore, MD 21205, USA; amcdermo@jhsph.edu
[4]   ICF International, 9300 Lee Highway, Fairfax, VA 22031, USA
[5]   Yale School of Forestry & Environmental Studies, Yale University, 195 Prospect Street, New Haven, CT 06511, USA; jesse.berman@yale.edu
[6]   Department of Epidemiology, Johns Hopkins Bloomberg School of Public Health, 615 N. Wolfe Street, Baltimore, MD 21205-2103, USA
*    Author to whom correspondence should be addressed; fcurrier@jhsph.edu; Tel.: +1-410-614-5817; Fax: +1-410-955-9334.
†    Alison Geyh deceased on 20 February 2011.

Received: 3 March 2014; in revised form: 22 April 2014; Accepted: 5 May 2014; Published: 19 May 2014

**Abstract:** In July 2002, lightning strikes ignited over 250 fires in Quebec, Canada, destroying over one million hectares of forest. The smoke plume generated from the fires had a major impact on air quality across the east coast of the U.S. Using data from the Medicare National Claims History File and the U.S. Environmental Protection Agency (EPA) National air pollution monitoring network, we evaluated the health impact of smoke exposure on 5.9 million elderly people (ages 65+) in the Medicare population in 81 counties in 11 northeastern and Mid-Atlantic States of the US. We estimated differences in the exposure to ambient $PM_{2.5}$—airborne particulate matter with aerodynamic diameter of $\leq 2.5$ μm—concentrations and hospitalizations for cardiovascular, pulmonary and injury outcomes, before and during the smoke episode. We found that there was an associated 49.6% (95% confidence interval (CI), 29.8, 72.3) and 64.9% (95% CI, 44.3–88.5) increase rate of hospitalization for respiratory and cardiovascular diagnoses, respectively, when the smoke plume was present compared to before the smoke plume had arrived. Our study suggests that rapid increases in $PM_{2.5}$ concentrations resulting from wildfire smoke can impact the health of elderly populations thousands of kilometers removed from the fires.

**Keywords:** air pollution; hospitalizations; $PM_{2.5}$; forest fires; global climate change

## 1. Introduction

Forest fires are known to be a major source of air pollutants [1] on a local and a global scale [2–6]. Each year, combustion products from local and distant wildfires impact large populations worldwide [5, 7–13]. The atmospheric pollutant that most consistently increases with biomass smoke from wildfires is suspended fine particulate matter (PM), which is commonly associated with increased mortality and morbidity [1,4,14–17]. The PM in biomass smoke consists mainly of black carbon (soot and charcoal particles), organic carbon, sulfates and/or nitrates, potassium carbonate and silica [12,18].

Short-term exposures to fine particulate matter, $PM_{2.5}$ (airborne particulate matter with aerodynamic diameter of $\leq 2.5$ μm) have been associated with increased mortality and morbidity in various communities around the world and in the United States [19–28]. Most studies addressing health impacts of short-term exposures are related to anthropogenically generated PM, which are commonly associated with automobile combustion and industrial practices. There has been a limited but growing body of literature addressing the impact of shorter-term exposure to smoke from forest and bush fires (referred to as wildfires in this paper) [8,9,11,17,29–33]. The majority of these studies have examined the impact on nearby local communities of exposures to wildfire aerosols. Associated health effects include increased emergency department visits and hospital admissions for chronic obstructive pulmonary disease (COPD), bronchitis, asthma and chest pain [7,13,15,34–37]. For example, the San Diego wildfire in October 2003 caused the daily 24 h average $PM_{2.5}$ concentrations to exceed 150 μg/m$^3$, and was associated with significant increases in hospital room emergency room visits for asthma, respiratory problems, eye irritation, and smoke inhalation [38], and increased eye and respiratory symptoms, medication use and physician visits in children living in the San Diego area [30]. In Canada, Moore *et al.* estimated that forest fire smoke in 2003 was associated with excess respiratory complaints in Kelowna (Kelowna, BC, Canada) area residents [31].

The wildfire aerosols have lifetimes on the order of many days [12], which allows transport over large distances [4,8,11,39]. While it is clear that local populations are affected by wildfire events, a growing concern is the potential health impact on geographically distant populations, specifically in susceptible groups such as the elderly. Epidemiologic research has identified the elderly, who are more likely to have pre-existing lung and heart diseases, as a population vulnerable to the effects of short-term exposures to air pollution including fine particles [24,40–50].

In July 2002, a dramatic increase in forest fire activity was registered in the province of Quebec, Canada [5]. Specifically, on 7 July 2002 at least 85 fires were burning out of control and destroyed approximately one million hectares of forest that month. The Canadian Forest Services indicated that the major causal factors contributing to these fires were a long period without precipitation and strong winds coming from the north. Lightning and dry conditions sparked fires on 2 July 2002 in two separate regions southeast of James Bay, which is between 200 and 400 miles north of the U.S. border.

The smoke plume generated from these forest fires had a major impact on air quality across the east coast of the United States during the first week in July 2002 [5,51]. The plume was carried by strong northerly winds from Quebec across the U.S. border covering a distance that extended from north of Montreal to northern Virginia and Maryland [5,51]. Satellite images show the plume on 7 July covering parts of New Hampshire, Vermont, and New York, and diffuse and patchy over the eastern seaboard down to Washington, DC (Figure 1: MODIS satellite image on 7 July 2002 [52]. An air quality study being conducted in Baltimore, MD at that time reported that the 24-hour $PM_{2.5}$ concentrations reached 86 μg/m$^3$ on 7 July, resulting in as much as a 30-fold increase in the daily ambient $PM_{2.5}$ concentrations [5] during the peak period of haze over the city. On that day, the highest $PM_{2.5}$ concentration (338 μg/m$^3$) was reported in New Hampshire [51]. In many cities in the region (such as New York City and Philadelphia), health advisories were issued for residents with espiratory conditions [53].

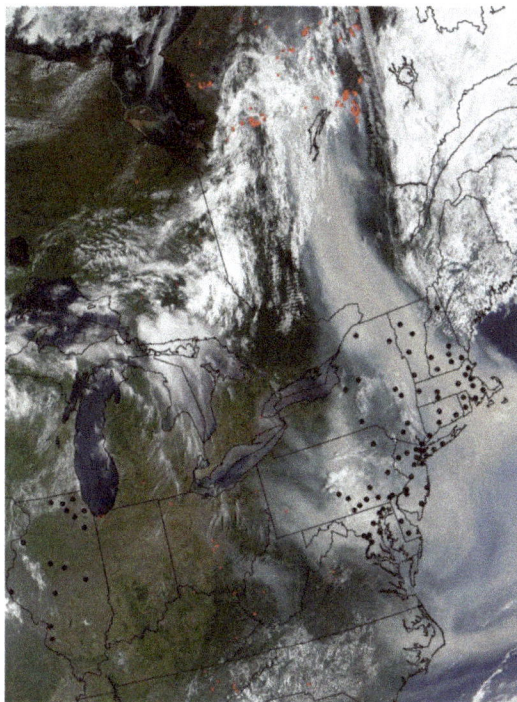

**Figure 1.** MODIS satellite image taken on 7 July 2002, 10:35 EDT. The red dots mark areas of high forest fire activity. The black dots represent the centroids of counties used in our analysis.

This transboundary wildfire smoke episode offered a unique opportunity to evaluate the vulnerability of a susceptible population in the United States exposed to smoke generated from a large-scale wildfires more than a thousand kilometers away. In this paper, we assess the relationship between hospital admissions for individuals 65 years and older and $PM_{2.5}$ concentrations during July 2002 for U.S. states impacted by the Canadian wildfire plume. Unlike previous research that focused on health impacts on populations living near or relatively close to wildfire events, this paper presents one of the first analyses of wildfire smoke health impacts to populations living at great distances, particularly within the U.S., from fires and not previously identified to be at significant risk.

## 2. Methods

### 2.1. Study Area

The study population includes 5.9 million Medicare enrollees aged 65 or older, residing in 81 U.S. counties in the northeastern and mid-Atlantic regions of the U.S. (11 states) and in Illinois. Medicare data included daily health care hospitalization information allowing us to look at changes in hospitalization rates associated with changes in PM air pollution. The study was restricted to 81 plume affected counties that could provide $PM_{2.5}$ data at least once every five days during our study period. These counties are identified in Figure 1. The state of Illinois was chosen as a reference area not affected by the plume for descriptive comparison to the forest fire induced levels of $PM_{2.5}$.

## 2.2. Data Sources

Daily counts of hospital admissions for June–July 2002 were obtained from billing claims for the Medicare National Claims History Files. Each billing claim contains the date of admission, age, sex, place of residence, and cause of hospitalization. Coding for hospitalization is based on the ninth revision of the International Classification of Diseases (ICD-9) [54]. The daily county-specific hospitalization rates were obtained by dividing the daily county-specific admission counts by the daily county-specific number of individuals at risk defined as the number of individuals enrolled in Medicare. The number of individuals at risk, the total Medicare population, varied each day. The daily counts of each health event within each county were obtained by summing the number of hospital admissions for each of the diseases considering both primary and secondary diagnoses. This study was exempt from the Johns Hopkins Bloomberg School of Public Health's Institutional Review Board because the data for this study did not involve individual identifiers.

The $PM_{2.5}$ air pollution monitoring data were obtained from the U.S Environmental Protection Agency Air Quality System [55]. Temperature and dew point temperature data were gathered from the National Climatic Data Center (NCDC) on the Earth-Info CD database [56]. The analysis was restricted to the 11 northeastern states affected by the plume [53]. Although, carbon monoxide (CO) is a gaseous pollutant commonly associated with forest fires, we did not include it in our analysis because only a quarter of the counties of interest reported CO monitoring data.

## 2.3. Statistical Analysis

County average $PM_{2.5}$ values for use in subsequent regression models were obtained using the geostatistical method known as block kriging. Kriging is a statistical method widely used in the environmental sciences that produces optimal spatial predictions and in the block kriging version produces statistically optimal aggregate level estimates based on point level (monitored) data [57,58]. To implement block kriging, a fine grid is placed over the study area and concentrations predicted at each grid cell using an ordinary kriging approach. Grid points within a specific county are then averaged to provide a single county measure. This allows the more sparsely point monitored $PM_{2.5}$ measurements to be optimally estimated at an aggregate county level by spatially borrowing information from proximal $PM_{2.5}$ measurements. To improve characterization of spatial dependence (a crucial step in the kriging process) and to address issues of edge effects in the block kriging procedure, air monitor values from all inclusive states in the affected region (Figure 1 and including Maine, Virginia, and West Virginia) were utilized. A similar approach was taken for block kriging county estimates in the chosen unaffected reference area of Illinois. Kriging was performed with the "gstat" package in the R Statistical Software [59]. The NCDC weather monitoring network is adequately dense so that county level temperature and dew point temperature estimates were obtained from averaging measurements observed at county specific monitoring stations.

In our primary analysis, we considered specific cardiovascular and respiratory outcomes that have been associated in the literature with short-term exposure to $PM_{2.5}$. This selection would also allow for comparability with previous studies. Outcomes considered were grouped in three broad categories: cardiovascular outcomes (ICD-9 codes (390–459)), respiratory outcomes (460–519) and injuries (800–849) (selected as our control outcome).

In our secondary analysis we targeted specific cardiovascular and respiratory outcomes, that represent endpoints more often associated with acute impacts of short-term exposures to $PM_{2.5}$ in the general population [26,60]: hypertensive disease (401–405), myocardial infarction (410), ischemic heart disease (410–414), acute pulmonary heart disease (415), acute heart disease (410–425), heart rhythm disturbances (426–427), heart failure (428), stroke (430–438), peripheral vascular disease (440–448), and asthma (493), chronic obstructive pulmonary disease (COPD) (490–492, 496), respiratory tract infections (RTI) (464–466, 480–487) and acute respiratory tract infections (460–466) (Table 1).

Satellite imagery and back-trajectory modeling showed that the smoke plume covering the northeastern states on 6–8 July 2002, a Saturday, Sunday, and Monday [5] originated from Province of

*ISPRS Int. J. Geo-Inf.* **2014**, *3*, 713–731

Quebec. Thus we defined 6–8 July 2002 as our haze period, when air pollution was also at its highest and compared it to a control non-haze period, corresponding to the same days of the preceding week (29, 30 June and 1 July 2002). We chose not to include the same days in the following week as part of the control because of potential persisting effects in the post-haze period [19] The same days of the week were used in this fashion to minimize potential time-varying confounding effects, such as might be expected for weekday *versus* weekend. Within our period of interest, we observed a total of 5772 for all respiratory hospital admissions, and 18,316 hospital admissions for all cardiovascular hospital admissions.

Table 1. Diagnosis Categories.

| Reason for Hospital Admission | ICD-9 Codes |
| --- | --- |
| Respiratory Outcomes | (460–519) |
| Chronic Obstructive Pulmonary Disease (COPD) | (490–492, 496) |
| Asthma | (493) |
| Respiratory Tract Infections | (464–466, 480–487) |
| Cardiovascular Outcomes | (390–492, 496) |
| Cerebrovascular Disease | (430–438) |
| Heart Failure | (440–448) |
| Stroke | (410) |
| Peripheral Vascular Disease | (410–414) |
| Myocardial Infarction | (426–427) |
| Ischemic Heart Disease | (426–428) |
| Heart Rhythm Disturbances | (46–427) |
| Heart Disease | (426–428) |
| Other Heart Disease | (420–425, 428) |
| Hypertension | (401–405) |
| Injury | (800–849) |

To assess the impact of exposure to $PM_{2.5}$ during the haze period on hospitalization rates for the outcomes of interest, we linked billing claims from Medicare with daily concentrations of $PM_{2.5}$ by county of residence for the Medicare enrollees [60,61] and employed Poisson regression. Regression inference was based on a generalized estimating equation approach (GEE) [62] to account for the possibility of residual temporal autocorrelation in the county-specific hospitalization rates. We assumed an independent working correlation structure for the county-specific daily hospitalization rates, which assumes that the correlation between daily rates decreases with time.

We let $Y_t^c$ and $\mu_t^c$ be the observed and expected daily number of cause-specific hospitalizations, $N_t^c$ be the number of people at risk on day t in county c, $X_t^c$ be the daily average $PM_{2.5}$ for county c, $T_t^c$ and $D_t^c$ the daily average temperature and dew point temperature for county c. We fit the following log-linear regression model:

$$\log(\mu_t^c) = \beta_0 + \beta_1 X_t^c + \beta_2 P_t^c + \beta_3 T_t^c + \beta_4 D_t^c + \log(N_t^c) \qquad (1)$$

with $P_t^c$ an indicator variable representing haze *versus* non-haze period and $\log(N_t^c)$ representing the regression offset. The parameter $\beta_1$ denotes the log relative risk of cause-specific hospitalization associated with one unit ($\mu g/m^3$) increase in daily average $PM_{2.5}$ (block kriged county estimates) adjusted for temperature and dew-point temperature. The parameter $\beta_2$ denotes the increase in log relative rate of hospitalizations during the haze period compared to that during the non-haze period.

To account for potential delays in disease incidence after exposure, we also explored single lag models, where we substitute $\beta_1 X_{t-l}^c$ in the county-specific model (1) for $\beta_1 X_t^c$, where $l$ = 0, 1, 2 days and $X^c_{t-l}$ is the PM$_{2.5}$ concentration for county $c$ on day $t$ at a lag of $l$ days [63]. A lag of 0 days corresponds to the association between PM$_{2.5}$ concentrations on a given day and the risk of hospitalization on the same day. We also applied distributed lag models [64,65] to estimate the relative risk of hospitalization associated with cumulative exposure over the current day and the previous two days. The distributed lag models are obtained by substituting $\sum^2_{i=0} \beta_1 X^c_{t-l}$ in the county-specific model (Equation (1)) for $\beta_1 X^c_{t-l}$. Models with a PM$_{2.5}$ by haze period interaction $\beta_5 X^c_t P^c_t$ were also considered allowing for the relative risk of hospitalizations associated with PM$_{2.5}$ exposure to change during the haze period compared to the non-haze period.

## 3. Results

Smoke from wildfires in the Quebec region of Canada drifted over the Northeastern and Mid-Atlantic region of the United States on 6–8 July 2002 blanketing much of the area (Figure 1). Figure 2 shows the region wide average daily PM$_{2.5}$ concentrations for the two months surrounding this event from both the affected area and unaffected reference area of Illinois, which are shown to have similar ambient PM$_{2.5}$ concentrations. However, during the identified haze period we observe a substantial spike in PM$_{2.5}$ only in the affected region. The daily averages shown Figure 2 were obtained using 10% trimmed mean to average across monitors after correcting for yearly averages for each monitor; an accepted approach for summarizing longer times series and larger geographic areas of air pollution data [55]. Missing concentrations were imputed using a natural spline interpolation method that accounts for the daily seasonality in PM$_{2.5}$.

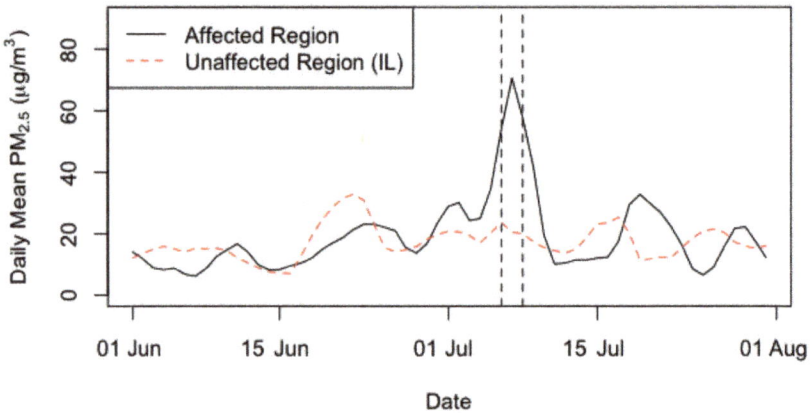

**Figure 2.** Countywide daily mean PM$_{2.5}$ ($\mu$g/m$^3$) in the affected and unaffected regions from 1 June to 31 Jul 2002.

Figure 3 displays for the affected region the spatial distribution in county specific average PM$_{2.5}$ concentrations (estimated via block kriging) as being substantially higher for days in the haze period compared to the same days in the preceding non-haze period. Table 2 presents summary statistics for these estimated county specific PM$_{2.5}$ concentrations in the affected states and the chosen Illinois reference area during the haze and non-haze periods. The average countywide PM$_{2.5}$ concentrations for 6–8 July 2002 were significantly higher (*p*-value < 0.001) in the haze period (mean 53.0 $\mu$g/m$^3$, standard deviation SD = 25.0), compared to non-haze period (mean 21.5 $\mu$g/m$^3$, SD = 10.3 for the affected states. No significant difference in average countywide PM$_{2.5}$ was found in the unaffected reference area of Illinois between the same haze and non-haze periods.

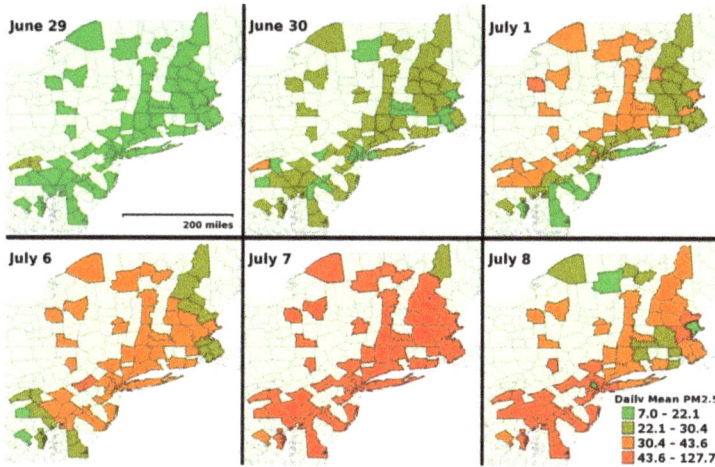

**Figure 3.** County level block kriged estimated mean $PM_{2.5}$ ($\mu g/m^3$) for the 81 affected counties in non-haze and haze periods. Top Row: Control period between 29 June–1 July designated as non-haze period. Bottom Row: Haze period between 6–8 July.

**Table 2.** Summary statistics for the block kriged estimated county average $PM_{2.5}$ ($\mu g/m^3$) for the plume affected area and Illinois reference area both stratified by the haze and non-haze periods. Data were pooled over the three days comprising both the haze and non-haze periods.

| Summary Statistics | Affected Region | | Unaffected Region (IL) | |
|---|---|---|---|---|
| | Haze | Non-Haze | Haze | Non-Haze |
| Min | 17.6 | 7.0 | 12.2 | 15.2 |
| 25th %tile | 35.6 | 13.7 | 16.0 | 19.0 |
| Median | 43.1 | 22.6 | 20.7 | 20.7 |
| Mean | 53.0 | 21.5 | 22.4 | 20.8 |
| 75th %tile | 69.1 | 25.3 | 29.7 | 22.1 |
| Max | 127.7 | 43.7 | 33.0 | 28.4 |
| SD | 25.0 | 10.3 | 7.2 | 5.0 |

Regression results for the parameter of interest, $\beta_2$ from model (1) representing the increase in log relative rate of hospitalization comparing the haze to non-haze period, are presented in Table 3. Results listed in Table 3 are for $(\exp(\beta_2) - 1) \times 100\%$ representing the percent change in admissions for the three primary outcome categories, all respiratory, all cardiovascular, and the selected control outcome injury. Compared to the non-haze period, this Medicare population had a 49.55% (95% confidence interval (CI): 29.82–72.29) significantly increased rate for respiratory related hospitalizations and a 64.93% (CI: 44.30–88.51) significantly increased rate for cardiovascular related hospitalizations during the haze period compared to the non-haze period, adjusting for weather and $PM_{2.5}$ on the same day (lag 0 model). For the chosen control outcome there was no significant increase in the rate of injury related hospitalizations between the haze and non-haze periods. Single and distributed lag model results show similar significant increases in respiratory and cardiovascular related hospitalizations although not as high as the lag 0 models.

**Table 3.** Percent change in hospital admissions for the haze period compared to the non-haze period in the affected region controlling for $PM_{2.5}$, temperature, and dew point. Bolded model lag types denote significant change in hospital admissions at the 0.05 level.

| Hospitalization Codes | PM$_{2.5}$ Model Lag | Percent Change | 95% Confidence Interval |
|---|---|---|---|
| | **Lag 0** | 49.55 | (29.82, 72.29) |
| All Respiratory | **Lag 1** | 21.14 | (2.69, 42.90) |
| | **Lag 2** | 23.36 | (10.20, 38.10) |
| | **Dlag1** | 30.20 | (7.66, 57.46) |
| | Dlag2 | 6.48 | (−14.63, 32.81) |
| | **Lag 0** | 64.93 | (44.30, 88.51) |
| | **Lag 1** | 36.06 | (18.43, 56.32) |
| All Cardiovascular | **Lag 2** | 49.28 | (34.39, 65.81) |
| | **Dlag1** | 33.85 | (12.45, 59.33) |
| | Dlag2 | 8.31 | (−13.89, 36.24) |
| | Lag 0 | 3.04 | (−21.31, 34.92) |
| | Lag 1 | 10.97 | (−14.04, 43.27) |
| Injury | Lag 2 | 13.23 | (−5.64, 35.89) |
| | Dlag1 | 0.37 | (−25.69, 35.55) |
| | Dlag2 | −6.04 | (−30.50, 27.02) |

The effect of the forest fire seemed to have a slightly greater impact on cardiovascular than respiratory admissions, a finding contrary to other studies [20,21], which found a higher impact on respiratory admissions than cardiovascular admissions. This may be because of over classification with the use of both primary and secondary discharge codes. However, an increase in cardiovascular and respiratory hospitalizations is consistent with the literature [8,16,24,50,66], though some literature has shown no increase in cardiovascular hospitalizations [2–21] or mortality [67].

Figure 4 displays the percent increase in hospital admissions for the haze period compared to the non-haze period in the affected region for all specific diagnoses of interest with single and distributed lag models. Percent increases in hospitalizations were significantly higher for same day models in all diagnosis groups except for respiratory tract infections, cerebrovascular disease, stroke, and myocardial infarction. For one-day lag models, COPD, heart rhythm disturbances, other heart disease, and hypertension were all significant. The largest change in hospitalization was observed on same day lag models; the magnitude of the effect decreases with increasing inclusion of lag effects. Some of the diagnostic codes, such as asthma, have a low prevalence in hospitalizations and may be more difficult to detect a change in rates of hospitalizations.

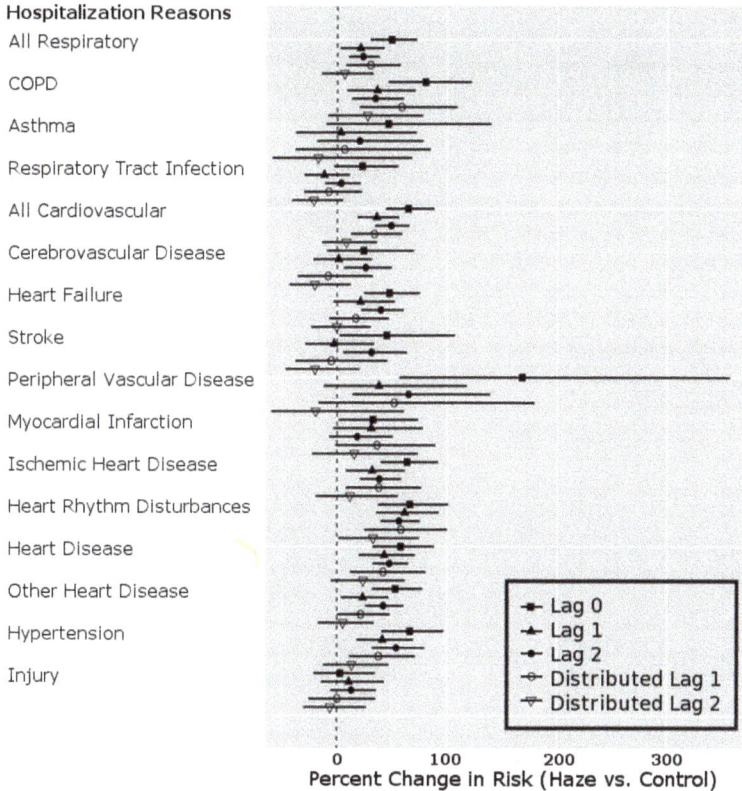

Note: Percent increase in admissions for the haze period compared to the non-haze period.

**Figure 4.** Percent change and 95% CI in hospital related admissions for the haze period compared to the non-haze period in the affected region controlling for $PM_{2.5}$, temperature, and dew point.

Models with the additional $PM_{2.5}$ by haze period interaction term did not reveal consistent or significant results for this interaction effect across many of the outcomes considered (results not shown) suggesting the $PM_{2.5}$ effect not to be the statistically different during the haze and non-haze periods. Results did however continue to reveal the strong significant increase in hospitalizations for the haze period compared to the non-haze period. The lack of evidence supporting a change in $PM_{2.5}$ associated relative risk of hospitalizations during the haze period compared to the non-haze period ($PM_{2.5}$ by haze period interaction) could be due to a combination of several factors including; $PM_{2.5}$ from wildfire sources not any more toxic than non-wildfire sources, the fact that the haze period happened to include both a weekday and the weekend possibly confounding exposure or that there are other drivers of hospitalizations during the haze period that is not entirely explained by $PM_{2.5}$ [68]. Although other studies have shown that wildfire PM is at least as toxic as urban PM [66,68]. Sensitivity analysis was performed increasing the haze period to five, seven, and nine days surrounding 7 July 2002 as well as lagging the haze period when considering models with a lagged $PM_{2.5}$ exposure. Results differed quantitatively, however the overall qualitative interpretations remained consistent. As such all reported and interpreted regressions results were based on model (1) with the predefined three-day haze period.

## 4. Discussion

With the Medicare National Claims History, National Climatic Data Center's weather data, and EPA's National Monitoring Network, we conducted an opportunistic study of the effects of wildfire air pollution from the Province of Quebec on the health of the elderly population stretching between New York and the District of Columbia.

The selection of our outcome categories was informed by several considerations. Acute exposure to PM may exacerbate existing pulmonary disease [69–72]. Because COPD is a substantial risk factor for cardiovascular mortality and morbidity [73–76], air pollution exposure may also contribute to cardiovascular risk through exacerbation of COPD symptoms.

We considered several specific cardiovascular and respiratory outcomes that are impacted by inflammatory processes, such as myocardial infarction, stroke, and asthma. The short term effects of exposure to high levels of air pollution are likely to cause inflammatory responses in the lung and release of cytokines with local and systemic consequences [24,77]. Acute effects of PM exposure have also been shown to increase plasma viscosity [25,78].

The log-linear model we used to estimate associations between day-to-day variations in $PM_{2.5}$ (at various lags) and day-to-day variations in the county-level hospitalization rates is typical of time-series analysis [79]. The advantage of the time-series approach is that confounding by individual-level covariates, such as smoking, is not an issue. However, factors that vary with daily pollution exposure, such as weather and co-pollutants, are likely to be confounders in these studies. Time-series analyses typically include nonlinear terms for weather and season [60,63]. One advantage of examining the effects of abrupt increases in $PM_{2.5}$ concentration over a short period of time is that it is unlikely that our analyses were confounded by any seasonal pattern unaccounted for in the model.

Our study focused on a population of interest, the elderly, using a nationally available database of health claims. In interpreting the findings of our analysis, considerations need to be given to the inherent limitations of the data analyzed. Information in the Medicare database is prone to bias due to inaccuracy of claims coding for specific diagnoses [80–83]. In an attempt to reduce misclassification for outcomes of interest, we used primary and secondary diagnosis codes to identify records for inclusion.

In addition, the ambient air pollution data from administrative databases such as EPA's National Monitoring Network, which have been created for regulatory purposes, only provide limited spatial and temporal coverage. This is an issue typical of air pollution studies relying on publicly available datasets. Also, during the short period when the plume affected the northeast U.S. (three peak days) the number of hospitalizations recorded was small compared to that observed in larger scale time-series studies. The small numbers of hospitalization counts and the limited exposure data reduces the power of our study.

In common with all studies examining the relationship between exposure to air pollution and health that depend on ambient air quality data, our study finding may be biased because of exposure misclassification. In our study all individuals were assumed to be exposed similarly to the corresponding ambient PM measured at EPA monitoring sites. However, some people may have listened to the health advisories (like the ones in New York and Pennsylvania, USA) and retreated indoors during the event. Although, as shown by Sapkota *et al.*, some indoor environments were substantially impacted by the elevated ambient $PM_{2.5}$ due to this event [5], the use air conditioning may also have ameliorated the indoor exposure [84].

Wildfires have rapid and substantial impacts on local air quality that elevate ambient PM concentrations well above the norm. The impact of this increased pollution on the health of local populations has been well documented [9,11,13,17,29,30,85–87]. For example, Duclos *et al.* showed a 40% and 30% increase in number of local emergency room visits for asthma and COPD respectively, during 1987 forest fires in northern California [13]. While composition of wildfire smoke has been shown to influence smoke related health outcomes [16], reliance on PM concentrations is common for regional studies because of the fire plume components it is most consistently elevated during smoke events, as opposed to other attributed components like carbon monoxide or nitric oxide [5,32].

*ISPRS Int. J. Geo-Inf.* **2014**, *3*, 713–731

Increases in wildfire activity have been linked to warmer spring and summer temperatures resulting from climate change and long-accumulated stocks of combustible vegetation [88,89]. In addition, climate models predict an increase in rising temperatures and regional drying will result in an increase in wildfire activity. Regions like the western U.S. can expect to see considerable increase in wildfire activity [90]. These wildfires may have significant impacts on future air quality and on the health of populations susceptible to the effects of air pollution even those living at a great distance. Our approach is replicable on a larger scale for estimating the health effects of large pollution events resulting from biomass burning over large distances.

## 5. Conclusions

This research adds to the growing body of literature demonstrating the significant impact of transboundary air pollution on public health. Short-term increases in $PM_{2.5}$ concentrations due to the Canadian forest fires in the Province of Quebec show a consistent significant increase in respiratory and cardiovascular hospital admissions for the elderly across the east coast of the U.S. as far south as Washington D.C. These results highlight the public health implications of long-range transport of wildfire-related air pollutants, and raise awareness of the health costs that may be associated with exposures to pollutants generated at remote locations. Preparedness planning by public health and clinical communities is needed to ensure a rapid and adequate response to the challenges that a large wildfire event poses to the health care emergency system. This is especially important as these events are expected to increase as a consequence of global climate change. Knowledge of the potential health risks of wildfire smoke can be translated into local action and used in risk prevention decisions and risk communication such as: educating vulnerable groups of taking their prescribed medication, preparing emergency rooms for an influx of patients with respiratory and cardiac conditions, and advising affected populations on how best to limit their exposures to the smoke.

**Acknowledgments:** This work was in part supported by the NIEHS (grant numbers P50ES015903, P01ES018176, and P30ES03819) and funding from the U.S. Environmental Protection Agency (grant numbers RD83213901, RD83451001, and RD83241701). Although the research described in this article has been funded wholly or in part by the United States Environmental Protection Agency it has not been subjected to the Agency's required peer and policy review and therefore does not necessarily reflect the views of the Agency and no official endorsement should be inferred. This manuscript is dedicated to Alison Geyh from the Department of Environmental Health Sciences, Johns Hopkins Bloomberg School of Public Health who passed away before this paper was completed. She was a great friend and outstanding colleague.

**Author Contributions:** Conception and design of the study: Patrick Breysse, Alison Geyh, Frank Curriero, and George Le. Analysis and interpretation of data: George Le, Jesse Berman, Frank Curriero.Collection and assembly of data: Patrick Breysse, Sorina Eftim, Jesse Berman, Aidan McDermott. Drafting of the article: George Le, Frank Curriero, Patrick Breysse, Jesse Berman, Sorina. Critical revision of the article for important intellectual content: Patrick Breysse, Frank Curriero, George Le. Final approval of the article: Frank Curriero, George Le, Patrick Breysse.

**Conflicts of Interest:** The authors declare no conflict of interest.

## References

1.   Ward, D. Health Guidelines for Vegetation Fire Events, Background Papers. In *Smoke from Wildland Fires*; WHO: Geneva, Switzerland, 1999; pp. 70–85.

2.   Miller, D.J. Assessing boreal forest fire smoke aerosol impacts on U.S. air quality: A case study using multiple data sets. *J. Geophys. Res.* **2011**, *116*, 1–19.

3.   Mathur, R. Estimating the impact of the 2004 Alaskan forest fires on episodic particulate matter pollution over the eastern Unites States through assimilation of satellite derived aerosol optical depths in a regional air quality model. *J. Geophys. Res.* **2008**, *113*, 1–14.

4.   Jalava, P.; Salonen, R.; Halinen, A.; Penttinen, P.; Pennanen, A.; Sillanpaa, M.; Sandell, E.; Hillamo, R.; Hirvonen, M.R. *In vitro* inflammatory and cytotoxic effects of size-segregated particulate samples collected during long-range transport of wildfire smoke to Helsinki. *Toxicol. Appl. Pharmacol.* **2006**, *215*, 341–353. [CrossRef]

5. Sapkota, A.; Symons, J.; Kleissl, J.; Wang, L.; Parlange, M.; Ondov, J.; Breysse, P.N.; Diette, G.B.; Eggleston, P.A.; Buckley, T.J. Impact of the 2002 Canadian forest fires on particulate matter air quality in Baltimore city. *Environ. Sci. Technol.* **2005**, *39*, 24–32. [CrossRef]

6. Morris, G.A.; Hershey, S.; Thompson, A.M.; Stohl, A.; Colarco, P.R.; McMillan, W.W.; Warner, J.; Johnson, B.J.; Witte, J.C.; Kucsera, T.L.; *et al.* Alaskan and Canadian forest fires exacerbate ozone pollution in Houston, Texas, on 19 and 20 July. *J. Geophys. Res.* **2004**, *111*, 1–10.

7. Ovadnevaite, J.; Kvietkus, K.; Marsalka, A. 2002 summer fires in Lithuania: Impact on the Vilnius city air quality and the inhabitants health. *Sci. Total Environ.* **2006**, *356*, 11–21. [CrossRef]

8. Mott, J.; Mannino, D.M.; Alverson, C.; Kiyu, A.; Hashim, J.; Lee, T.; Falter, K.; Redd, S.C. Cardiorespiratory hospitalizations associated with smoke exposure during the 1997, southeast Asian forest fires. *Int. J. Hyg. Environ. Health* **2005**, *208*, 75–85. [CrossRef]

9. Heil, A.; Goldammer, J. Smoke-haze pollution: A review of the 1997 episode in southeast Asia. *Reg. Environ. Chang.* **2001**, *2*, 24–37. [CrossRef]

10. Wotawa, G.; Trainer, M. The influence of Canadian forest fires on pollutant concentrations in the United States. *Science* **2000**, *288*, 324–328. [CrossRef]

11. Health Guidelines for Vegetation Fire Events. In *Technical Report*; World Health Organization: Geneva, Switzerland, 1999.

12. Andreae, M. Biomass burning in the tropics: Impact on environmental quality and global climate. *Popul. Dev. Rev.* **1990**, *16*, 268–291. [CrossRef]

13. Duclos, P.; Sanderson, L.M.; Lipsett, M. The 1987 forest fire disaster in California: Assessment of emergency room visits. *Arch. Environ. Health* **1990**, *45*, 53–58. [CrossRef]

14. Emmanuel, S.C. Impact to lung health of haze from forest fires: The Singapore experience. *Respirology* **2000**, *5*, 175–182. [CrossRef]

15. Johnston, F.H.; Webby, R.J.; Pilotto, L.S.; Bailie, R.S.; Parry, D.L.; Halpin, S.J. Vegetation fires, particulate air pollution and asthma: A panel study in the Australian monsoon tropics. *Int. J. Environ. Health Res.* **2006**, *16*, 391–404. [CrossRef]

16. Rappold, A.G.; Stone, S.L.; Cascio, W.E.; Neas, L.M.; Kilaru, V.J.; Carraway, M.S.; Szykman, J.J.; Ising, A.; Cleve, W.E.; Meredith, J.T.; *et al.* Peat bog wildfire smoke exposure in rural north Carolina is associated with cardiopulmonary emergency department visits assessed through syndromic surveillance. *Environ. Health Perspect.* **2011**, *119*, 1415–1420. [CrossRef]

17. Sastry, N. Forest fires, air pollution and mortality in southeast Asia. *Demography* **2002**, *39*, 1–23. [CrossRef]

18. Larson, T.V.; Koenig, J.Q. Wood smoke: Emissions and noncancer respiratory effects. *Annu. Rev. Public Health* **1994**, *15*, 133–156. [CrossRef]

19. Delfino, R.J.; Brummel, S.; Wu, J.; Stern, H.; Ostro, B.; Lipsett, M.; Winer, A.; Street, D.H.; Zhang, L.; Tjoa, T.; *et al.* The relationship of respiratory and cardiovascular hospital admissions to the southern California wildfires of 2003. *Occup. Environ. Med.* **2009**, *66*, 189–197. [CrossRef]

20. Henderson, S.B.; Brauer, M.; MacNab, Y.C.; Kennedy, S.M. Three measures of forest fire smoke exposure and their associations with respiratory and cardiovascular health outcomes in a population-based cohort. *Environ. Health Perspect.* **2011**, *119*, 1266–1271. [CrossRef]

21. Johnston, F.H.; Bailie, R.S.; Pilotto, L.S.; Hanigan, I.C. Ambient biomass smoke and cardio-respiratory hospital admissions in Darwin, Australia. *BMC Public Health* **2007**, *7*. [CrossRef]

22. Kunii, O.; Kanagawa, S.; Yajima, I.; Hisamatsu, Y.; Yamamura, S.; Amagai, T.; Ismail, I.T. The 1997 haze disaster in Indonesia: Its air quality and health effects. *Arch. Environ. Health* **2002**, *174*, 16–22.

23. Ostro, B.; Broadwin, R.; Green, S.; Feng, W.; Lipsett, M. Fine particulate air pollution and mortality in nine California counties: Results from Calfine. *Environ. Health Perspect.* **2006**, *114*, 29–33.

24. Pope, C.A., III; Dockery, D.W. Health effects of particulate air pollution: Lines that connect. *J. Air Waste Manag. Assoc.* **2006**, *56*, 709–742. [CrossRef]

25. Rückerl, R.; Schneider, A.; Breitner, S.; Cyrys, J.; Peters, A. Health effects of particulate air pollution: A review of epidemiological evidence. *Inhal. Toxicol.* **2011**, *23*, 555–592. [CrossRef]

26. Samet, J.M.; Domnici, F.; Curriero, F.; Coursac, I.; Zeger, S. Fine particulate air pollution and mortality in 20 U.S. cities: 1987–1994. *N. Engl. J. Med.* **2000**, *342*, 1742–1757. [CrossRef]

27. Schwartz, J.; Dockery, D.W.; Neas, L. Is daily mortality associated specifically with fine particles? *J. Air Waste Manag. Assoc.* **1996**, *46*, 927–939. [CrossRef]

28. Zanobetti, A.; Schwartz, J. Air pollution and emergency admissions in Boston, MA, USA. *J. Epidemiol. Community Health* **2006**, *60*, 890–896. [CrossRef]

29. Aditama, T. Impact of haze from forest fire to respiratory health: Indonesian experience. *Respirology* **2000**, *5*, 169–174. [CrossRef]

30. Künzli, N.; Avol, E.; Wu, J.; Gauderman, W.J.; Rappaport, E.; Millstein, J.; Bennion, J.; McConnell, R.; Gilliland, F.D.; Berhane, K.; *et al.* Health effects of the 2003 Southern California wildfire on children. *Amer. J. Respir. Crit. Care Med.* **2006**, *174*, 1221–1228. [CrossRef]

31. Moore, D.; Copes, R.; Fisk, R.; Joy, R.; Chan, K.; Brauer, M. Population health effects of air quality changes due to forest fires in British Columbia in 2003: Estimates from physician-visit billing data. *Can. J. Public Health* **2006**, *97*, 105–108.

32. Naeher, L.P.; Brauer, M.; Lipsett, M.; Zelikoff, J.T.; Simpson, C.D.; Koenig, J.Q.; Smith, K.R. Woodsmoke health effects: A review. *Inhal. Toxicol.* **2007**, *19*, 67–106.

33. Vedal, S. Ambient particles and health: Lines that divide. *J. Air Waste Manag. Assoc.* **1997**, *37*, 551–581. [CrossRef]

34. Mott, J.; Meyer, P.; Mannino, D.M.; Redd, S.; Smith, E.; Gotway-Crawford, C.; Chase, E. Wildland forest fire smoke: Health effects and intervention evaluation, Hoopa, California, 1999. *West. J. Med.* **2002**, *176*, 157–162. [CrossRef]

35. Schreuder, A.; Larson, T.; Sheppard, L.; Claiborn, C. Ambient woodsmoke and associated respiratory emergency department visits in Spokane, Washington. *Int. J. Occup. Environ. Health* **2006**, *12*, 147–153. [CrossRef]

36. Shusterman, D.; Kaplan, J.; Canabarro, C. Immediate health effects of an urban wildfire. *West. J. Med.* **1993**, *158*, 133–138.

37. Sutherland, E.; Make, B.; Vedal, S.; Zhang, L.; Dutton, S.; Murphy, J.; Silkoff, P.E. Wildfire smoke and respiratory symptoms in patients with chronic obstructive pulmonary disease. *J. Allergy Clin. Immunol.* **2005**, *115*, 420–422. [CrossRef]

38. Viswanathan, S.; Eria, L.; Diunugala, N.; Johnson, J.; McClean, C. An analysis of effects of San Diego wildfire on ambient air quality. *J. Air Waste Manag. Assoc.* **2006**, *56*, 56–67. [CrossRef]

39. Report of the Biregional Workshop on the Health Impacts of Haze Related Air Pollution. In *Technical Report RS/98/GE/17(MAA)*; Regional Office for the Western Pacific, World Health Organization: Manila, Philippines, 1998.

40. American Lung Association. State of the Air: 2004. In *Technical Report*; American Lung Association: New York, NY, USA, 2004.

41. Anderson, H.; Atkinson, R.; Peacock, J.; Marston, L.; Konstantinou, K. Meta-Analysis of Time-Series Studies and Panel Studies of Particulate Matter (PM) and Ozone ($O_3$). In *Technical Report*; World Health Organization (WHO): Geneva, Switzerland, 2004.

42. Annesi-Maesano, I.; Agabiti, N.; Pistelli, R.; Couilliot, M.; Forastiere, F. Subpopulations at increased risk of adverse health outcomes from air pollution. *Eur. Respir. J.Suppl.* **2003**, *40*, S57–S63.

43. Barnett, A.; Williams, G.; Schwartz, J.; Best, T.; Neller, A.; Petroeschevsky, A.; Simpson, R. The effects of air pollution on hospitalizations for cardiovascular disease in elderly people in Australian and New Zealand cities. *Environ. Health Perspect.* **2006**, *114*, 1018–1023. [CrossRef]

44. Boezen, H.; Vonk, J.; van der Zee, S.; Gerritsen, J.; Hoek, G.; Brunekreef, B.; Schouten, J.P.; Postma, D.S. Susceptibility to air pollution in elderly males and females. *Eur. Respir. J.* **2005**, *25*, 1018–1024. [CrossRef]

45. Brook, R.; Franklin, B.; Cascio, W.; Hong, Y.; Howard, G.; Lipsett, M.; Luepker, R.; Mittleman, M.; Samet, J.; Smith, S.C., Jr.; *et al.* Air pollution and cardiovascular disease: A statement for healthcare professionals from the Expert Panel on Population and Prevention Science of the American Heart Association. *Circulation* **2004**, *109*, 2655–2671. [CrossRef]

46. Chen, Y.; Yang, Q.; Krewski, D.; Burnett, R.; Shi, Y.; McGrail, K. The effect of coarse ambient particulate matter on first, second, and overall hospital admissions for respiratory disease among the elderly. *Inhal. Toxicol.* **2005**, *17*, 649–655. [CrossRef]

47. Goldberg, M.S.; Burnett, R.; Bailar, J., III; Brook, J.; Bonvalot, Y.; Tamblyn, R.; Singh, R.; Valois, M. Identifying Subgroups of 19 the General Population that may be Susceptible to Short-Term Increases in Particulate Air Pollution: A Time-Series Study in Montreal, Quebec. In *Research Report 097*; Health Effects Institute: Cambridge, MA, USA, 2000.

48. Pope, C.A., III. Epidemiology of fine particulate air pollution and human health: Biologic mechanisms and who's at risk? *Environ. Health Perspect.* **2000**, *108*, 713–723. [CrossRef]

49. Pope, C.A., III; Burnett, R.; Thun, M.; Calle, E.; Krewski, D.; Ito, K.; Thurston, G.D. Lung cancer, cardiopulmonary mortality and long-term exposure to particulate air pollution. *JAMA* **2002**, *287*, 1132–1141. [CrossRef]

50. Villeneuve, P.; Chen, L.; Stieb, D.; Rowe, B. Associations between outdoor air pollution and emergency department visits for stroke in Edmonton, Canada. *Eur. J. Epidemiol.* **2006**, *21*, 689–700. [CrossRef]

51. DeBell, L.J.; Talbot, R.; Dibb, J. A major regional air pollution event in the northeastern United States caused by extensive forest fires in Quebec, Canada. *J. Geophys. Res.* **2004**, *109*, D19305. [CrossRef]

52. NASA: Visible Earth. Wildfires in Quebec and Smoke over United States East Coast. Available online: http://visibleearth.nasa.gov/view.php?id=60610 (accessed on 1 February 2011).

53. GFMC (Global Fire Monitoring Center)/Fire Ecology Research Group. 2007 UN International Strategy for Disaster Reduction (ISDR). Available online: http://www.fire.uni-freiburg.de/ (accessed on 3 December 2010).

54. *International Classification of Diseases, Ninth Revision (ICD-9)*; World Health Organization: Geneva, Switzerland, 1977.

55. U.S. Environmental Protection Agency. Air Quality System Data Mart (Internet Database). Available online: http://www.epa.gov/ttn/airs/aqsdatamart (accessed on 1 February 2011).

56. National Climatic Data Center. *NCDC Surface Airways, CD Database*; EarthInfo Inc.: Boulder, CO, USA, 2011.

57. Cressie, N.A.; Cassie, N.A. *Statistics for Spatial Data*; Wiley: New York, NY, USA, 1993.

58. Young, L.J.; Gotway, C.A.; Yang, J.; Kearney, G.; DuClos, C. Linking health and environmental data in geographical analysis: It's so much more than centroids. *Spat. Spatiotemporal Epidemiol.* **2009**, *1*, 73–84. [CrossRef]

59. R Core Team. R: A Language and Environment for Statistical Computing, 2014. Available online: Availableonline:www.r-project.org (accessed on 1 February 2014).

60. Dominici, F.; Peng, R.; Bell, M.; Pham, L.; McDermott, A.; Zeger, S.; Samet, J.M. Fine particulate air pollution and hospital admission for cardiovascular and respiratory diseases. *JAMA* **2006**, *295*, 1127–1134. [CrossRef]

61. Eftim, S.E.; Janes, H.; McDermott, A.; Samet, J.; Dominici, F. A comparative analysis of the chronic effects of fine particulate matter. *Epidemiology* **2008**, *19*, 209–216. [CrossRef]

62. Liang, K.-Y.; Zeger, S. Longitudinal data analysis using generalized linear models. *Biometrika* **1986**, *73*, 13–22. [CrossRef]

63. Peng, D.; Dominici, F. *Statistical Methods for Environmental Epidemiology with R*; Springer Science, Business Media, LLC: New York, NY, USA, 2008.

64. Schwartz, J. The distributed lag between air pollution and daily deaths. *Epidemiology* **2000**, *1*, 320–326. [CrossRef]

65. Welty, L.; Zeger, S. Are the acute effects of particulate matter on mortality in the national morbidity, mortality, and air pollution study the result of inadequate control for weather and season? A sensitivity analysis using flexible distributed lag models. *Amer. J. Epidemiol.* **2005**, *162*, 80–88. [CrossRef]

66. Morgan, G.; Sheppard, V.; Khalaj, B.; Ayyar, A.; Lincoln, D.; Jalaludin, B.; Beard, J.; Corbett, S.; Lumley, T. Effects of brushfire smoke on daily mortality and hospital admissions in Sidney, Australia. *Epidemiology* **2010**, *21*, 47–55. [CrossRef]

67. Vedal, S.; Steven, J.D. Wildfire air pollution and daily mortality in a large urban area. *Environ. Res.* **2006**, *102*, 29–35. [CrossRef]

68. Dennekamp, M.; Abramson, M.J. The effects of bushfire smoke on respiratory health. *Respirology* **2011**, *16*, 198–209. [CrossRef]

69. Dockery, D.; Ware, J.; Ferris, B.J.; Speizer, F.; Cook, N.; Herman, S. Change in pulmonary function in children associated with air pollution episodes. *J. Air Pollut. Control Assoc.* **1982**, *32*, 937–942. [CrossRef]

70. Roemer, W.; Hoek, G.; Brunekreef, B. Effect of ambient winter air pollution on respiratory health of children with chronic respiratory symptoms. *Amer. Rev. Resp. Dis.* **1993**, *147*, 118–124. [CrossRef]

71. Hoek, G.; Brunekreef, B. Effect of photochemical air pollution on acute respiratory symptoms in children. *Amer. J. Resp. Crit. Care Med.* **1995**, *151*, 27–32. [CrossRef]

72. Hoek, G.; Dockery, D.; Pope, C.A.; Neas, L.; Roemer, W.; Brunekreef, B. Association between $PM_{10}$ and decrements in peak expiratory flow rates in children: Reanalysis of data from five panel studies. *Eur. Resp. J.* **1998**, *11*, 1307–1311. [CrossRef]

73. Brook, R.D.; Rajagopalan, S.; Pope, C.A.; Brook, J.R.; Bhatnagar, A.; Diez-Roux, A.V.; Holguin, F.; Hong, Y.; Luepker, R.V.; Mittleman, M.A.; *et al.* Particulate matter air pollution and cardiovascular disease: An update to the scientific statement from the American Heart Association. *Circulation* **2010**, *121*, 2331–2378. [CrossRef]

74. Sin, D.; Man, S. Chronic obstructive pulmonary disease as a risk factor for cardiovascular morbidity and mortality. *Proc. Amer. Thorac. Soc.* **2005**, *2*, 8–11. [CrossRef]

75. Sin, D.; Wu, L.; Man, S. The relationship between reduced lung function and cardiovascular mortality: A population-based study and a systematic review of the literature. *Chest* **2005**, *127*, 1952–1959. [CrossRef]

76. Van Eeden, S.; Yeung, A.; Quinlam, K.; Hogg, J. Systemic response to ambient particulate matter: Relevance to chronic obstructive pulmonary disease. *Proc. Amer. Thorac. Soc.* **2005**, *2*, 61–67. [CrossRef]

77. Swiston, J.R.; Davidson, W.; Attridge, S.; Li, G.T.; Brauer, M.; van Eeden, S.F. Wood smoke exposure induces a pulmonary and systemic inflammatory response in fire fighters. *Eur. Resp. J.* **2008**, *32*, 129–138. [CrossRef]

78. Peters, A.; Döring, A.; Wichmann, H.E.; Koenig, W. Increased plasma viscosity during an air pollution episode: A link to mortality? *Lancet* **1997**, *349*, 1582–1587. [CrossRef]

79. Dominici, F.; Sheppard, L.; Clyde, M. Statistical review of air pollution. *Int. Statist. Rev.* **2003**, *71*, 243–276. [CrossRef]

80. Baicker, K.; Chandra, A.; Skinner, J.S.; Wennberg, J.E. Who you are and where you live: How race and geography affect the treatment of Medicare beneficiaries. *Health Affair.* **2004**, *23*. [CrossRef]

81. Fisher, E.; Whaley, F.; Krushat, W.; Malenka, D.; Fleming, C.; Baron, J.A.; Hsia, D.C. The accuracy of Medicare's hospital claims data: Progress has been made, but problems remain. *Amer. J. Public Health* **1992**, *82*, 243–248. [CrossRef]

82. Havranek, E.; Wolfe, P.; Masoudi, F.; Rathore, S.; Krumholz, H.; Ordin, D. Provider and hospital characteristics associated with geographic variation in the evaluation and management of elderly patients with heart failure. *Arch. Intern. Med.* **2004**, *164*, 1186–1191.

83. Kiyota, Y.; Schneeweiss, S.; Glynn, R.; Cannuscio, C.; Avorn, J.; Solomon, D. Accuracy of Medicare claims-based diagnosis of acute myocardial infarction: Estimating positive predictive value on the basis of review of hospital records. *Amer. Heart J.* **2004**, *148*, 99–104. [CrossRef]

84. Janssen, N.A.H.; Schwartz, J.; Zanobetti, A.; Suh, H.H. Air conditioning and source-specific particles as modifiers of the effect of $PM_{10}$ on hospital admissions for heart and lung disease. *Environ. Health* **2002**, *110*, 43–49.

85. Analitis, A.; Georgiadis, I.; Katsouyanni, K. Forest fires are associated with elevated mortality in a dense urban setting. *Occup. Environ. Med.* **2012**, *69*, 158–162.

86. Hanigan, I.C.; Johnston, F.H.; Morgan, G.G. Vegetation fire smoke, indigenous status and cardio-respiratory hospital admissions in Darwin, Australia, 1996–2005: A time-series study. *Environ. Health* **2008**, *7*. [CrossRef]

87. See, S.; Balasubramanian, R.; Rianawati, E.; Karthikevyan, S.; Streets, D. Characterization and source apportionment of particulate matter ≤2.5 μm in Sumatra, Indonesia, during a recent peat fire episode. *Environ. Sci. Technol.* **2007**, *41*, 3488–3494. [CrossRef]

88. *IPCC 2007: Climate Change: Synthesis Report. Contribution of Working Groups I, II and III to the Fourth Assessment Report of the Intergovernmental Panel on Climate Change*; Core Writing Team; Pachauri, R.K.; Reisinger, A. (Eds.) IPCC: Geneva, Switzerland, 2007; p. 104.

89. Westerling, A.L.; Hidalgo, H.G.; Cayan, D.R.; Swetnam, T.W. Warming and earlier spring increases western U.S. forest wildfire activity. *Science* **2006**, *313*, 940–943. [CrossRef]

90. Pechony, O.; Shindell, D.T. Driving forces of global wildfires over the past millennium and the forthcoming century. *Proc. Nat. Acad. Sci. USA* **2010**, *107*, 19167–19170. [CrossRef]

ISPYS International Journal of
Geo-Information

MDPI

*Article*

# Nexus of Health and Development: Modelling Crude Birth Rate and Maternal Mortality Ratio Using Nighttime Satellite Images

**Koel Roychowdhury [1],* and Simon Jones [2]**

[1]  Institute for Sustainability and Peace (UNU-ISP), United Nations University, Tokyo 150-8925, Japan
[2]  School of Mathematical and Geospatial Sciences, RMIT University, Melbourne, VIC 3001, Australia;
    simon.jones@rmit.edu.au
*  Author to whom correspondence should be addressed; roychowdhury@unu.edu;
    Tel.: +81-354-671-212; Fax: +81-334-992-828.

Received: 26 February 2014; in revised form: 14 April 2014; Accepted: 25 April 2014; Published: 8 May 2014

**Abstract:** Health and development are intricately related. Although India has made significant progress in the last few decades in the health sector and overall growth in GDP, there are still large regional differences in both health and development. The main objective of this paper is to develop techniques for the prediction of health indicators for all the districts of India and examine the correlations between health and development. The level of electrification and district domestic product (DDP) are considered as two fundamental indicators of development in this research. These data, along with health metrics and the information from two nighttime satellite images, were used to propose the models. These successfully predicted the health indicators with less than a 7%–10% error. The chosen health metrics, such as crude birth rate (CBR) and maternal mortality rate (MMR), were mapped for the whole country at the district level. These metrics showed very strong correlation with development indicators (correlation coefficients ranging from 0.92 to 0.99 at the 99% confidence interval). This is the first attempt to use Visible Infrared Imaging Radiometer Suite (VIIRS) (satellite) imagery in a socio-economic study. This paper endorses the observation that areas with a higher DDP and level of electrification have overall better health conditions.

**Keywords:** health GIS; health atlas; development; geospatial technology; nighttime lights; satellite earth observation; remote sensing; modelling; public health

## 1. Introduction

Health and economic development go hand in hand. Health planning has been an integral part of socio-economic planning in India since its independence [1]. There have been recommendations from health committees over the years, which were translated into health policies [2]. Along with this, India has also progressed in the world economy. The per capita GDP of the country has increased from around U.S. $142 in 1950–1951 to around U.S. $720 in 2010–2011 (values at 2004–2005 constant prices) [3].

For the past 30 years, the country has made significant improvement in the infrastructure in the health sector, with achievements, such as the eradication of communicable diseases, including polio [3]. However, the health sector has often been one of the neglected areas in India's planning scenario [4]. India is one of the countries with the lowest public spending in the health sector [5]. At the beginning of the last decade (2000), only one percent of the GDP in India was dedicated to the health sector [4]. In 2013, the public expenditure in health sector stands at 1.1% of the GDP [5]. There is a rise of unregulated private healthcare facilities in the country [4], and over 70% of the healthcare expenditure is out-of-pocket expenditures [5]. All these factors have led India to lose around 6% of its

GDP every year to premature deaths and preventable diseases [5], with the country accounting for 21% of the global burden of diseases.

In order to use economic development to improve the health of its population, there is a need for disaggregated data to support planning and policy in the health sector. State-level data often mask out the districts deserving special attention for healthcare development. As a result, the Registrar General of India formulated the plan for Annual Health Survey in 2005. This was planned to be conducted from 2010–2013 in nine selected states of India with the highest rate of population growth. These are: Bihar, Jharkhand, Uttar Pradesh, Uttarakhand, Madhya Pradesh, Chhattisgarh, Orissa, Rajasthan and Assam. However, this survey was mainly undertaken by means of household interviews across 284 districts (administrative units that make up an Indian state) by enumerators. Secondly, due to the constraints of the vastness of the states along with the differences in the socio-economic conditions, it was conducted on a sample of a population of 20.1 million from 4.1 million households.

The main objective of this paper is to predict health indicators at the district level for the whole of India. Correlations were examined between health indicators (obtained from the Annual Health Survey (AHS) Reports) and development metrics (percentage of houses with access to electricity and district domestic product (DDP)). The information from two nighttime satellite images were used along with these data to propose models. Unlike previous works by the authors [6,7], this research takes into account the development indicators along with the information from satellite images as predictor variables to prepare the models. The models successfully predict the health indicators with high accuracy (less than 7%–10% error). Health metrics, such as crude birth rate and maternal mortality rate, demonstrated strong correlations with development metrics, with correlation coefficients ranging from 0.92 to 0.99 at the 99% confidence interval.

The paper is divided into three parts. The first part examines the correlations between DDP and the level of electrification with the health indicators. Part two of the paper focuses on building the models, and the final part prepares the health atlas of India and establishes the correlation between health and development.

## 2. Study Area

The first part of this study was based on all the districts of Bihar, Jharkhand, Uttar Pradesh, Uttarakhand, Madhya Pradesh, Chhattisgarh, Odisha, Rajasthan and Assam. In the second part, models were proposed to predict the health indicators using the data from the five states of Assam, Bihar, Jharkhand, Odisha and Rajasthan (142 districts). Data on DDP was freely available for these five states. The models were finally used to predict health indicators for all the districts of India.

## 3. Data Used

The current study used satellite images, as well as data from the Indian census, Indian Annual Health Survey and Indian Economic Survey.

Two kinds of satellite images captured at night were used in this research. The first image was obtained from F18 DMSP-OLS satellite in 2010. The stable light product of the satellite was used in this paper. These products were created by compositing the cloud-free orbital sections in the centre half of the swath, with no sun and moonlight present [8]. This was followed by a process of the filtering of noise and transient light sources, such as fires. The grid cell size of these images was 30 arc seconds or 1 square kilometre at the equator [8]. However, the bright areas, such as urban cores, were saturated in these images, and variations in the distribution of radiance within these areas were indiscernible. The second nighttime image was captured in 2011 by the Visible Infrared Imaging Radiometer Suite (VIIRS) instrument on board the Suomi National Polar Partnership (SNPP) satellite. NASA and NOAA launched this satellite jointly in 2011 [8]. The data composites were generated using VIIRS day/night band (DNB) data collected on nights with zero moonlight. Cloud screening was done based on the detection of clouds in the VIIRS M15 thermal band [8]. The original radiance values have been multiplied by a billion, and thus, the units of radiance obtained from the images are nano-Watts/cm$^2$/sr [9].

Data on the percentage of households with access to electricity and the total population of the districts were obtained from the Indian census, 2011. The data on health indicators was obtained from the second round of Annual Health Survey Bulletins, 2011–2012. These data were collected for the nine states of Assam, Bihar, Chhattisgarh, Jharkhand, Orissa, Madhya Pradesh, Uttar Pradesh, Rajasthan and Uttarakhand. District Domestic Product information was obtained from the Economic Survey reports of the respective states. The DDP data at the base price of 2004–2005 was used in this study.

The DMSP-OLS sensor has captured images of the Earth at night for more than 40 years. The data has been widely applied in different research studies. It has been used to estimate the population of the countries of the world [10–12] and in estimating population without access to electricity [13]. The first global map of the GDP purchasing power parity (GDP-PPP) [14] and gas flares [15] used radiance calibrated DMSP-OLS data to map urban and suburban extent around cities [16,17]. At the smaller spatial scale, the DMSP-OLS dataset was used to estimate population and other socio-economic metrics for the state of Maharashtra in India [7,18–21] and GDP estimates [22].

The VIIRS satellite sensor has a similar polar orbit to the DMSP-OLS image and a swath of 3000 km. It also has the same spectral band pass as the DMSP-OLS low-light imaging band (0.5–0.9 um). The DMSP-OLS, however, has an overpass time of around 19:30, while the VIIRS passes after midnight around 01:30 [9].

The VIIRS data is relatively new [23] and has been mainly used for nocturnal detection and characterization of combustion sources, such as emissions from gas flares, biomass burning, industrial sites and volcanoes [24,25].

## 4. Method

### 4.1. Step I: Satellite Image Processing

The DMSP–OLS satellite image for 2010 is an eight-bit image with pixel values ranging from 0–63, while the VIIRS image is a 14-bit image. In both of the images, the areas with persistent lights have higher pixel values. These correspond with higher population areas on the surface and act as surrogate maps of cities and towns. Subsets of each of the nine states were taken from the satellite images. This was followed by image geometric correction and image enhancement. The DMSP-OLS images were re-projected from the World Geographic System (WGS) Projection to the Lambert Conformal Conic Projection with the datum, WGS 1984. The DMSP-OLS images were contrast-enhanced, and a standard deviation stretch was applied. The average stable light from the DMSP-OLS image and the mean radiance from the VIIRS image for each district in the states was calculated using the Arc Info software package.

### 4.2. Step II: Processing of Secondary Data Sources

Analyses began with the calculation of the mean of the health metrics, the percentage of houses with electricity and the DDP for each state. The Annual Health Survey includes crude birth rate (CBR), crude death rate, natural growth rate, infant mortality rate (IMR), neo-natal mortality rate, post-neo-natal mortality rate, under five mortality rate, the gender ratio at birth, the gender ratio (0–4 years), maternal mortality rate (MMR) and the gender ratio of maternal and non-maternal mortality rates. These maternal and non-maternal mortality rates were available only for regional subdivisions within a state and not for each district [26].

In order to examine the nature of the distribution of the health indicators over the states, statistical tests for a normal distribution were conducted. Histograms were plotted to graphically assess the distribution of these metrics. In addition to this, skewness and kurtosis were also calculated. Skewness is a measure of symmetry in the data, while kurtosis refers to the peakiness of the data distribution. In order to calculate whether the value of zero for both skewness and kurtosis lie within the 95%

confidence interval, the z-scores of skewness and kurtosis were calculated. In a perfectly normal distribution, the values of skewness and kurtosis are zero. An absolute value of these scores greater than 1.96 is significantly different from the normal distribution at more than the 95% confidence interval. Therefore, all the variables with z-scores less than 1.96 are normally distributed. The results of the distribution helped in the selection of health metrics for modelling.

Table 1 shows the mean, standard deviation, skewness, kurtosis and the z-scores of skewness and kurtosis. It was observed that only IMR and under 5 mortality rate (U5MR) have a normal distribution (z-scores less than 1.96).

**Table 1.** Skewness and kurtosis of health metrics. CBR, crude birth rate; IMR, infant mortality rate; MMR, maternal mortality rate.

| | | CBR | CDR | IMR | Neo-Natal Mortality Rate | Under 5 Mortality Rate | MMR | Gender Ratio |
|---|---|---|---|---|---|---|---|---|
| | | | | | Statistics | | | |
| N | Vaild | 284 | 284 | 284 | 284 | 284 | 284 | 284 |
| | Missing | 0 | 0 | 0 | 0 | 0 | 0 | 0 |
| Mean | | 23.992 | 7.571 | 58.458 | 39.479 | 78.595 | 279.557 | 954.197 |
| Std. Deviation | | 3.8667 | 1.5036 | 13.9609 | 11.2543 | 20.0106 | 62.9007 | 66.6693 |
| Skewness | | 0.384 | 0.376 | 0.039 | 0.382 | 0.087 | 0.339 | 1.113 |
| Std. Error of Skewness | | 0.145 | 0.145 | 0.145 | 0.145 | 0.145 | 0.145 | 0.145 |
| Kutosis | | 0.941 | −0.084 | 0.409 | 0.145 | 0.432 | −0.010 | 2.037 |
| Std. Error of Kutosis | | 0.288 | 0.288 | 0.288 | 0.288 | 0.288 | 0.288 | 0.288 |
| Z score_skewness | | 2.65619438 | 2.59710546 | 0.2691124 | 0.639865406 | 0.604767764 | 2.34790618 | 7.695079608 |
| Z score_kutosis | | 3.26451241 | −0.2899202 | 1.41850337 | 0.503650003 | 1.500358571 | −0.03550822 | 7.069284246 |

**Table 2.** Skewness and kurtosis of health metrics (log normal).

| | | CBR | CDR | IMR | Neo-Natal Mortality Rate | Under 5 Mortality Rate | MMR | Gender Ratio |
|---|---|---|---|---|---|---|---|---|
| | | | | | Statistics | | | |
| N | Vaild | 284 | 284 | 284 | 284 | 284 | 284 | 284 |
| | Missing | 0 | 0 | 0 | 0 | 0 | 0 | 0 |
| Mean | | 3.165 | 2.005 | 4.037 | 3.633 | 4.328 | 5.608 | 6.859 |
| Std. Deviation | | 0.1623 | 0.1999 | 0.2628 | 0.3029 | 0.2814 | 0.2298 | 0.0678 |
| Skewness | | −0.198 | −0.133 | −1.044 | −0.665 | −1.026 | −0.271 | 0.845 |
| Std. Error of Skewness | | 0.145 | 0.145 | 0.145 | 0.145 | 0.145 | 0.145 | 0.145 |
| Kutosis | | 0.300 | −0.269 | 2.381 | 1.294 | 2.204 | -0.148 | 1.288 |
| Std. Error of Kutosis | | 0.288 | 0.288 | 0.288 | 0.288 | 0.288 | 0.288 | 0.288 |
| Z score_skewness | | −1.365517241 | −0.917241379 | −7.2 | −4.586206897 | −7.075862069 | −1.868965517 | 5.827586207 |
| Z score_kutosis | | 1.041666667 | −0.934027778 | 8.267361111 | 4.493055556 | 7.652777778 | −0.513888889 | 4.472222222 |

In order to reduce the peakiness and to increase the symmetry of the data, the metrics were converted to logarithmic values. Logarithmic values to base "e" or "log-normal" values were calculated. This significantly improved the results. After converting to log values, it was observed that CBR, CDR and MMR were normally distributed (Table 2).

In the next stage, correlations between the health metrics and the development indicators (percentage of households with electricity and DDP) were obtained. The correlations of these metrics with the average stable light and the average radiance were also calculated. On the basis of the results from these correlations, CBR and MMR were selected for further analyses and modelling.

*4.3. Step III: Models*

In order to prepare the health atlas of India (maps of health metrics for the districts of the country), multiple regression models were proposed. Due to the availability of DDP information over five states, models were calculated on the data from those states. They include the states of Assam, Bihar,

Jharkhand, Odisha and Rajasthan (142 districts). One hundred districts were randomly chosen to build the models, and the results were validated over 42 withheld districts. Models were also prepared to predict DDP. This model was used to predict DDP for the districts of India for which the data was not available.

### 4.4. Step IV: Preparing the Atlas

Using the models, CBR and MMR were predicted for all the districts of India. A strong correlation was noted between predicted data and the percentage of households with electricity and DDP for all the districts. Correlation coefficients of more than 0.92 at the 99% confidence between the health metrics and the development indicators were observed.

## 5. Health and Development

### 5.1. Health Status in India

India displays a complex picture in the health sector. There are large regional variations of health indicators within the country. The health indicators of some of the southern states, such as Kerala and Tamil Nadu, are comparable to those of developed countries [27]. However, states, like Uttar Pradesh, Madhya Pradesh, Bihar, Jharkhand, Chhattisgarh and Odisha, have lagged behind in overall progress in health sector [27]. These are the most populated states in the country, with more than 50% of the total population living in these areas. Moreover, 70% of infant deaths and 62% of maternal deaths for the country are recorded from these regions [26].

The Millennium Development Goals (MDGs) proposed by United Nations in the UN Millennium Summit [28] helped in formulating the template health infrastructure of the country. The health-related MDGs include poverty and hunger (MDG 1), child health and immunization (MDG 4), maternal and reproductive health (MDG 5), HIV/AIDS, malaria and other diseases (MDG 6), environment sustainability (MDG 7) and global partnership for development (MDG 8) [29]. The Indian government has asserted its commitment to the achievement of these goals [27]. The progress of the country towards the MDGs is tracked by a system of sample registrations in India. This is a state- and national-level system designed to collect data on fertility and mortality [30].

IMR is a measure of the number of deaths of infants less than one year of age. The highest IMR in India is found in the state of Madhya Pradesh, while the lowest is found in Kerala. Kerala (12) and Tamil Nadu (28) have already achieved the Millennium Development Goal (MDG) IMR target of 28 [30]. Closely following are the states of Delhi (33), Maharashtra (31) and West Bengal (33). The IMR in India has declined by 30 points in the last 20 years, with an annual average decline of 1.5 points [30]. The maternal mortality ratio (MMR) measures the number of women aged 15–49 years dying due to maternal causes per 100,000 live births [30]. There is a decline in MMR of about 17% for the whole of India over the last decade. The states of Tamil Nadu, Kerala and Maharashtra have reached the MDG MMR target of 109. The states of Andhra Pradesh, West Bengal, Gujarat and Haryana are close to the target [30].

Table 3 shows the health indicators for the nine states of India as recorded in the AHS. CBR, defined as the ratio of the number of live births in a year to mid-year population multiplied by 1000 [26], is highest for the state of Bihar (26.3). IMR and the under-five mortality rate is highest for the state of Uttar Pradesh and lowest for Jharkhand. Uttarakhand has the best gender ratio, with 995 females per 1000 males.

There are also intra-state variations of these metrics. For example, in Uttar Pradesh, the highest CBRs (more than 40) are found in the districts of Shrawasti, Siddharthnagar and Balrampur. In Assam, 15 districts have CBRs ranging from 15–22. In Bihar, the CBR ranges from 21–32, with the majority of the districts having 24–26 CBRs. The district of Bageshwar of Uttarakhand has the lowest CBR. IMR is the highest for the districts of Shrawasti and Faizabad of Uttar Pradesh, Balangir of Odisha and Panna of Madhya Pradesh. These districts have IMR values of more than 90. The lowest IMR (20) is noted in the districts of Pithoragarh, Almora and Rudraprayag of Uttarakhand. MMR is the highest (more than 435) for the districts of Faizabad, Barabanki and Ambedkar Nagar of Uttar Pradesh. The lowest MMR is recorded in the districts of Rudraprayag, Chamoli and Dehra Dun of Uttarakhand. Overall, the state of Uttarakhand displays better heath indicators compared to the other states considered in the AHS survey.

**Table 3.** Health indicators of nine states as obtained from the Annual Health Survey (AHS) Second Round Survey, 2011–2012.

| State | CBR | CDR | IMR | Neo-Natal Mortality Rate | Under 5 Mortality Rate | MMR | Gender Ratio |
|---|---|---|---|---|---|---|---|
| Assam | 21.3 | 7.1 | 57 | 38 | 75 | 347 | 959 |
| Bihar | 26.3 | 7 | 52 | 34 | 73 | 294 | 951 |
| Chhatisgarh | 23.5 | 7.4 | 50 | 35 | 66 | 263 | 970 |
| Jharkhand | 23.3 | 5.8 | 38 | 24 | 55 | 267 | 944 |
| Madhya Pradesh | 24.8 | 7.8 | 65 | 43 | 86 | 277 | 916 |
| Odisha | 19.8 | 8.2 | 59 | 39 | 79 | 237 | 995 |
| Rajasthan | 24.4 | 6.4 | 57 | 38 | 76 | 264 | 929 |
| Uttar Pradesh | 25 | 8.4 | 70 | 50 | 92 | 300 | 944 |
| Uttarakhand | 18.2 | 6.4 | 41 | 29 | 50 | 162 | 995 |

## 5.2. Rate of Electrification in India

Access to modern sources of energy is one of the key factors of development. In 2011, the major source of energy in the Indian states was electricity. It accounted for 51% of the total energy consumption [31], followed by coal and lignite (25%) and crude petroleum (20%). The domestic sector accounted for the highest contributor to the electricity consumption. The growth in the consumption of electricity in this sector was 15.91% between 2009–2010 and 2010–2011. However, marked regional disparities in the level of electrification are noted within the country. More than 90% of the households have access to electricity in the states of Gujarat, Haryana, Sikkim, Punjab, Himachal Pradesh, Chandigarh and the southern states of Kerala, Andhra Pradesh, Tamil Nadu and Karnataka [32]. Almost all the households (more than 99%) in the National Capital Territory of Delhi and the Union Territories of Lakshadweep and Daman and Diu use electricity as the major source of energy for lighting [32]. In contrast, a state, like Bihar, has only 16.4% of its households with access to electricity. The distribution of electricity is slightly better in the states of Uttar Pradesh (36.8%), Assam (37.1%), Odisha (43%), Jharkhand (45.8%), Rajasthan (67%) and Madhya Pradesh (67.1%) [32].

Figure 1 shows the nature of the distribution of electricity in the states of: (i) Assam; (ii) Bihar; (iii) Chhattisgarh; (iv) Jharkhand; (v) Odisha; (vi) Rajasthan; (vii) Madhya Pradesh; (viii) Uttar Pradesh; and (ix) Uttarakhand. Of these nine states, on an average, Uttarakhand has 85.13% of households with access to electricity followed by the states of Chhattisgarh (71.68%), Madhya Pradesh (66.26%) and Rajasthan (64.95%). Odisha has just about 40% of households using electricity as the main source of power supply. The lowest level of electrification is found in the state of Bihar, where only 16% of the households use electricity as their domestic power source.

District-level variations in the rate of electrification are also observed within these nine states. In Assam, the districts of Dibrugarh, Jorhat and Tinsukia have more than 50% of households with electricity. The districts of North Cachar Hills, Sibsagar, Nalbari, Kamrup, Goalpara, Cachar and Golaghat have a level of electrification of more than 35%. The district of Dhubri alone has less than 20% of households with electricity access. In Bihar, more than 57% of the households have access to electricity, only in the capital district of Patna. The rest of the districts have a less than 35% electrification rate. On the contrary, more than 50% of the districts of Chhattisgarh have higher than 80% of households using electricity as the main source of power. The district of Janjgir-Champa in Chhattisgarh has an electrification rate of around 90%. A similar kind of electrification is also found in the state of Uttarakhand. Ten out of 13 districts have more than 80% of households with access to electricity, the highest being found in the capital of Dehradun (96%). Although the districts of Indore and Bhopal in Madhya Pradesh have more than 90% electrification, most of the districts have electrification of around 60%–85%. In Rajasthan, the highest rate of electrification is found in the districts of Kota, Jaipur, Chittaurgarh and Ajmer (more than 80%). Most of the districts of the state, however, have 50%–75% of households with access to electricity. In Uttar Pradesh, 30%–70% of the households have electricity access in most of the districts, with the highest concentration (around 80%) is found in the districts of Ghaziabad, Gautam Budhha Nagar and Agra.

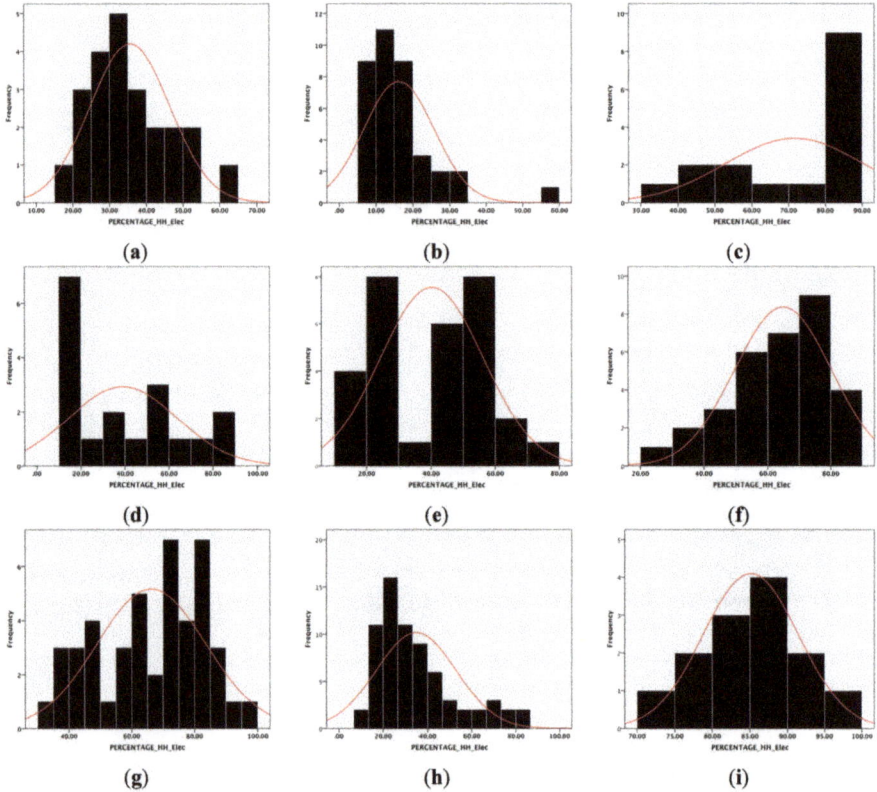

**Figure 1.** Level of electrification in nine states of India: (**a**) Assam; (**b**) Bihar; (**c**) Chhattisgarh; (**d**) Jharkhand; (**e**) Odisha; (**f**) Rajasthan; (**g**) Madhya Pradesh; (**h**) Uttar Pradesh; (**i**) Uttarakhand.

*5.3. Contribution of DDP of the States*

Domestic product has always been considered as an indicator of development [33]. According to the Ministry of Statistics and Program Implementation, Government of India [33], gross domestic product (GDP) can be defined using three approaches: production approach, income approach and expenditure approach. In the National Accounts Statistics of India, the production approach is considered to be a firmer estimate. "The production approach GDP measures the sum of value added of all economic activities within the country's territory (sum of output minus intermediate consumption) plus indirect taxes minus subsidies on products" [33]. The state accounts are an extension of the national accounts at the regional level. At the state level, the most important aggregate is the state domestic product (SDP) and that at the district level is the district domestic product (DDP) [33]. They are also calculated in exactly the same manner as the domestic product for the country.

Out of the nine states, DDPs at constant prices (2004–2005) were obtained for the states of Assam [34], Bihar [35], Rajasthan [36], Odisha [37] and Jharkhand [38]. This data was used to propose models and predict the DDP and health indicators for the other states at the district level.

Figure 2 shows the distribution of DDP in the districts of the five states. The mean DDP at 2004–2005 constant prices is the highest for the state of Rajasthan (INR 57,878.46 million) followed by the state of Odisha (INR 38,617.18 million). The state of Bihar has the lowest DDP (INR 19,845.17 million), while both Assam and Jharkhand have mean DDPs of INR 26,154.86 million and INR 26,212.42 million, respectively.

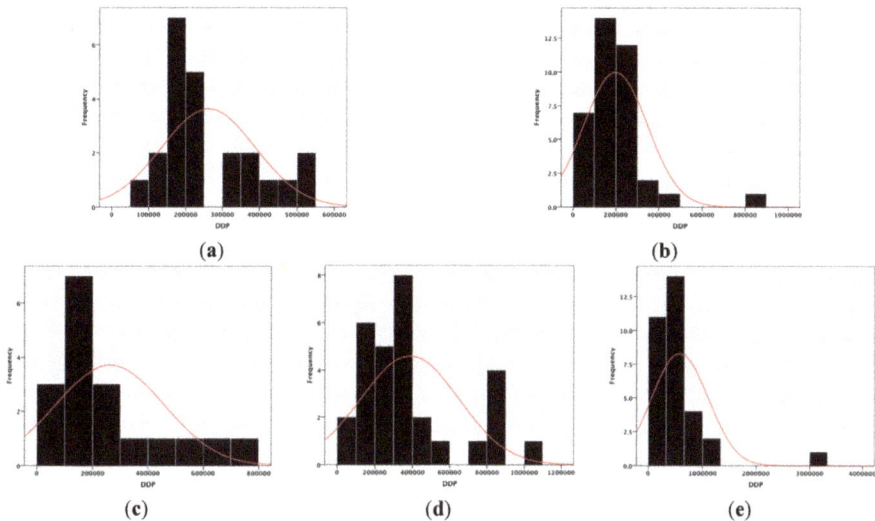

**Figure 2.** Distribution of DDP in nine states of India: (**a**) Assam; (**b**) Bihar; (**c**) Jharkhand; (**d**) Odisha; (**e**) Rajasthan.

The intra-state variation in the contribution of DDP shows that the highest DDP is recorded in the state of Jaipur in Rajasthan (more than INR 300,000 million), followed by the states of Alwar and Jodhpur in Rajasthan. Sundargarh district is the highest contributor in Odisha (more than INR 100,000 million). Lakhisarai district of Bihar has the highest DDP (more than INR 82,000 million). The lowest DDP figures are recorded for the states of Khagaria, Nalanda, Bhojpur, Sheohar and Begusarai of Bihar (less than INR 4000 million).

## 5.4. Correlating the Indicators

In the next stage, CBR, IMR, MMR and DDP were correlated with the percentage of households with access to electricity, the average radiance obtained from the VIIRS image and the average stable lights. Due to the availability of DDP data for the five states of Assam, Bihar, Jharkhand, Odisha and Rajasthan, the correlations were carried out in those areas only. The results from the correlations are shown in Table 4.

**Table 4.** Results of correlations between health metrics and rate of electrification, average stable lights and average radiance. VIIRS, Visible Infrared Imaging Radiometer Suite.

| | Percentage of Households with Electricity | Total Population | Average Radiance VIIRS | Average Stable Light | DDP |
|---|---|---|---|---|---|
| CBR 2011–2012 | −0.412 ** | 0.204 * | −0.249 ** | −0.390 ** | −0.220 ** |
| IMR 2011–2012 | −0.084 | −0.137 | −0.180 * | −0.317 ** | −0.019 |
| Maternal mortality rate 2011–2012 | −0.322 ** | −0.04 | 0.157 | −0.287 ** | −0.265 ** |
| DDP | 0.547 ** | 0.428 ** | 0.152 | 0.454 ** | 1 |

** Correlation is significant at the 0.01 level (two-tailed); * correlation is significant at the 0.05 level (two-tailed).

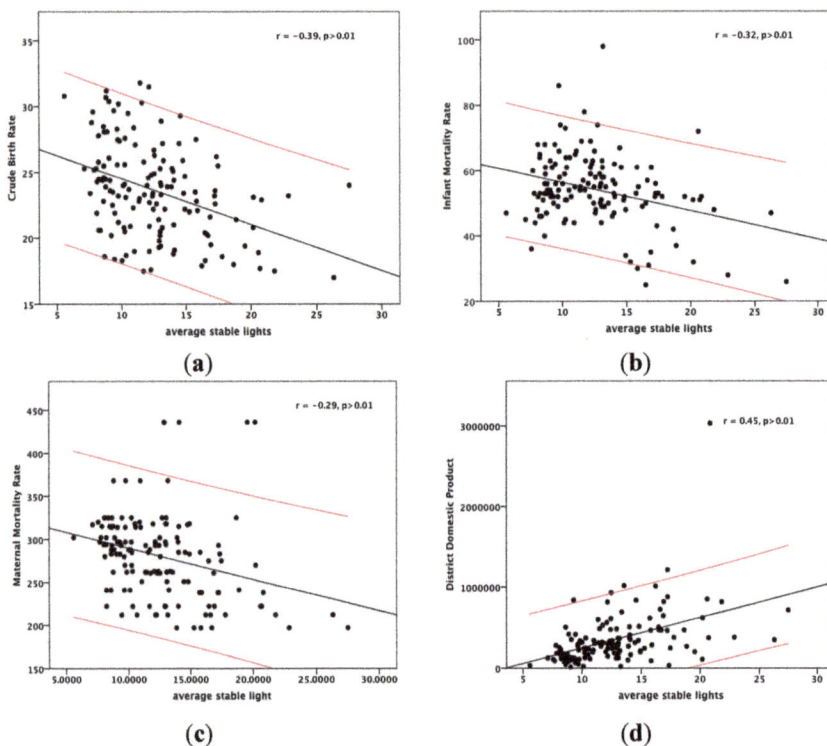

**Figure 3.** Correlations between health indicators and average stable lights, average radiance and DDP: (a) CBR; (b) IMR; (c) MMR; (d) DDP.

The percentage of households with electricity and DDP has a significant correlation with CBR and MMR. There are negative correlations with both these variables. This suggests that with an increase in the percentage of electrification and DDP, there is a fall in the CBR and MMR of these areas. The correlations are significant at the 99% confidence level. IMR however, does not show any significant correlation with these indicators.

Correlations were also examined with the mean stable lights and radiance obtained from the VIIRS images of 2010 and 2011, respectively. The nature of the correlations is shown in Figure 3. Correlation coefficients ranged from ±0.2 to ±0.6 at both the 99% and 95% confidence intervals. CBR has significant correlations with the percentage of households with electricity, average radiance (weak correlation), average stable lights and DDP (weak correlation) at the 99% confidence level. It also shows significant positive correlation with the total population at the 95% confidence interval. IMR shows significant correlation with both average stable light ($r = -0.32$, $p > 0.01$) and average radiance ($r = -0.18$, $p > 0.05$). MMR is significantly correlated with average stable lights ($r = -0.23$, $p > 0.01$), the percentage of households with electricity ($r = -0.32$, $p > 0.01$) and DDP ($r = -0.27$, $p > 0.01$). From Table 3, it was observed that all of the health indicators show a significant correlation with average stable light.

It was also observed that CBR and MMR had significant correlations with the percentage of households with access to electricity, average stable lights and average radiance at the 99% confidence interval. As a result, these two health metrics were chosen for further analyses and modelling.

## 6. Models for Predicting Health Indicators

### 6.1. Models

In order to propose the models, 100 out of 142 districts were chosen using a random sample. The log to the base "e" (log normal) values were calculated for all the indicators. Multiple regression models were proposed to predict DDP, as well as health indicators for each district. DDP was proposed for the districts for which the data was not readily available.

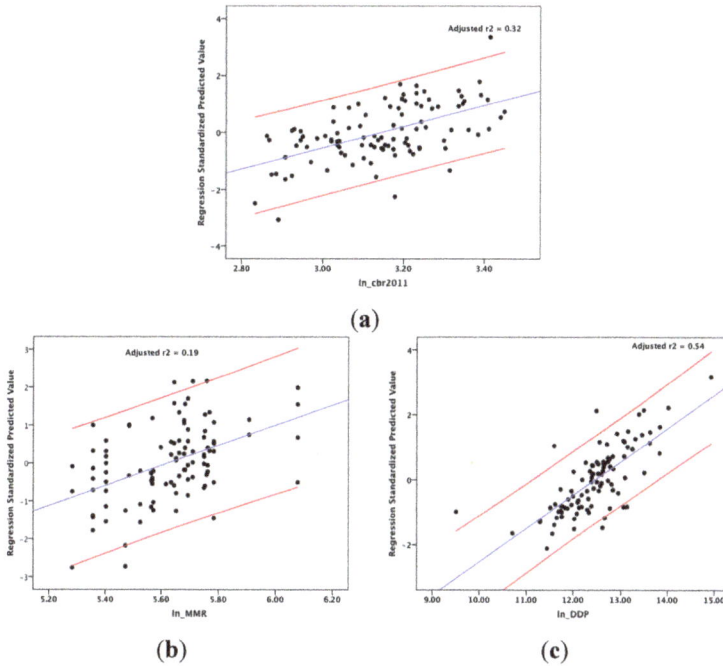

**Figure 4.** Actual and predicted values of CBR, IMR and DDP: (a) CBR; (b) MMR; (c) DDP.

Multiple regression models were calculated in this study using the method of backward elimination. In this method, to begin with, all the independent variables were used to fit in the regression equation. The null hypothesis was tested for each of the partial correlation coefficients. When the null hypothesis was rejected, all of the independent variables were considered for the model. However, if the null hypothesis was true, the independent variable with the lowest |t| value was eliminated from the equation, and the remaining variables were considered so as to fit to the multiple regression equation. The step continued, until all the partial regression coefficients were significant [39].

### 6.1.1. Models to Propose DDP

Average stable lights, total population, percentage of households with access to electricity and average radiance from VIIRS image were used as the input variables for calculating the model to propose DDP for the districts (Figure 4). Average radiance from VIIRS image was pooled in the model calculation. The multiple regression model has an adjusted $r^2$ value of 0.54.

The model proposed to calculate DDP is shown in Table 5.

**Table 5.** Model to predict DDP.

| Metrics | Model |
|---------|-------|
| Ln_DDP | 4.054 + 0.563 (ln_average stable lights) + 0.450 (ln_population) + 0.15 (percentage of households with electricity) |

### 6.1.2. Models to Propose Health Indicators

For predicting health indicators, multiple regression models were proposed. The method of backward elimination was also used in this case. The variables that are pooled and included in the models are shown in Table 5.

All of the models were significant at the level of $p > 0.05$.

From Table 6, it is observed that DDP and average stable lights were used as inputs in both of the models. Radiance from VIIRS image was only used as an input in the model predicting MMR. The adjusted $r^2$ was the highest for the model predicting CBR (0.30).

**Table 6.** Variables pooled and included in the models to predict CBR and MMR.

| Independent Variables in the Models | Variables Pooled in Stepwise Regression | Variables Included in Multiple Regression | Model $r^2$ | Adjusted $r^2$ ($p > 0.05$) |
|---|---|---|---|---|
| Crude birth rate | Radiance from VIIRS image, percentage of households with electricity | DDP, population, average stable light | 0.32 | 0.30 |
| Maternal mortality rate | Total population, percentage of households with electricity | DDP, radiance from VIIRS image, average stable light | 0.21 | 0.19 |

The proposed models to predict the health indicators are shown in Table 7. The states for which the AHS survey was undertaken are the least developed states of India. They have the highest population and are predominantly rural. There are no major urban centres in these states. The capital cities and towns are smaller compared to the mega cities of the country. As a result, the overall radiance and stable lights captured by nighttime images over these states are comparatively low when compared to the Mumbai and Delhi conurbations. Therefore, the model adjusted $r^2$ values, though significant, are low compared to other parts of the country, as observed by the authors in their previous publications [6,7,18,20]. Furthermore, the health data was calculated on a sample population from the districts. This sample data was used in the models with light information obtained for the whole area of districts. This was probably another reason for the low adjusted $r^2$ of the models.

**Table 7.** Models to predict CBR and MMR.

| Health Metrics | Models |
|---|---|
| Ln.(CBR) | 3.078 − 0.019(Ln. average stable lights) + 0.098(Ln. population) − 0.08(Ln. DDP) |
| Ln. (MMR) | 6.937 + 0.106(Ln. average radiance from VIIRS image) − 0.266(Ln. average stable lights) − 0.064(Ln. DDP) |

*6.2. Model Validation*

The proposed models were validated over the withheld 42 districts from the states of Chhattisgarh, Uttar Pradesh, Madhya Pradesh and Uttarakhand.

The actual and predicted values of CBR and MMR are shown in Figure 4. Most of the districts have predicted CBR within the 95% confidence band. For MMR, there are a few outliers. Furthermore, in the case of MMR, some of the districts displayed had different predicted values for the same actual values. For example the districts of Hazaribagh and Lakhisarai had an actual MMR of 368.00, while their predicted MMRs were 269.79 and 291.47, respectively. A similar case was also found for the districts of Nabarangapur, Godda and Jalore. All of these districts have AHS recorded MMR of 297.00, while the predicted MMRs were 284.89, 288.31 and 299.03, respectively.

The results of model validation are shown in Figure 5. The mean error of the prediction of CBR was 0.59%, and MMR was −0.24%. For IMR, the mean percentage of error was −1.59. The difference between the actual and predicted values was less than ±10% for both CBR and MMR.

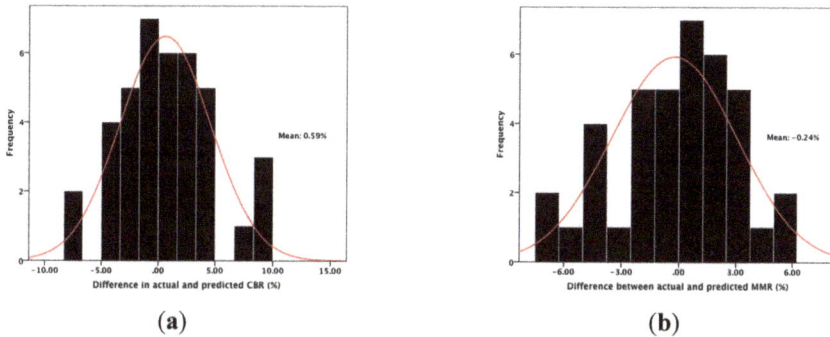

**Figure 5.** Results of model validation. (**a**) CBR; (**b**) MMR.

## 7. Health Atlas of India

The models were used to predict CBR and MMR for all the districts of India. The maps showing the distribution of CBR and MMR are presented in Figure 6.

The maps show the following trends in distribution of CBR and MMR:

- CBRs ranging from 7.00 to 10.00 are found in patches over some districts of Himachal Pradesh, Punjab, Andhra Pradesh, Goa, Lakshadweep, Nagaland, Pondicherry, Daman and Diu and the National Capital Territory of Delhi.
- The highest CBRs (more than 25) are found in the districts of Bihar, Uttar Pradesh, Jharkhand and Odisha.
- MMRs less than 70 are found in the National Capital Territory of Delhi and in the districts of Andhra Pradesh, Kerala, Maharashtra and Punjab.
- The highest values (more than 250) are found in Bihar, Arunachal Pradesh, Jharkhand, Orissa and Uttar Pradesh.

   The predicted results of CBR and MMR show a strong correlation with DDP and the percentage of households with access to electricity. The correlation coefficients range from −0.92 to −0.99 at a significance level of 0.01. The correlations are shown in Figure 7. The results show that health metrics and development indicators go hand in hand and are very strongly correlated with each other.

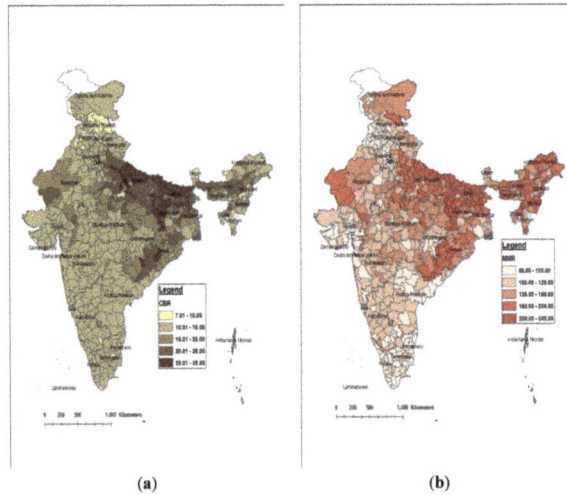

**(a)**                                          **(b)**

**Figure 6.** CBR and MMR for all the districts of India. (**a**) Crude birth rate; (**b**) Maternal mortality rate.

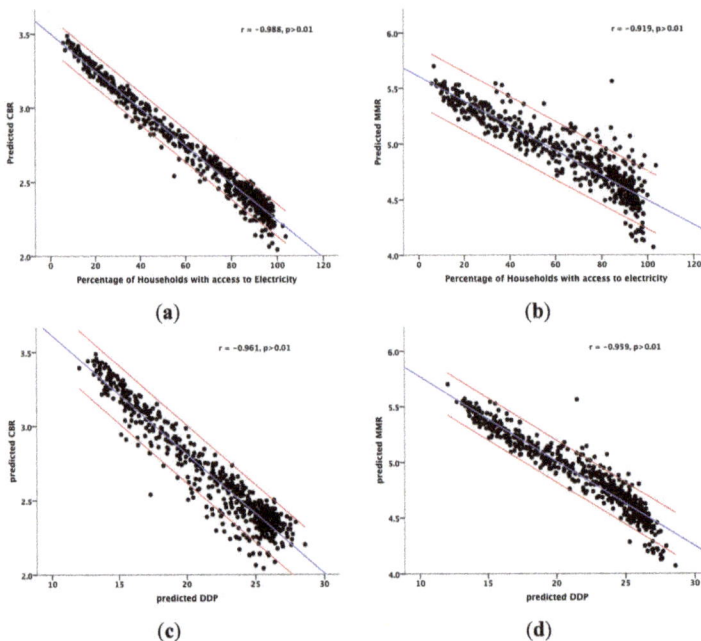

**(a)**                                          **(b)**

**(c)**                                          **(d)**

**Figure 7.** Correlations between health metrics (CBR and MMR) and development indicators (rate of electrification and DDP) for India. (**a**) CBR and electricity; (**b**) MMR and electricity; (**c**) CBR and DDP; (**d**) MMR and DDP.

## 8. Conclusions

The paper demonstrates that better health is correlated with the overall development of an area. The study is based on the health metrics available from the second round of the Annual Health Survey Report of India, 2011–2012, and census data, 2011. Due to logistical constraints, health data in India was surveyed for only nine states. However, the importance of the availability of disaggregated health data cannot be overlooked. The paper used remote sensing techniques to model the health indicators at the sub-national level. The models were proposed using information from two nighttime images. In addition to the variations amongst states, district level differences in the health metrics within the states were also highlighted in the paper. Models were proposed using the information from satellite images and census metrics as predictor variables. The chosen health indicators were mapped to show the spatial variations over the country as a whole. Correlations between health indicators and development were examined for all of the states. The results show:

- A very strong correlation between health and development exists, with more developed areas, such as the National Capital Territory of Delhi, urban areas and states, such as Kerala, having better health metrics than the other areas.
- Nighttime images captured by both the DMSP-OLS and VIIRS sensors are useful for mapping health indicators at the sub-national level.

The method proposed in this paper is useful for predicting the metrics at a smaller spatial scale, such as taluks and villages, as well. These disaggregated predicted data will be invaluable for micro-level health planning and formulating rural development policies. Moreover, the health metrics can be mapped more frequently than those available from the census, which is carried out every 10 years. The metrics can be predicted and mapped as often as the satellite data is available. This is also useful in recording the trend of change in the health indicators. For a country like India, where manual data collection is a cumbersome, time-consuming and expensive process, the method proposed in this study has the potential to be of immense use.

**Acknowledgments:** The authors would like to thank everyone in the National Geophysical Data Centre for the nighttime datasets and also the officials in the Office of the Registrar General and Census Commissioner, Government of India, for the census data.

**Author Contributions:** This paper is a combined effort by Koel Roychowdhury and Simon Jones. The author order reflects that the major contribution came from Koel Roychowdhury.

**Conflicts of Interest:** There are no conflicts of interest.

## References

1. Mishra, C. Nexus of poverty, energy balance and health. *Ind. J. Commun. Med.: Off. Publ. Ind. Assoc. Prev. Soc. Med.* **2012**, *37*. [CrossRef]
2. Government of India. *Economic Survey of India*; Government of India: New Delhi, India, 2012.
3. UNICEF Polio Eradication. Available online: http://www.unicef.org/india/health_3729.htm (accessed on 3 February 2014).
4. Duggal, R. Health Planning in India. Available online: http://www.cehat.org/go/uploads/Publications/a168.pdf (accessed on 3 February 2014).
5. World Health Organization Country Cooperation Strategy at a Glance: India. Available online: http://www.who.int/countryfocus/cooperation_strategy/ccsbrief_ind_en.pdf?ua=1 (accessed on 3 February 2014).
6. Roychowdhury, K.; Jones, S.; Arrowsmith, C.; Reinke, K. A Comparison of high and low gain DMSP/OLS satellite images for the study of socio-economic metrics. *J. Sel. Top. Earth Observ. Remote Sens.* **2010**. [CrossRef]
7. Roychowdhury, K.; Jones, S.; Arrowsmith, C.; Reinke, K. Indian Census Using Satellite Images: Can DMSP-OLS Data be Used for Small Administrative Regions? In Processings of 2011 Joint Urban Remote Sensing Event, Munich, German, 11–13 April 2011.

8.   Baugh, K.; Elvidge, C.D.; Ghosh, T.; Ziskin, D. Development of a 2009 Stable Lights Product Using DMSP-OLS Data. In Proceedings of Asia-Pacific Advanced Network (APAN) Workshop, Hanoi, Vietnam, 25–27 November 2010.

9.   Elvidge, C.D.; Baugh, K.; Zhizhin, M.; Hsu, F.C. Why VIIRS data are superior to DMSP for mapping nighttime lights. *Proc. Asia Pac. Adv. Netw.* **2013**. [CrossRef]

10.  Elvidge, C.D.; Baugh, K.E.; Hobson, V.R.; Kihn, E.A.; Kroehl, H.W.; Davis, E.R.; Cocero, D. Satellite inventory of human settlements using nocturnal radiation emissions: A contribution for the global toolchest. *Glob. Chang. Biol.* **1997**, *3*, 387–395.

11.  Elvidge, C.D.; Baugh, K.E.; Kihn, E.A.; Kroehl, H.W.; Davis, E.R. Mapping city lights with nighttime data from the DMSP operational linescan system. *Photogramm. Eng. Remote Sens.* **1997**, *63*, 727–734.

12.  Imhoff, M.L.; Lawrence, W.T.; Stutzer, D.C.; Elvidge, C.D. A technique for using composite DMSP/OLS "City Lights" satellite data to map urban area. *Remote Sens. Environ.* **1997**, *61*, 361–370. [CrossRef]

13.  Doll, C.N.H.; Pachauri, S. Estimating rural populations without access to electricity in developing countries through night-time light satellite imagery. *Energy Policy* **2010**. [CrossRef]

14.  Doll, C.N.H.; Muller, J.P.; Elvidge, C.D. Night-time imagery as a tool for global mapping of socioeconomic parameters and greenhouse gas emissions. *AMBIO J. Hum. Environ.* **2000**, *29*, 157–162.

15.  Elvidge, C.; Ziskin, D.; Baugh, K.; Tuttle, B.; Ghosh, T.; Pack, D.; Erwin, E.; Zhizhin, M. A fifteen year record of global natural gas flaring derived from satellite data. *Energies* **2009**, *2*, 595–622. [CrossRef]

16.  Roychowdhury, K.; Jones, S.D.; Arrowsmith, C. Mapping Urban Areas of India from DMSP/OLS Night-Time Images. In Proceeedings of the Surveying & Spatial Sciences Institute Biennial International Conference, Adelaide, SA, Australia, 28 September 2009.

17.  Roychowdhury, K.; Taubenbock, H.; Jones, S. Delineating Urban, Suburban and Rural Areas Using Landsat and DMSP-OLS Night-Time Images. In Proceedings of 2011 Joint Urban Remote Sensing Event, Munich, German, 11–13 April 2011.

18.  Roychowdhury, K.; Jones, S.; Arrowsmith, C. Assessing the Utility of DMSP/OLS Night-Time Images for Characterizing Indian Urbanization. In Proceedings of 2009 Joint Urban Remote Sensing Event, Shanghai, China, 20–22 May 2009; pp. 1–7.

19.  Roychowdhury, K.; Jones, S.J.; Arrowsmith, C.; Reinke, K. Night-Time Lights and Levels of Development: A Study Using DMSP-OLS Night-Time Images at the Sub-National Level. In Proceedings of the XXII ISPRS Congress, Melbourne, VIC, Australia, 25 August–1 September 2012; pp. 93–98.

20.  Roychowdhury, K.; Jones, S.D.; Arrowsmith, C.; Reinke, K. A Comparison of high and low gain DMSP/OLS satellite images for the study of socio-economic metrics. *IEEE J. Sel. Top. Appl. Earth Observ. Remote Sens.* **2011**, *4*, 35–42. [CrossRef]

21.  Global Adaptation Institute. *Global Adaptation Index (GaIn): Measuring What Matters: Building Resilience to Cliate Change and Global Forces*; Global Adaptation Institute: Washington, DC, USA, 2011.

22.  Bhandari, L.; Roychowdhury, K. Night lights and economic activity in India: A study using DMSP-OLS night time images. *Proc. Asia Pac. Adv. Netw.* **2012**. [CrossRef]

23.  Baugh, K.; Hsu, F.C.; Elvidge, C.D.; Zhizhin, M. Nighttime lights compositing using the VIIRS day-night band: Preliminary results. *Proc. Asia Pac. Adv. Netw.* **2013**. [CrossRef]

24.  Elvidge, C.D.; Baugh, K.; Zhizhin, M.; Hsu, F.C. What is so great about nighttime VIIRS data for the detection and characterization of combustion sources? *Proc. Asia Pac. Adv. Netw.* **2013**. [CrossRef]

25.  Zhizhin, M.; Elvidge, C.D.; Hsu, F.C.; Baugh, K. Using the short-wave infrared for nocturnal detection of combustion sources in VIIRS data. *Proc. Asia Pac. Adv. Netw.* **2013**. [CrossRef]

26.  Office of the Registrar General. *Annual Health Survey (AHS) in 8 EAG States and Assam-Release of District Level Factsheet 2010–2011*; Ministry of Home Affairs, Government of India: New Delhi, India, 2012.

27.  Hazarika, I. India at the crossroads of millennium development goals 4 and 5. *Asia-Pac. J. Public Health* **2012**, *24*, 450–463.

28.  United Nations Development Program the Millenium Development Goals: Eight Goals for 2015. Available online: http://www.undp.org/content/undp/en/home/mdgoverview/ (accessed on 19 November 2013).

29.  World Health Organization Global Health Observatory Data Repository. Available online: http://apps.who.int/gho/data/node.main.519?lang=en (accessed on 3 February 2014).

30. Office of the Registrar General Maternal and Child Mortality and Total Fertility Rates: Sample Registration System (SRS). Available online: http://www.censusindia.gov.in/vital_statistics/SRS_Bulletins/MMR_release_070711.pdf (accessed on 21 November 2013).

31. Central Statistics Office. *Energy Statistics*; Government of India: New Delhi, India, 2012.

32. Office of the Registrar General Source of Lighting (2001–2011). Available online: http://www.censusindia.gov.in/2011census/hlo/Data_sheet/India/Source_Lighting.pdf (accessed on 4 February 2014).

33. Ministry of Statistics and Program Implementation. *National Accounts Statistics: Manual on Estimation of State and District Income*; Government of India: New Delhi, India, 2008.

34. Planning and Development Department. *Economic Survey of Assam 2011–2012*; Government of India: New Delhi, India, 2012.

35. Department of Finance. *Economic Survey, Bihar 2011–2012*; Government of India: New Delhi, India, 2012.

36. Planning and Development Department. *Rajasthan Economic Survey*; Government of India: New Delhi, India, 2012.

37. Planning and Coordination Department. *Economic Survey, Odisha 2011–2012*; Government of Odisha: Bhubaneshwar, India, 2013.

38. Directorate of Economics and Statistics. *GSDP & NSDP 2009–2010, 2010–2011(P), 2011–2012 (Q) & 2012–2013 (A)*; Government of India: New Delhi, India, 2012.

39. Zar, J.H. *Biostatistical Analysis*, 5th ed.; Pearson Education Inc: Upper Saddle River, NJ, USA, 2010.

MDPI

St. Alban-Anlage 66

4052 Basel

Switzerland

Tel. +41 61 683 77 34

Fax +41 61 302 89 18

www.mdpi.com

*ISPRS International Journal of Geo-Information* Editorial Office

E-mail: ijgi@mdpi.com

www.mdpi.com/journal/ijgi

www.ingramcontent.com/pod-product-compliance
Lightning Source LLC
Chambersburg PA
CBHW051729210326
41597CB00032B/5658

9 783038 971726